Compact Stars in the QCD Phase Diagram

Compact Stars in the QCD Phase Diagram

Special Issue Editors

David Blaschke
Alexander Ayriyan
Alexandra Friesen
Hovik Grigorian

MDPI • Basel • Beijing • Wuhan • Barcelona • Belgrade

The cover depicts a poem by Prof. Qun Wang (USTC Hefei, China) with its calligraphy by Prof. Dong-pei Zhu (USTC Hefei, China), October 2019.

中子星辰密
夸克物態豐
宇宙初創時
剎那藏永恆

Its English translation reads:
The neutron star is dense,
the state of quark matter is abundant,
at the birth moment of the universe,
the eternity is held.

Special Issue Editors
David Blaschke
University of Wroclaw
Poland

Alexander Ayriyan
Joint Institute for Nuclear Research
Russia

Alexandra Friesen
Bogoliubov Laboratory of Theoretical Physics
Russia

Hovik Grigorian
Joint Institute for Nuclear Research
Russia

Editorial Office
MDPI
St. Alban-Anlage 66
4052 Basel, Switzerland

This is a reprint of articles from the Special Issue published online in the open access journal *Universe* (ISSN 2218-1997) from 2017 to 2018 (available at: https://www.mdpi.com/journal/universe/special_issues/Compact_Stars).

For citation purposes, cite each article independently as indicated on the article page online and as indicated below:

LastName, A.A.; LastName, B.B.; LastName, C.C. Article Title. *Journal Name* **Year**, *Article Number*, Page Range.

ISBN 978-3-03921-958-2 (Pbk)
ISBN 978-3-03921-959-9 (PDF)

© 2019 by the authors. Articles in this book are Open Access and distributed under the Creative Commons Attribution (CC BY) license, which allows users to download, copy and build upon published articles, as long as the author and publisher are properly credited, which ensures maximum dissemination and a wider impact of our publications.

The book as a whole is distributed by MDPI under the terms and conditions of the Creative Commons license CC BY-NC-ND.

Contents

About the Special Issue Editors vii

Preface to "Compact Stars in the QCD Phase Diagram" ix

Toru Kojo
QCD Equations of State in Hadron–Quark Continuity
Reprinted from: *Universe* **2018**, *4*, 42, doi:10.3390/universe4020042 1

Niels-Uwe F. Bastian, David Blaschke, Tobias Fischer and Gerd Röpke
Towards a Unified Quark-Hadron-Matter Equation of State for Applications in Astrophysics and Heavy-Ion Collisions
Reprinted from: *Universe* **2018**, *4*, 67, doi:10.3390/universe4060067 11

Hermann Wolter
The High-Density Symmetry Energy in Heavy-Ion Collisions and Compact Stars
Reprinted from: *Universe* **2018**, *4*, 72, doi:10.3390/universe4060072 46

Sanjin Benić
Equation of State for Dense Matter with a QCD Phase Transition
Reprinted from: *Universe* **2018**, *4*, 45, doi:10.3390/universe4030045 62

Konstantin A. Maslov, Evgeny E. Kolomeitsev and Dmitry N. Voskresensky
Charged ρ Meson Condensate in Neutron Stars within RMF Models
Reprinted from: *Universe* **2018**, *4*, 1, doi:10.3390/universe4010001 68

Mateusz Cierniak, Thomas Klähn, Tobias Fischer, Niels–Uwe F. Bastian
Vector-Interaction-Enhanced Bag Model
Reprinted from: *Universe* **2018**, *4*, 30, doi:10.3390/universe4020030 76

Vahagn Abgaryan, David Alvarez-Castillo, Alexander Ayriyan, David Blaschke, Hovik Grigorian
Two Novel Approaches to the Hadron-Quark Mixed Phase in Compact Stars
Reprinted from: *Universe* **2018**, *4*, 94, doi:10.3390/universe4090094 85

Stefan Typel and David Blaschke
A Phenomenological Equation of State of Strongly Interacting Matterwith First-Order Phase Transitions and Critical Points
Reprinted from: *Universe* **2018**, *4*, 32, doi:10.3390/universe4020032 100

Yuri B. Ivanov
Directed Flow in Heavy-Ion Collisions andIts Implications for Astrophysics
Reprinted from: *Universe* **2017**, *3*, 79, doi:10.3390/universe3040079 110

Sylvain Mogliacci, Isobel Kolbé and W. A. Horowitz
From Heavy-Ion Collisions to Compact Stars: Equation of State and Relevance of the System Size
Reprinted from: *Universe* **2018**, *4*, 14, doi:10.3390/universe4010014 119

Ludwik Turko
Looking for the Phase Transition—Recent NA61/SHINE Results
Reprinted from: *Universe* **2018**, *4*, 52, doi:10.3390/universe4030052 129

Amir Ouyed, Rachid Ouyed and Prashanth Jaikumar
Hadron–Quark Combustion as a Nonlinear, Dynamical System
Reprinted from: *Universe* 2018, 4, 51, doi:10.3390/universe4030051 137

Alessandro Drago, Giuseppe Pagliara, Sergei B. Popov, Silvia Traversi and Grzegorz Wiktorowicz
The Merger of Two Compact Stars: A Tool for Dense Matter Nuclear Physics
Reprinted from: *Universe* 2018, 4, 50, doi:10.3390/universe4030050 145

Manjari Bagchi
Prospects of Constraining the Dense Matter Equation of State from Timing Analysis of Pulsars in Double Neutron Star Binaries: The Cases of PSR J0737−3039A and PSR J1757−1854
Reprinted from: *Universe* 2018, 4, 36, doi:10.3390/universe4020036 160

Hovik Grigorian, Evgeni E. Kolomeitsev, Konstantin A. Maslov and Dmitry N. Voskresensky
On Cooling of Neutron Stars with a Stiff Equation of State Including Hyperons
Reprinted from: *Universe* 2018, 4, 29, doi:10.3390/universe4020029 174

Francesco Tonelli and Massimo Mannarelli
Cracking Strange Stars by Torsional Oscillations
Reprinted from: *Universe* 2018, 4, 41, doi:10.3390/universe4020041 180

Rodrigo Negreiros, Cristian Bernal, Veronica Dexheimer and Orlenys Troconis
Many Aspects of Magnetic Fields in Neutron Stars
Reprinted from: *Universe* 2018, 4, 43, doi:10.3390/universe4030043 187

Enping Zhou, Antonios Tsokaros, Luciano Rezzolla, Renxin Xu and Kōji Uryū
Rotating Quark Stars in General Relativity
Reprinted from: *Universe* 2018, 4, 48, doi:10.3390/universe4030048 209

Prashanth Jaikumar, Thomas Klähn and Raphael Monroy
Non-Radial Oscillation Modes of Superfluid Neutron Stars Modeled with CompOSE
Reprinted from: *Universe* 2018, 4, 53, doi:10.3390/universe4030053 217

Efrain J. Ferrer, Vivian de la Incera
Anomalous Electromagnetic Transport in Compact Stars
Reprinted from: *Universe* 2018, 4, 54, doi:10.3390/universe4030054 225

William M. Spinella, FridolinWeber, Gustavo A. Contrera, and Milva G. Orsaria
Neutrino Emissivity in the Quark-Hadron Mixed Phase
Reprinted from: *Universe* 2018, 4, 64, doi:10.3390/universe4050064 239

Vinzent Steinberg, Dmytro Oliinychnko, Jan Staudenmaier, Hannah Petersen
Strangeness Production in Nucleus-Nucleus Collisions at SIS Energies
Reprinted from: *Universe* 2018, 4, 37, doi:10.3390/universe4020037 254

About the Special Issue Editors

David Blaschke obtained his PhD in theoretical physics from Rostock University in 1987 and habilitated in 1995. During 2001–2007 he served as Vice Director of the Bogoliubov Laboratory of Theoretical Physics at the Joint Institute for Nuclear Research in Dubna. He has been Professor at the University of Wroclaw since 2006. His works are mainly devoted to topics in quantum field theory at finite temperature, dense hadronic matter and QCD phase transitions, quark matter in heavy-ion collisions and in compact stars, as well as pair production in strong fields with applications to high-intensity lasers. He has published more than 380 articles, most of them in peer-reviewed international journals, edited more than 12 books, and currently has an h-index of 44. He was awarded honorary doctorates from Dubna State University (2017) and Russian-Armenian University in Yerevan (2019).

Alexander Ayriyan obtained his master's degree in Computer Science at MIREA—Russian Technological University in 2007 and is a Researcher at the Laboratory of Information Technologies at the Joint Institute for Nuclear Research in Dubna. During 2010–2014, he was chairman of the Association of Young Scientists and Specialists of the JINR. He has been part of the Computational Physics and IT Division of the A. Alikhanyan National Science Laboratory since joining in 2019. His work is devoted to numerical simulations of complex physical phenomena, like structure and evolution of compact stars, equation of state for nuclear matter under extreme conditions, and heavy-ion collisions. He has published 41 articles in peer-reviewed journals. He won an award from the Open Scientific Research Competition of Students from the Russian Federation and CIS (2007) and from the Moscow Region Governor for Young Scientists (2014).

Alexandra Friesen obtained her PhD in theoretical physics from the Bogoliubov Laboratory of Theoretical Physics at the Joint Institute for Nuclear Research in 2016, where she continues to work as a Scientific Researcher. Her scientific interests are centered around topics in quantum field theory at finite temperature, dense hadronic matter, QCD phase transitions, and quark matter in heavy-ion collisions. She has published more than 20 articles in peer-reviewed journals. She was awarded a scholarship for Young Scientists and Specialists named after D.I. Blokhintsev (2015, 2016) and L.D. Soloviev (2017).

Hovik Grigorian obtained his PhD in Mathematics and Physics from Yerevan State University (YSU) in 1989 and was a Senior Researcher at the Chair of Theoretical Physics at YSU from 1986. During 2006–2009, as well as from 2013 until now, he is a Senior Researcher at the Laboratory for Information Technologies of JINR Dubna. He is also a Senior Researcher at A. Alikhanyan National Science Laboratory since his appointment in 2019. His works are mainly devoted to topics in gravitational field theory (including general relativity theory and its applications in compact stars physics), the theory of dense hadronic and quark matter equations of state and phase transitions of stellar matter in compact stars, as well as the mathematical modeling and applications of the Bayesian analysis method in theoretical physics. He has published more than 80 articles, most of them in peer-reviewed international journals.

Preface to "Compact Stars in the QCD Phase Diagram"

Back in 2001 at NORDITA Copenhagen, the Conference "Compact Stars in the QCD Phase Diagram" was organized by Rachid Oyed and Francesco Sannino for the first time, and it was not obvious before 2009, when Renxin Xu organized a sequel at the recently established Kavli Institute for Astronomy and Astrophysics at Peking University in Beijing, that this would become the start of a series of biannual meetings organized by a growing community worldwide. Rodrigo Negreiros and Jorge Horvath organized the 2012 meeting in Guaruja (Brazil), Tobias Fischer and Jochen Wambach the 2014 meeting in Prerow (Germany), Ignazio Bombaci and Massimo Mannarelli the 2016 meeting in Gran Sasso (Italy), and Vivian Incera and Efrain Ferrer the 2018 meeting in New York (USA). The meeting 2017 in Dubna was an extraordinary one. It was squeezed in to the scheme because of the great interest and relevance of the topic in view of the first gravitational wave being detected from merging black holes and the rapid progress in constructing the new accelerator complex, NICA, at JINR Dubna for discovering signals of a possible first-order phase transition to quark matter in heavy-ion collisions at not-too-high beam energies.

Therefore, the special aim of the CSQCD-VI conference was to bring together the experts in fields of compact stars, their mergers, and their involved explosive astrophysical phenomena with those studying the QCD phase diagram with heavy-ion collision experiments. Consequently, the conference covered the following main topics:

- QCD phase diagram for HIC vs. astrophysics;
- Quark deconfinement in HIC vs. supernovae, neutron stars, and their mergers;
- Strangeness in HIC and in compact stars;
- Equation of state and QCD phase transitions.

We are grateful to the RSF for supporting this research under grant number 17-12-01427. We also want to express our thanks to the MDPI journal *Universe* which supported a Special Issue under the theme of our conference series, with the beneficial conditions of open access publishing in addition to being free of article processing charges. We acknowledge the COST Actions MP1304 "NewCompStar" and CA16214 "PHAROS" for supporting the networking activities of the participants. We thank the directorate of the JINR Dubna and the Bogoliubov Laboratory of Theoretical Physics for hosting the conference and providing the atmosphere for inspiring and fruitful discussions as well as for all their ongoing efforts to construct a major infrastructure with NICA, home to the BM@N and MPD heavy-ion collisions experiments for exploring the structure of the QCD phase diagram.

We hope that the book edition of this Special Issue of *Universe* "Compact Stars in the QCD Phase Diagram" will serve as a useful guide in the education of young and senior scientists in this emerging field that represents an intersection of the communities of strongly interacting matter theory, heavy-ion collision physics, and compact star astrophysics.

David Blaschke, Alexander Ayriyan, Alexandra Friesen, Hovik Grigorian
Special Issue Editors

Article

QCD Equations of State in Hadron–Quark Continuity

Toru Kojo

Institute of Particle Physics (IOPP) and Key Laboratory of Quark and Lepton Physics (MOE),
Central China Normal University, Wuhan 430079, China; torujj@mail.ccnu.edu.cn

Received: 3 January 2018; Accepted: 13 February 2018; Published: 19 February 2018

Abstract: The properties of dense matter in quantum chromodynamics (QCD) are delineated through equations of state constrained by the neutron star observations. The two solar mass constraint, the radius constraint of $\simeq$11–13 km, and the causality constraint on the speed of sound, are used to develop the picture of hadron–quark continuity in which hadronic matter continuously transforms into quark matter. A unified equation of state at zero temperature and β-equilibrium is constructed by a phenomenological interpolation between nuclear and quark matter equations of state.

Keywords: hadron–quark continuity; neutron stars; QCD phase diagram

1. Introduction

The study of the phase structure in quantum chromodynamics (QCD) at large baryon density has been a difficult problem, partly because the lattice Monte Carlo simulations based on the QCD action are not at work, and partly because many-body problems with strong interactions are very complex in theoretical treatments. Currently, the best source of information for dense QCD is the physics of neutron stars from which one can extract useful insights into QCD equations of state [1], as well as the transport properties in matter. Since the domain relevant for these physics is the baryon density of $n_B \sim 1 - 10 n_0$ ($n_0 \simeq 0.16\,\mathrm{fm}^{-3}$: nuclear saturation density) or baryon chemical potential of $\mu_B \sim 1 - 2$ GeV, we can use the neutron star constraints to explore the properties of matter beyond the nuclear regime.

There have been remarkable progress in observations that constrain our understanding on the nature of dense QCD matter. They include the discoveries of two-solar mass ($2M_\odot$) neutron stars [2,3], the constraints for the neutron star radii from X-ray analyses [4,5], and, most remarkably, the detection of the gravitational waves (GW170817) [6] and the electromagnetic signals [7] from the neutron star merger found on 17 August. While the GW170817 was announced only after this meeting, we include this topic in this article because of its significance.

Of particular concern in this article are the constraints on equations of state through the neutron star mass–radius (M-R) relations. In principle, a precisely determined M-R relation can be used to directly reconstruct the neutron star equations of state [8], even without any knowledge about microscopic properties of the matter. Actually, the current precision of M-R relations is not good enough to pin down the unique equation of state. Nevertheless, the current constraints are already significant for us to derive qualitative and semi-quantitative understanding about the nature of dense QCD matter.

Based on equations of state supposed from the M-R and causality constraints, we will develop the picture of hadron–quark continuity in which hadronic matter continuously transforms into quark matter without experiencing thermodynamic phase transitions. Such continuity picture was developed in the context of the crossover from the superfluid hadronic phase to the color-flavor-locked superconducting phase [9]. This scenario was revisited in [10,11] where the role of $U_A(1)$ anomaly is emphasized. The previous studies are based on theoretical considerations and model calculations, while, in our approach, we reach the continuity picture from the demand to satisfy the neutron star constraints.

2. The Neutron Star Constraints and the Implications for QCD Equations of State

To begin with, we first define some terminology in this article. "Stiff" equations of state mean equations of state with large pressure P at given energy density ε. The stiffer equations of state generally lead to larger maximum masses and larger radii for neutron stars. We will not use the speed of sound $c_s = (\partial P/\partial \varepsilon)^{1/2}$ as the measure of the stiffness, as even ideal gas equations of state with the relatively small sound velocity (compared to what we will consider) can generate very large maximum masses.

Secondly, we should specify at which region of density the equations of state are stiff. We will use the terminology such as "soft-stiff", by which we mean that equations of state is soft at low density, $n_B \leq 2n_0$, and stiff at high density, $n_B \geq 5n_0$. For the reasons described below, equations of state leading to $R_{1.4} \leq 13$ km for $1.4 M_\odot$ stars will be called "soft at low density", and equations of state leading to $M \geq 2 M_\odot$ will be called "stiff at high density'. Then, the soft-stiff equations of state generate the M-R curves with the typical radii of $R_{1.4} \leq 13$ km and the maximum mass $\geq 2 M_\odot$.

The classification of equations of state by the baryon density is useful because it has been known [12] that the shapes of the M-R curves have strong correlations with equations of state at several fiducial densities (see Figure 1). At very low density, the material is loosely bound by the gravity, but, as M increases, R rapidly decreases because of the stronger gravity. Around $\sim 2n_0$, the matter starts to observe the repulsive forces in microscopic dynamics; then, the M-R curve starts to go vertically. Eventually, the curve reaches the maximum in M at $n_B \geq 5n_0$. Using these correlations between M-R and n_B, one can focus on the radius constraint in the studies of low density equations of state, or one can focus on the maximum mass when studying high density equations of state.

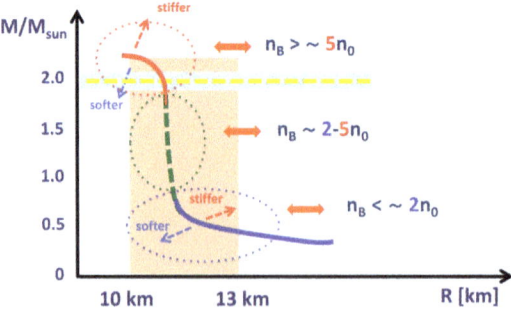

Figure 1. The correlation between the M-R relation and equations of state.

The existence of two-solar mass ($2M_\odot$) neutron stars [2,3] tells us that high density equations of state at $n_B \geq 5n_0$ must be stiff. Meanwhile the estimate of $R_{1.4}$ is relatively uncertain. There have been many theoretical predictions for $R_{1.4}$ which range from $\simeq 10$ km to $\simeq 16$ km. The observational constraints on $R_{1.4}$, which have been based on spectroscopic analyses of the X-rays from the neutron star surface, include several systematic uncertainties, but the current trend converges toward the estimate R = 11–13 km [1]. In addition, the analyses of gravitational waves from GW170817 favors

---

[1] The exception can be found in [13], where the authors (Suleimanov et al.) estimate $R_{1.4} > 13.9$ km using the X-ray burst from 4U 1724-307. The paper was published in 2011. Later, further analyses were done by two of the authors and their collaborators. In a recent paper [14], they discussed that the event used to extract $R_{1.4} > 13.9$ km is not suitable for reliable analyses due to large contaminations in the neutron star atmosphere. The newer analyses include more samples and cleaner events than the previous ones, and yield the estimate 11 km < R < 13 km for neutron stars with the masses ranging from 1.1–2.1$M_\odot$ [15]. The author appreciates Dr. David Blaschke for mentioning these papers.

equations of state with the radii smaller than ~13 km. More precisely, the actual constraint is on the dimensionless tidal deformability, $\Lambda = \frac{2}{3}k_2(R/G_N M)^5$ (G_N: Newton constant; k_2: Love number [16]), of each star before the coalescence; clearly Λ is very sensitive to the compactness and radius of the star.

Therefore, the QCD equation of state is likely to be the soft-stiff type. For the left over region $n_B = 2 - 5n_0$, there is also a causality constraint on the speed of sound $c_s^2 = \partial P/\partial \varepsilon$, i.e., c_s must be less than the light velocity [2]. This constraint becomes significant for soft-stiff equations of state because $P(\varepsilon)$ is small at low density but must be large at high density, meaning that in between there must be a region where $\partial P/\partial \varepsilon$ must be large. The difficulty is even more signified if there are the first order phase transitions, see Figure 2; during such transitions, $P(\varepsilon)$ is constant for increasing ε, and, after the phase transitions, even larger $\partial P/\partial \varepsilon$ is necessary to get connected to $P(\varepsilon)$ at high density.

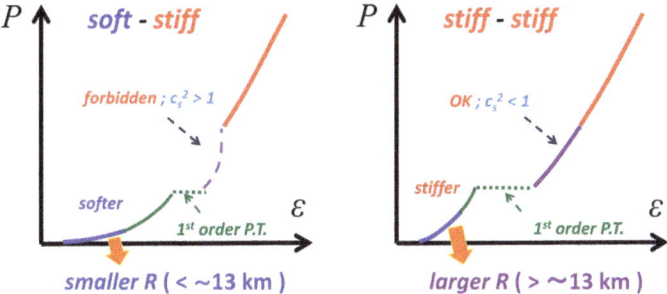

Figure 2. The pressure vs. energy density for soft-stiff (**left**) and stiff-stiff (**right**) equations of state. The slope is given by $\partial P/\partial \varepsilon = c_s^2$, the sound speed square, which must be smaller than the light speed, 1. The soft-stiff equations of state have smaller radii than the stiff-stiff ones, and disfavor the strong 1st order phase transitions.

If we assume the neutron star radii to be large > 13 km, then the equations of state at low density is so stiff that, even after strong 1st order phase transitions, the low density equations of state have the causal connection to $P(\varepsilon)$ at high density [17]. Thus, the determination of the neutron star radii is crucial for our understanding of the QCD phase structure. It should be evident that if the strength of transitions is sufficiently weak, the soft-stiff equations of state is still possible even with the 1st order transitions. For more quantitative and systematic analyses, we refer to Ref. [18].

These considerations for soft-stiff equations of state motivate us to consider the picture of hadron–quark continuity in which equations of state at $2n_0$ and $5n_0$ are continuously connected.

3. The 3-Window Modeling

Now, we turn to the discussions about the microscopic nature of matter. We consider the matter by decomposing it into 3-windows [19–22]; the nuclear regime at $n_B \leq 2n_0$; the crossover regime for $2n_0 - 5n_0$; and the quark matter regime at $n_B \geq 5n_0$. The picture we have is illustrated in Figure 3. At low density, $n_B \leq 2n_0$, the matter is dilute and baryons remain well-defined objects, so the equations of state are described by nuclear ones. Beyond ~$2n_0$, it is unlikely that nucleons are effective degrees of freedom; many-body forces become increasingly important as seen from microscopic nuclear calculations, which include nuclear interaction up to 3-body forces [23,24], and, in addition, typical calculations indicate that baryonic excitations other than nucleons are no longer negligible. Even though the matter is presumably not dense enough to consider quark matter,

---

[2] Some people postulated that the c_s^2 should be smaller than the conformal limit $c^2/3$ (c: light velotiy). As argued by Bedaque and Steiner, this hypothesis is in tension with the neutron star observations.

the above-mentioned problems demand us to think of matter based on microscopic quark degrees of freedom. At $n_B \sim 5n_0$, baryons with the radii of ~ 0.5 fm start to touch one another. If we assume a 3-flavor quark matter, the density $5n_0$ corresponds to the quark Fermi momentum of $p_F \sim 400$ MeV (for 2-flavor matter p_F is even larger), reasonably large compared to the QCD non-perturbative scale, $\Lambda_{QCD} \sim 200$ MeV.

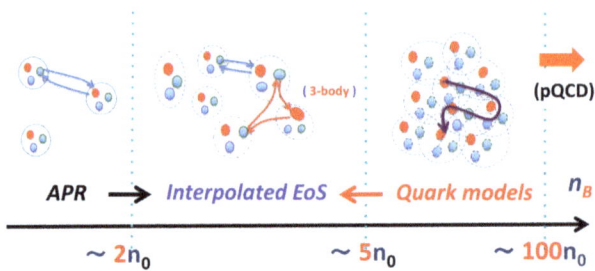

Figure 3. The 3-window modeling of the QCD matter.

One might think that, since some phenomenological hadronic equations of state have been made consistent with the $2M_\odot$ constraint (e.g., [23]), there is no need to introduce the quark matter descriptions for neutron star matter. However, to pass the $2M_\odot$ constraint is the necessary but not sufficient condition to validate the hadronic models; the construction of equations of state must be reasonable from the microscopic point of view, but, at this point, we have problems in extrapolating purely hadronic descriptions beyond $5n_0$, for the reasons already discussed above. This motivates us to start with quark matter picture at high density side and approach the hadronic side by including hadronic correlations. This approach, even when $\sim 5n_0$ happens to not be high for the quark matter formation, at least will shed light on the nature of hadronic matter in terms of quark descriptions.

We will construct equations of state based on this 3-window picture. For the nuclear regime, we use the Akmar–Phandheripande–Ravenhall (APR) equation of state as a representative [3] [23]. For the quark matter regime, we use a schematic quark model that concisely expresses microscopic interactions relevant in hadron and nuclear physics. In between, neither purely hadronic nor quark matter descriptions are appropriate, so here we use the hadron–quark continuity picture to smoothly interpolate the APR and quark model equations of state. Specifically, our interpolation is done with polynomials [21]

$$P(\mu_B) = \sum_{n=0}^{5} c_n \mu_B^n. \quad (1)$$

To determine the coefficients c_ns, we first compute $n_B = \partial P / \partial \mu_B$, and then demand, at $n_B = 2n_0$ and $5n_0$, the interpolating function to match with the APR and quark equations of state up to the second order derivatives of $P(\mu_B)$.

4. A Model for Quark Matter

In our phenomenological modeling, we need to choose a quark model for $n_B \geq 5n_0$. Guided by the continuity picture, the form of effective models is exported from those for hadron physics. Here, semi-long range interactions, relevant for the energy scale of 0.2–1 GeV or distance scale ~ 0.2–1 fm^{-1} [25], should remain important from low to high densities because the quark matter regime

---

[3] Actually, we also need to use some crust equations of state for $n_B < 0.2 - 0.5n_0$. We use the Togashi equation of state [24], which is based on the microscopic calculations with techniques similar to the APR, and is consistent with the regime of laboratory nuclei below the neutron drip regime.

observes the contents inside of hadrons. Meanwhile, due to the overlap of baryon wavefunctions, the confining forces that try to neutralize the color are expected to be less important at higher densities, except for any excitations that break the local color neutrality. The confining force is a long range interaction relevant for the energy scale $\Lambda_{QCD} \sim 0.2\,\text{GeV} \sim 1\,\text{fm}^{-1}$.

Our effective Hamiltonian is ($\mu_q = \mu_B/3$) [1]

$$\mathcal{H} = \bar{q}(i\gamma_0\vec{\gamma}\cdot\vec{\partial} + m - \mu_q\gamma_0)q - G_s\sum_{i=0}^{8}\left[(\bar{q}\tau_i q)^2 + (\bar{q}i\gamma_5\tau_i q)^2\right] + 8K(\det_f \bar{q}_R q_L + \text{h.c.}) \\ + \mathcal{H}_{\text{conf}}^{3q\to B} - H\sum_{A,A'=2,5,7}(\bar{q}i\gamma_5\tau_A\lambda_{A'}C\bar{q}^T)(q^T Ci\gamma_5\tau_A\lambda_{A'}q) + g_V(\bar{q}\gamma^\mu q)^2. \quad (2)$$

The first line is the standard Nambu-Jona-Lasinio (NJL) model with u,d,s-quarks and responsible for the chiral symmetry breaking. We use the Hatsuda–Kunihiro parameter set [26] with which the constitutent quark masses are $M_{u,d} \simeq 336\,\text{MeV}$ and $M_s \simeq 528\,\text{MeV}$. The first term in the second line includes the confining interactions which trap 3-quarks into a baryon. The second term is the color-magnetic interaction for color-flavor-antisymmetric S-wave channel; they play very important roles in the level splitting in the hadron spectra, e.g., N-Δ splitting. The last term is the phenomenological vector repulsive interactions, which are inspired from the ω-meson exchange in nuclear physics. In actual calculations, the confining term is not explicitly included as we do not know a good modeling for it. Therefore, we restrict the use of this model to $n_B \geq 5n_0$ where we expect that confining effects are not significant.

While the form of the Hamiltonian is obtained by extrapolating the description of hadron and nuclear physics, in principle the range of parameters (G_s, K, g_V, H) at $n_B \geq 5n_0$ can be considerably different from those used in hadron physics due to, e.g., medium screening effects. In a strongly correlated region, the estimate of medium modifications is difficult; for instance, screening masses in 2-color QCD, measured in lattice QCD [27], are qualitatively different from the perturbative behaviors [28]. For 3-color QCD, no reliable estimates on medium modifications are available, so here we use the neutron star constraints to examine the range of these parameters, and then use them to delineate the properties of QCD matter at $n_B \geq 5n_0$. Below, we vary (g_V, H), while assuming that (G_s, K) do not change from the vacuum values appreciably; this assumption will be checked posteriori. More elaborated treatment is to explicitly determine the medium running coupling $g_V(\mu_B)$, as demonstrated in Ref. [29].

Our Hamiltonian for quarks, together with the contributions from leptons, is solved within the mean field (MF) approximation. We impose the neutrality conditions for electric and color charges as well as the β-equilibrium condition. In the MF treatments, we find that the chiral and diquark condensates coexist at $n_B \geq 5n_0$. For the range of parameters that we have explored, the diquark pairing always appears to be the color-flavor-locked (CFL) type at $n_B \geq 5n_0$; other less symmetric pairings such as the 2SC type appear only at lower density, thus we will not take their appearance at face value.

Now, we examine the roles of effective interactions by subsequently adding g_V and then H to the standard NJL model [21]. First of all, in order to make equations of state stiff, $(G_s, K)_{@5n_0}$ should remain comparable to the size of its vacuum values; the large reduction of these parameters accelerates the chiral restoration that yields contributions similar to the bag constant, i.e., the positive (negative) contributions to energy (pressure). As a result, the significant softening takes place in equations of state. Actually, even if we fix $(G_s, K)_{@5n_0}$ to the vacuum values, the strong 1st order chiral transition takes place at $n_B \sim 2$–$3\,n_0$ in the standard NJL model, so the equations of state at $n_B \geq 5n_0$ is too soft to pass the $2M_\odot$ constraint.

This situation is changed by adding g_V. It stiffens the equations of state in two-fold ways. Firstly, the repulsive interactions obviously contribute to the stiffening. Secondly, it delays the chiral restoration by tempering the growth of baryon density as a function of μ_B, so that there is no radical softening associated with the chiral restoration. In fact, the 1st order transition turns into a crossover in the range of g_V we explored. The value of g_V large enough to pass the $2M_\odot$ constraint, however, causes another kind of problem in connecting the APR and quark model pressure; see the left panel of Figure 4; with larger g_V quark pressure, $P(\mu_B)$ tends to appear at higher μ_B with less slope, and, as a consequence, the pressure curve in the interpolation region tends to contain an inflection point at which $\partial^2 P/(\partial \mu_B)^2$ is negative. Such region is thermodynamically unstable and so must be excluded. Therefore, while a larger value of g_V is favored to pass the $2M_\odot$ constraint, it generates more mismatch between the APR and quark pressure in the μ_B direction.

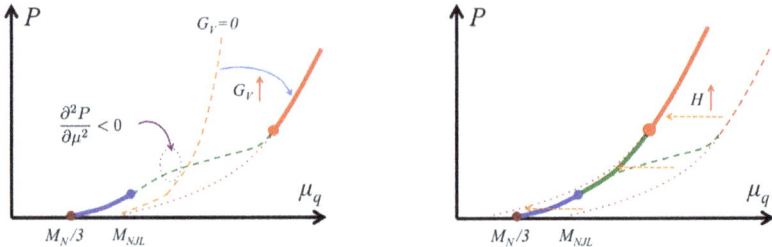

Figure 4. The impacts of the vector and color-magnetic interactions.

Here, the color-magnetic interactions improve the situation; see the right panel of Figure 4. We note that the onset chemical potential of the APR pressure is the nucleon mass $\mu_B \simeq 939$ MeV, while, for the NJL pressure, it is $\mu_B \simeq 3M_{u,d} \simeq 1018$ MeV. In a conventional picture of quark models, the nucleon and Δ masses are split by the color-magnetic interaction, and the nucleon mass is reduced from $3M_{u,d}$. From this viewpoint, the color-magnetic interactions naturally induce the overall shift of the NJL pressure toward the lower chemical potential, thus making the matching between the APR and quark pressure curves much better.

The M-R relations are shown in Figure 5 for the parameter sets $(g_V, H)/G_s = (0.5, 1.4), (0.8, 1.5)$, and $(1.0, 1.6)$. For all these sets, the radius of a neutron star at the canonical mass $1.4M_\odot$ is 11.3–11.5 km, mainly determined by our APR equations of state. In these sets, only the set $(0.8, 1.5)$ fulfills all of the constraints; the set $(0.5, 1.4)$ is slightly below the $2M_\odot$ constraint, while $(1.0, 1.6)$ slightly violates the causality bound. More exhaustive parameter surveys [1] show that g_V should be $>\sim 0.7\ G_s$, and $H >\sim 1.4\ G_s$ which are comparable to the vacuum scalar coupling. For given g_V, the value of H is fixed to $\sim 10\%$; in fact, we do not have much liberty in our choice when we connect the APR and quark matter pressures.

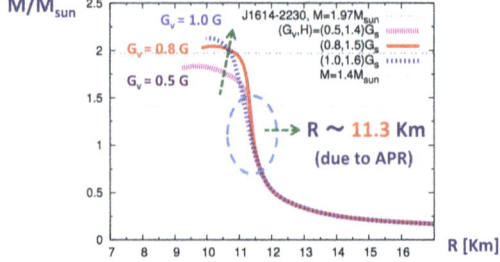

Figure 5. The mass–radius relations from the 3-window equations of state for sets of parameters, $(g_V, H)/G_s = (0.5, 1.4), (0.8, 1.5), (1.0, 1.6)$. Only the set $(0.8, 1.5)$ satisfies the $2M_\odot$ and causality constraints.

We note that the couplings (G_s, g_V, H) as large as the vacuum coupling of G_s are necessary to fulfill the constraints from neutron star observations and causality. With such strong effective couplings, we expect that gluons in the non-perturbative regime still survive in spite of the presence of quark matter. In addition, substantial amounts of chiral and diquark condensates coexist [21]. It is also important to emphasize that the quark matter contains the strange-quarks as much as up- and down-quarks.

5. Discussion and Conclusion

We first mention the difference between the finite temperature crossover and low temperature crossover (see Figure 6). The relevant thermodynamic relations are $P = Ts - \varepsilon$ (s: entropy) and $P = \mu_B n_B - \varepsilon$, respectively. In the finite temperature crossover, which has been established by the lattice Monte Carlo calculations [30,31], the QCD matter changes from a hadron resonance gas to a quark gluon plasma as the temperature increases. The transition is smooth, but radical changes take place in the thermodynamic quantities. In particular, there is radical growth in the entropy and energy densities as a consequence of liberation of quarks and gluons, which in turn lead to a dip in the speed of sound c_s. In contrast, this feature is not present in the low temperature crossover; the sound velocity should have a peak, rather than a dip, in the crossover region [1]. Neither the baryon density nor energy density radically change; instead, as the matter approaches the crossover region, the strong interactions among baryons temper the growth of the baryon density at increasing μ_B. In this respect, the distinction between strongly interacting hadronic and quark matter is more difficult than that between a hadron resonance gas and a quark gluon plasma. It may be appropriate to characterize the hadron–quark crossover in terms of the quark–hadron duality, or in the context of quarkyonic matter [32–35] that has the quark Fermi sea but baryonic Fermi surface; hence, it naturally interpolates the hadronic and quark matter. To get qualitative insights for the quarkyonic matter, we refer to the studies of QCD in (1+1) dimensions [36] where analytic insights are available.

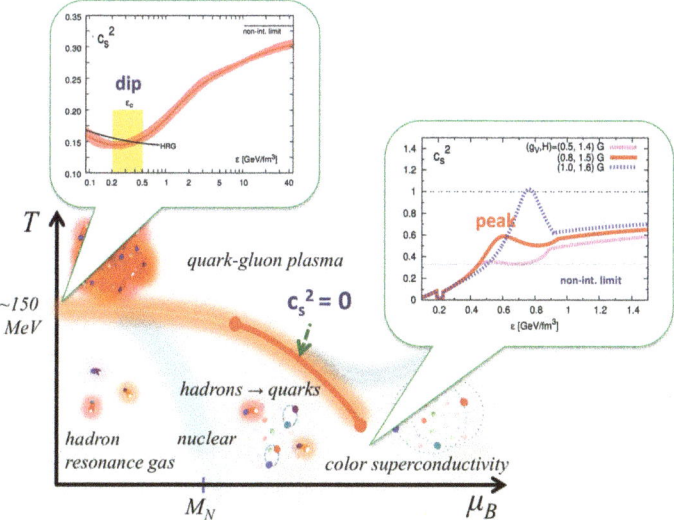

Figure 6. The speed of sound square c_s^2 around the finite temperature crossover from hadron resonance gas to quark gluon plasma, on the possible first order chiral restoration line, and around the possible low temperature crossover from hadron to quark matter. While the finite temperature crossover has a dip in c_s^2, the low temperature crossover has a peak.

Finally, we present a conjecture concerning the crossover in the gauge dynamics, namely from a confining phase to a Higgs phase with colored diquark condensates $\sim\phi = |\phi|e^{i\theta}$. This question must be addressed when we consider the crossover from hadronic to quark matter with diquark condensates. In the presence of matter fields in the fundamental representations, there are no strict order parameters based on symmetries, and in fact the two phases can be smoothly connected [37]. On the other hand, the symmetry concepts are not necessary conditions for phase transitions, as can be seen in liquid–gas phase transition. Thus, we need to discuss the dynamical aspects. As for the confining vs. Higgs phases, there are two important elements to distinguish them in qualitative terms. The first is the strength of the gauge coupling, g_s, and the other is the size of Higgs field (diquark) amplitudes, $|\phi|$. Two extreme limits are relatively easy to imagine: in the strong coupling limit $g_s \gg 1$ and small Higgs amplitudes, the strong color fluctuations disfavor the colored objects and the confinement takes place [38]. Meanwhile, in the weak coupling limit $g_s \ll 1$ and large Higgs amplitudes, the matter should look like a Higgs phase as in textbook examples. The important question is how they can be connected depending on the trajectories of the g_s and $|\phi|$ as functions of μ_B.

This question is hard to answer for dense QCD, but some insights can be obtained from the gauge Higgs models with the fixed Higgs amplitude $|\phi|$ [37]. There are two characteristic paths from the confining to the Higgs phase (see Figure 7). In the first path, we move along the small $|\phi|$ region in the confining phase; move from $(g_s \gg 1, |\phi| \sim 0)$ to $(g_s \ll 1, |\phi| \sim 0)$ domain, and then go to the domain of $(g_s \ll 1, |\phi| \gg 1)$. In this path, we hit the phase transition increasing the value of $|\phi|$ at the weak coupling region. Indeed, it is difficult to imagine that the Higgs phase at weak coupling, which apparently looks very different from the confining phase, continuously transforms into the confining phase. The other path, however, allows the crossover transition: starting again from $g_s \gg 1, |\phi| \sim 0$, one can move along the $g_s \gg 1$ region with increasing $|\phi|$, and reaches the confining phase at large Higgs fields, or Higgs phase at strong coupling. This regime was not studied as much as the weak coupling regime in quark matter.

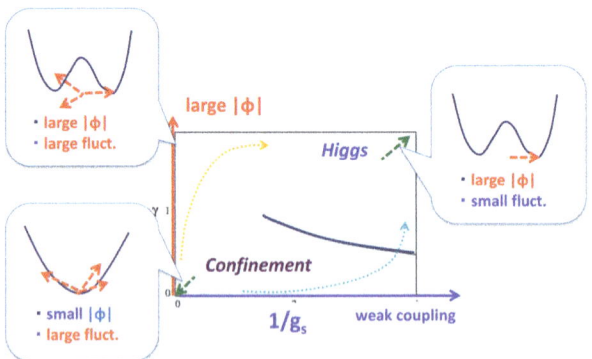

Figure 7. The phase diagram for gauged-Higgs model with the fixed Higgs amplitudes, in the $1/g_s - |\phi|$ plane. At large coupling, the confining and Higgs phases are smoothly connected.

From this example and neutron star constraints, we conjecture that the matter remains strongly coupled from hadronic to quark matter regimes, so that the Higgs fields develop within the confining regime and then the system gradually relaxes to the Higgs phase at weak coupling. Gluons remain non-perturbative until the weak coupling regime is reached at sufficiently high density.

More observational constraints will come in the next 10 years through the timing analyses of X-rays in the NICER program [39] and the GW detection by currently operating aLIGO, Virgo, GEO [40], and also KAGRA [41] under construction, which will be ready soon. The electromagnetic counterparts associated with the GWs give the information about the ejecta, from which one can learn the dynamics at the coalescence regime. It is desirable to utilize all this information to improve our understanding of dense QCD matter.

Acknowledgments: The author thanks Gordon Baym, Kenji Fukushima, Tetsuo Hatsuda, Phillip Powell, Yifan Song, Tatsuyuki Takatsuka for collaboration, and D. Blaschke for information about references for neutron star radii. T.K. is grateful for the workshop organizers and participants who made this workshop very enjoyable. This work is supported by an NSFC grant 11650110435.

Conflicts of Interest: The author declares no conflicts of interest.

References

1. Baym, G.; Hatsuda, T.; Kojo, T.; Powell, P.D.; Song, Y.; Takatsuka, T. From hadrons to quarks in neutron stars: A review. *arXiv* **2017**, arXiv:1707.04966.
2. Demorest, P.; Pennucci, T.; Ransom, S.; Roberts, M.; Hessels, J. Shapiro Delay Measurement of A Two Solar Mass Neutron Star. *Nature* **2010**, *467*, 1081–1083.
3. Antoniadis, J.; Freire, P.C.C.; Wex, N.; Tauris, T.M.; Lynch, R.S.; van Kerkwijk, M.H.; Kramer, M.; Bassa, C.; Dhillon, V.S.; Driebe, T.; et al. A Massive Pulsar in a Compact Relativistic Binary. *Science* **2013**, *340*, 1233232.
4. Ozel, F.; Freire, P. Masses, Radii, and Equation of State of Neutron Stars. *Ann. Rev. Astron. Astrophys.* **2016**, *54*, 401–440.
5. Steiner, A.W.; Lattimer, J.M.; Brown, E.F. Neutron Star Radii, Universal Relations, and the Role of Prior Distributions. *Eur. Phys. J. A* **2016**, *52*, 18.
6. Abbott, B.P.; Abbott, R.; Abbott, T.D.; Acernese, F.; Ackley, K.; Adams, C.; Adams, T.; Addesso, P.; Adhikari, R.X.; Adya, V.B.; et al. GW170817: Observation of Gravitational Waves from a Binary Neutron Star Inspiral. *Phys. Rev. Lett.* **2017**, *119*, 161101.
7. Abbott, B.P.; Abbott, R.; Abbott, T.D.; Acernese, F.; Ackley, K.; Adams, C.; Adams, T.; Addesso, P.; Adhikari, R.X.; Adya, V.B.; et al. Multi-messenger observations of a binary neutron star merger. *Astrophys. J. Lett.* **2017**, *848*, L1–L59.
8. Lindblom, L. Determining the nuclear equation of state from neutron-star masses and radii. *Astrophys. J.* **1992**, *398*, 569–573.
9. Schäfer, T.; Wilczek, F. Continuity of quark and hadron matter. *Phys. Rev. Lett.* **1999**, *82*, 3956–3959.
10. Hatsuda, T.; Tachibana, M.; Yamamoto, N.; Baym, G. New critical point induced by the axial anomaly in dense QCD. *Phys. Rev. Lett.* **2006**, *97*, 122001.
11. Zhang, Z.; Fukushima, K.; Kunihiro, T. Number of the QCD critical points with neutral color superconductivity. *Phys. Rev. D* **2009**, *79*, 014004.
12. Lattimer, J.M.; Prakash, M. Neutron Star Observations: Prognosis for Equation of State Constraints. *Phys. Rep.* **2007**, *442*, 109–265.
13. Suleimanov, V.; Poutanen, J.; Revnivtsev, M.; Werner, K. Neutron star stiff equation of state derived from cooling phases of the X-ray burster 4U 1724-307. *Astrophys. J.* **2011**, *742*, 122.
14. Nättilä, J.; Steiner, A.W.; Kajava, J.J.E.; Suleimanov, V.F.; Poutanen, J. Equation of state constraints for the cold dense matter inside neutron stars using the cooling tail method. *Astron. Astrophys.* **2016**, *591*, A25.
15. Nättilä, J.; Miller, M.C.; Steiner, A.W.; Kajava, J.J.E.; Suleimanov, V.F.; Poutanen, J. Neutron star mass and radius measurements from atmospheric model fits to X-ray burst cooling tail spectra. *Astron. Astrophys.* **2017**, *608*, A31.
16. Hinderer, T.; Lackey, B.D.; Lang, R.N.; Read, J.S. Tidal deformability of neutron stars with realistic equations of state and their gravitational wave signatures in binary inspiral. *Phys. Rev. D* **2010**, *81*, 123016.
17. Benic, S.; Blaschke, D.; Alvarez-Castillo, D.E.; Fischer, T.; Typel, S. A new quark-hadron hybrid equation of state for astrophysics—I. High-mass twin compact stars. *Astron. Astrophys.* **2015**, *577*, A40.
18. Alford, M.G.; Han, S.; Prakash, M. Generic conditions for stable hybrid stars. *Phys. Rev. D* **2013**, *88*, 083013.
19. Masuda, K.; Hatsuda, T.; Takatsuka, T. Hadron-Quark Crossover and Massive Hybrid Stars with Strangeness. *Astrophys. J.* **2013**, *764*, 12.
20. Masuda, K.; Hatsuda, T.; Takatsuka, T. Hadron-quark crossover and massive hybrid stars. *Prog. Theor. Exp. Phys.* **2013**, *2013*, 073D01.
21. Kojo, T.; Powell, P.D.; Song, Y.; Baym, G. Phenomenological QCD equation of state for massive neutron stars. *Phys. Rev. D* **2015**, *91*, 045003.
22. Kojo, T. Phenomenological neutron star equations of state: 3-Window modeling of QCD matter. *Eur. Phys. J. A* **2016**, *52*, 51.

23. Akmal, A.; Pandharipande, V.R.; Ravenhall, D.G. The Equation of state of nucleon matter and neutron star structure. *Phys. Rev. C* **1998**, *58*, 1804.
24. Togashi, H.; Nakazato, K.; Takehara, Y.; Yamamuro, S.; Suzuki, H.; Takano, M. Nuclear equation of state for core-collapse supernova simulations with realistic nuclear forces. *Nucl. Phys. A* **2017**, *961*, 78–105.
25. Manohar, A.; Georgi, H. Chiral Quarks and the Nonrelativistic Quark Model. *Nucl. Phys. B* **1984**, *234*, 189–212.
26. Hatsuda, T.; Kunihiro, T. QCD phenomenology based on a chiral effective Lagrangian. *Phys. Rep.* **1994**, *247*, 221–367.
27. Hajizadeh, O.; Boz, T.; Maas, A.; Skullerud, J.I. Gluon and ghost correlation functions of 2-color QCD at finite density. *arXiv* **2017**, arXiv:1710.06013.
28. Kojo, T.; Baym, G. Color screening in cold quark matter. *Phys. Rev. D* **2014**, *89*, 125008.
29. Fukushima, K.; Kojo, T. The Quarkyonic Star. *Astrophys. J.* **2016**, *817*, 180.
30. Aoki, Y.; Endrodi, G.; Fodor, Z.; Katz, S.D.; Szabo, K.K. The Order of the quantum chromodynamics transition predicted by the standard model of particle physics. *Nature* **2006**, *443*, 675–678.
31. Ding, H.T.; Karsch, F.; Mukherjee, S. Thermodynamics of strong-interaction matter from Lattice QCD. *Int. J. Mod. Phys. E* **2015**, *24*, 1530007.
32. McLerran, L.; Pisarski, R.D. Phases of cold, dense quarks at large N(c). *Nucl. Phys. A* **2007**, *796*, 83.
33. Kojo, T.; Hidaka, Y.; McLerran, L.; Pisarski, R.D. Quarkyonic Chiral Spirals. *Nucl. Phys. A* **2010**, *843*, 37–58.
34. Kojo, T.; Pisarski, R.D.; Tsvelik, A.M. Covering the Fermi Surface with Patches of Quarkyonic Chiral Spirals. *Phys. Rev. D* **2010**, *82*, 074015.
35. Kojo, T.; Hidaka, Y.; Fukushima, K.; McLerran, L.D.; Pisarski, R.D. Interweaving Chiral Spirals. *Nucl. Phys. A* **2012**, *875*, 94–138.
36. Kojo, T. A (1+1) dimensional example of Quarkyonic matter. *Nucl. Phys. A* **2012**, *877*, 70.
37. Fradkin, E.H.; Shenker, S.H. Phase Diagrams of Lattice Gauge Theories with Higgs Fields. *Phys. Rev. D* **1979**, *19*, 3682–3697.
38. Wilson, K.G. Confinement of Quarks. *Phys. Rev. D* **1974**, *10*, 2445–2459.
39. Gendreau, K.C.; Arzoumanian, Z.; Adkins, P.W.; Albert, C.L.; Anders, J.F.; Aylward, A.T.; Baker, C.L.; Balsamo, E.R.; Bamford, W.A.; Benegalrao, S.S.; et al. The Neutron star Interior Composition Explorer (NICER): Design and development. *Proc. SPIE* **2016**, *9905*, 99051H.
40. Hough, J.; Meers, B.J.; Newton, G.P.; Robertson, N.A.; Ward, H.; Leuchs, G.; Niebauer, T.M.; Rüdiger, A.; Schilling, R.; Schinupp, L.; et al. Proposal for a Joint German-British Interferometric Gravitational Wave Detector. Available online: eprints.gla.ac.uk/114852/7/114852.pdf (accessed on 3 September 2017).
41. Aso, Y.; Michimura, Y.; Somiya, K.; Ando, M.; Miyakawa, O.; Sekiguchi, T.; Tatsumi, D.; Yamamoto, H. Interferometer design of the KAGRA gravitational wave detector. *Phys. Rev. D* **2013**, *88*, 043007.

© 2018 by the author. Licensee MDPI, Basel, Switzerland. This article is an open access article distributed under the terms and conditions of the Creative Commons Attribution (CC BY) license (http://creativecommons.org/licenses/by/4.0/).

Article

Towards a Unified Quark-Hadron-Matter Equation of State for Applications in Astrophysics and Heavy-Ion Collisions

Niels-Uwe F. Bastian [1,*], David Blaschke [1,2,3], Tobias Fischer [1] and Gerd Röpke [3,4]

1. Institute of Theoretical Physics, University of Wroclaw, 50-137 Wroclaw, Poland; david.blaschke@gmail.com (D.B.); tobias.fischer@ift.uni.wroc.pl (T.F.)
2. Bogoliubov Laboratory for Theoretical Physics, Joint Institute for Nuclear Research, 141980 Dubna, Russia
3. National Research Nuclear University (MEPhI), 115409 Moscow, Russia; gerd.roepke@uni-rostock.de
4. Institute of Physics, University of Rostock, 18051 Rostock, Germany
* Correspondence: niels-uwe@bastian.science

Received: 26 April 2018; Accepted: 16 May 2018; Published: 25 May 2018

Abstract: We outline an approach to a unified equation of state for quark-hadron matter on the basis of a Φ−derivable approach to the generalized Beth-Uhlenbeck equation of state for a cluster decomposition of thermodynamic quantities like the density. To this end we summarize the cluster virial expansion for nuclear matter and demonstrate the equivalence of the Green's function approach and the Φ−derivable formulation. As an example, the formation and dissociation of deuterons in nuclear matter is discussed. We formulate the cluster Φ−derivable approach to quark-hadron matter which allows to take into account the specifics of chiral symmetry restoration and deconfinement in triggering the Mott-dissociation of hadrons. This approach unifies the description of a strongly coupled quark-gluon plasma with that of a medium-modified hadron resonance gas description which are contained as limiting cases. The developed formalism shall replace the common two-phase approach to the description of the deconfinement and chiral phase transition that requires a phase transition construction between separately developed equations of state for hadronic and quark matter phases. Applications to the phenomenology of heavy-ion collisions and astrophysics are outlined.

Keywords: cluster virial expansion; quark-hadron matter; Mott dissociation; Beth-Uhlenbeck equation of state; heavy-ion collisions; supernova explosions; mass-twin stars

---

1. Introduction

The development of a unified equation of state (EoS) for quark-hadron matter, where the hadrons are not elementary degrees of freedom but rather appear as composites of their quark constituents (bound and scattering states of effective quark interactions) is a formidable task because it requires the implementation of dynamical mechanisms of quark confinement as well as chiral symmetry breaking which characterise the hadronic phase of matter and mechanisms for deconfinement and chiral symmetry restoration which determine the transition to the quark(-gluon) phase of strongly interacting matter.

The aspect of bound state formation and dissociation has been well studied in warm dense nuclear matter, where a cluster virial expansion has been developed as a most effective means of description of a system that contains clusters of different sizes with internal quantum numbers like in the nuclear statistical equilibrium picture, but generalizes it by including residual binary interactions among them (second cluster virial coefficient) and their dissociation due to compression and heating within the Mott effect. The consistent description of bound and scattering states on the same footing is provided by the generalized Beth-Uhlenbeck equation of state that uses phase shifts in order to describe correlations and their modifications in a hot, dense environment. It turns out that for the short-ranged strong interactions it is not the screening of the interaction which drives the dissociation of the bound states but rather the Pauli blocking effect that inhibits cluster formation when the phase space is densely populated. Being based on pure symmetry arguments, this effect is sufficiently general to apply to any fermionic many-particle system where—as a function of the density—the formation of bound state gets first favored (principle of Le Chatelier and Brown) until the Pauli principle inhibits scattering processes that would lead to cluster formation and a homogeneous phase of fermionic quasiparticles emerges: The dense nuclear matter.

Since nucleons themselves can be considered as clusters of quarks, the more fundamental degrees of freedom of quantum chromodynamics (QCD) as the underlying gauge field theory of strong interactions, we want to develop in this work the basics of a corresponding approach that will describe the transition from hadronic to quark matter as a Mott transition driven by the Pauli blocking effect. Despite this striking analogy there are also specifics for the quark degrees of freedom which need to be taken into account. These are mainly the internal quantum numbers (color, flavor, spin) that cause the confinement of colored states (quarks, gluons, diquarks etc.), the dynamical chiral symmetry breaking and combined symmetry requirements as well as relativistic kinematics.

In Figure 1 we show the phase diagram of QCD according to the present state of knowledge. In particular it addresses the aspect of cluster formation and dissociation in dense quark-hadron matter, which is the main goal of the approach to be outlined in this work.

To this end in Section 2 we first review the cluster virial approach for nuclear matter in the form of a generalized Beth-Uhlenbeck EoS. Then in Section 3 we suggest that it might be obtained from the Φ−derivable approach in its field theoretic formulation when for the Φ functional the class of all two-loop Feynman diagramns is chosen that can be constructed from cluster Green's functions and cluster T-matrices and consequently take the form of generalized "sunset" type diagrams.

In Section 4 we develop the approach for quark matter with mesons and baryons as clusters dominating the hadronic phase of matter since quarks and diquarks are suppressed by confining interactions that give rise to diverging selfenergies for those states in the low-density (confinement) phase. This mechanism is realized here within the so-called string-flip model that is generalized in relativistic form. As a first step, a selfconsistent mean-field approximation is performed which results in a new quark matter equation of state with confining features, thus superior to previous NJL model approaches and their Polyakov-loop generalized versions. We demonstrate how the quark Pauli-blocking effect for hadrons is already inherent in the Φ−derivable approach and can be made apparent by a perturbation expansion with respect to cluster selfenergies on top of the quasiparticle approximation. The Pauli-blocking effect can be mimicked by a hadronic excluded volume which should therefore be taken into account when a hadronic EoS is extrapolated from a low-density limit (hadron resonance gas, nuclear statistical equilibrium) to the vicinity of the deconfinement transition. In the quark matter phase, we implement higher order repulsive interactions which may be justified by multi-pomeron exchange interactions, thus non-perturbative effects of the gluon sector which we are not treating dynamically at this stage. These effects cause a stiffening of the high-density quark matter EoS that are essential for hybrid compact star applications.

In Section 5 we discuss the role of nuclear clusters for the astrophysics of supernovae and the deconfinement transition in compact stars while in Section 6 we consider the potential of a unified approach to quark-hadron matter for applications in heavy-ion collisions. We point out that nuclear cluster formation may occur directly at the hadronisation transition and thus result in their (sudden) chemical freeze-out together with other hadronic species as indicated by the ALICE experiment at the CERN LHC.

In Section 7 we draw conclusions for the further development of the approach towards a unified description of quark-hadron matter and its phenomenology in heavy-ion collisions and astrophysics.

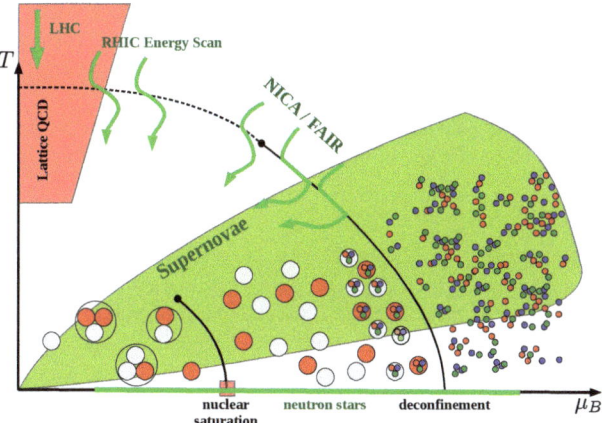

Figure 1. Schematic phase diagram of strongly interacting matter with the solid lines indicating first order liquid-gas and hadronic-quark matter phase transitions ending in dots symbolizing the critical endpoints. While for the nuclear liquid-gas phase transition the critical point is at a temperature of $T_c^{LG} = 16.6 \pm 0.9$ MeV [1], the position of the critical endpoint of the deconfinement transition is not known. The dashed line indicates the pseudocritical temperature $T_c(\mu_B)$ of the crossover transition from hadronic to quark matter which has been determined at vanishing baryochemical potential $\mu_B = 0$ in lattice QCD simulations to $T_c(0) = 154 \pm 9$ MeV [2]. The red and white colored circles stand for neutrons and protons which at low densities can form nuclear clusters that get dissociated in the liquid phase. Similar to that case, nucleons themselves can be viewed as clusters of quarks symbolized by colored dots that get dissociated in the quark matter phase. The green shaded area is the domain of the phase diagram accessible by supernova explosions. The red shaded area is the region accessible by ab-initio simulations of QCD at finite temperatures and chemical potentials on space-time lattices. The green arrowed lines symbolize trajectories of heavy-ion collisions at different center of mass energies ranging form LHC over RHIC to the planned NICA/FAIR experiments. The green line on the chemical potential axis depicts the range of values accessible in compact star interiors. In this case, the third dimension of the charge (or isospin) chemical potential relevant for isospin asymmetric systems in supernovae, compact stars and their mergers is suppressed.

2. Quantum Statistical Approach and the Cluster Virial Expansion

A systematic approach to the cluster expansion of thermodynamic properties is obtained from quantum statistics. We consider a system consisting of species i with the (conserved) particle number N_i and the corresponding chemical potential μ_i, described by the hamiltonian H at equilibrium with temperature T. The index i will be dropped in the following. The grand canonical thermodynamic potential is given by

$$\Omega = -PV = -T \ln \mathrm{Tr}\ e^{-(H-\mu N)/T}, \tag{1}$$

where P is the pressure and V the volume. It can be represented by diagrams within a perturbation expansion [3,4], see also [5]. We have

$$P = \frac{1}{V}\mathrm{Tr}\ln[-G_1^{(0)}] - \frac{1}{2V}\int_0^1 \frac{d\lambda}{\lambda}\mathrm{Tr}\Sigma_\lambda G_\lambda, \tag{2}$$

or

$$P = P_0 - \frac{1}{2V}\int_0^1 \frac{d\lambda}{\lambda}\left\{ \begin{array}{c}\end{array} + \begin{array}{c}\end{array} + \begin{array}{c}\end{array} + \begin{array}{c}\end{array} + \begin{array}{c}\end{array} + \begin{array}{c}\end{array} + \ldots \right\} \tag{3}$$

where λ is a scaling factor substituting the two-particle interaction V_2 by $\lambda\, V_2$. $G_1^{(0)}$ is the free single-particle propagator that gives the ideal part of the pressure P_0. The full single-particle Green function G_λ and the self-energy Σ_λ are taken with the coupling constant λ. Depending on the selected diagrams, different approximations can be found. In particular, the second virial coefficient for charged particle systems has been investigated, see Ref. [5].[1]

An alternative way to derive the equation of state is to start from the expression for the total nucleon density

$$n_{\tau_1}(T, \mu_p, \mu_n) = \frac{2}{V} \sum_{\mathbf{p}_1} \int_{-\infty}^{\infty} \frac{d\omega}{2\pi} f_1(\omega) S_1(1, \omega), \qquad (4)$$

where V is the system volume, the variable $1 = \{\mathbf{p}_1, \sigma_1, \tau_1\}$ denotes the single particle states (here nucleons in momentum representation), $\tau_1 = n, p$, and summation over spin direction is collected in the factor 2. Both the Fermi distribution function and the spectral function depend on the temperature and the chemical potentials μ_p, μ_n, which are not given explicitly. The spectral function $S_1(1,\omega)$ of the single-particle Green function $G_1(1, iz_\nu)$ is related to the single-particle self-energy $\Sigma(1,z)$ according to

$$S_1(1,\omega) = \frac{2\mathrm{Im}\,\Sigma_1(1, \omega - i0)}{(\omega - E^{(0)}(1) - \mathrm{Re}\,\Sigma_1(1,\omega))^2 + (\mathrm{Im}\,\Sigma_1(1, \omega - i0))^2}, \qquad (5)$$

where the imaginary part has to be taken for a small negative imaginary part in the frequency; $E^{(0)}(1) = p_1^2/(2m_1)$.

Both approaches are equivalent. The perturbation expansion can be represented by Feynman diagrams, see Equation (3). However, a finite order perturbation theory will not produce bound states, and partial summations of an infinite number of e.g., ladder diagrams must be performed to get bound states. As shown by Baym and Kadanoff [3,4], self-consistent approximations to the one-particle Green function can be given based on a functional Φ so that

$$\Sigma_1(1, 1') = \frac{\delta \Phi}{\delta G_1(1, 1')}. \qquad (6)$$

Different approximations for the generating functional Φ are discussed in the following Sections 3 and 4. The self-consistent Φ−derivable approximations not only lead to a fully-conserving transport theory. In the equilibrium case they also have the property that different methods to obtain the grand partition function such as integrating the expectation value of the potential energy with respect to the coupling constant λ, Equation (3), or integrating the density n with respect to the chemical potential μ, lead to the same result [6,7]. In particular, with

$$\Omega = -\mathrm{Tr}\,\ln(-G_1) - \mathrm{Tr}\Sigma_1 G_1 + \Phi \qquad (7)$$

also

$$n = -\frac{\partial \Omega}{\partial \mu} \qquad (8)$$

holds in the considered approximation.

The latter approach (using Equation (4)) has been extensively used in many-particle systems [8–12], in particular in connection with the chemical picture. The main idea of the chemical picture is to treat bound states on the same footing as "elementary" single particles. This way one describes correctly

---

[1] Note that the interaction in nuclear systems is strong. However, the perturbation expansion is performed with respect to the imaginary part of the self-energy that is assumed to be small. Most of the interaction is already taken into account in the self-consistent determination of the quasiparticle energies. With increasing density, the Fermi energy will dominate the potential energy so that the correlations are suppressed. A quasiparticle description can be used to calculate the nuclear structure.

the low-temperature, low-density region of many-body systems where bound states dominate. Within a quantum statistical approach, the chemical picture is realized considering the A-particle propagator. In ladder approximation, see Figure 2, the Bethe-Salpeter equation (BSE)

$$G_A^{\text{ladder}}(1\ldots A;1'\ldots A';z_A) = G_A^0(1\ldots A;z_A)\delta_{\text{ex}}(1\ldots A;1'\ldots A') \\ + \sum_{1''\ldots A''} G_A^0(1\ldots A;;z_A)V_A(1\ldots A;1''\ldots A'')G_A^{\text{ladder}}(1''\ldots A'';1'\ldots A';z_A), \quad (9)$$

is obtained, where z_A is the A-particle Matsubara frequency, G_A^0 is the product of single-particle propagators, $V_A(1\ldots A;1''\ldots A'') = \frac{1}{2}\sum_{i,j} V_2(ij,i''j'')\prod_{k\neq i,j}\delta(k,k'')$ is the A-particle interaction. $\delta_{\text{ex}}(1\ldots A;1'\ldots A')$ describes the antisymmetrization of the A-particle state.

The BSE is equivalent to the A-particle wave equation. Neglecting all medium effects we have

$$[E^{(0)}(1)+\cdots+E^{(0)}(A)]\psi_{A\nu P}(1\ldots A)+\sum_{1'\ldots A'}V_A(1\ldots A;1'\ldots A')\Psi_{A\nu P}(1'\ldots A') \\ = E_{A,\nu}^{(0)}(P)\Psi_{A\nu P}(1\ldots A), \quad (10)$$

where ν indicates the internal quantum number of the A-particle eigen states, including the channel c describing spin and isospin state. Different excitations are possible in each channel c, in particular bound states and scattering states.

Figure 2. Ladder approximation G_A^{ladder} for $A=2$. Iteration gives the infinite sum of ladder diagrams.

The chemical picture uses the eigen-representation of the A-particle propagator. In ladder approximation, neglecting medium effects,

$$G_A^{\text{ladder},0}(1\ldots A;1'\ldots A';z_A) = \sum_\nu \langle 1\ldots A|\psi_{A\nu P}\rangle \frac{1}{z_A - E_{A,\nu}^{(0)}(P)}\langle \psi_{A\nu P}|1'\ldots A'\rangle. \quad (11)$$

In particular, it contains the contribution of A-particle bound states (nuclei) similar to the "elementary" single particle propagator

$$G_1^{(0)}(1,z) = \frac{1}{z-E_1^{(0)}(p_1)}. \quad (12)$$

Within the representation of the perturbative expansion by Feynman diagrams, the chemical picture implies that diagrams containing single-particle propagators should be completed by adding A-particle propagators, at least for the bound states ν. However, also the scattering parts ν have to be considered for a full description.

Coming back to Equation (5), different approximations for the self-energy $\Sigma_1(1,z)$ can be considered. The main issues are the implementation of bound state formation which is of relevance in the low-density region, and the account for density effect if considering higher densities, such as mean-field approximations. To give some systematics with respect to the different approximations calculating the self-energy and the corresponding approximation for the EoS, in Ref. [13] the following overview was presented.

As seen in Table 1, to consider density effects, the ideal gas approximation is improved by the mean-field approximation. Quasiparticle shifts are introduced in a semi-empirical way within the relativistic mean-field (RMF) approximation[2] or considering Skyrme parametrizations.

---

[2] A Padé approximation of the nucleon quasiparticle shifts, applicable in a wide temperature range, can be found in Ref. [14].

This approximation is assumed to give an adequate description of dense matter near and above the saturation density but fails to describe correlations, in particular bound state formation, in the low density region.

Table 1. Aspects of the quantum statistical description of a many-particle system with bound and scattering states in the low-density limit (left column) and its modification due to medium effects at high densities (right column).

Low Density Limit	High Density Modification (Medium Effects)
(1) elementary particles	
Ideal Fermi gas: neutrons, protons (electrons, neutrinos,...)	*Quasiparticle quantum liquid:* mean-field approximation Skyrme, Gogny, RMF
(2) bound state formation	
Nuclear statistical equilibrium: ideal mixture of all bound states chemical equilibrium, mass action law	*Chemical equilibrium of quasiparticle clusters:* medium modified bound state energies self-energy and Pauli blocking
(3) continuum contributions	
Second virial coefficient: account of continuum correlations ($A = 2$) scattering phase shifts, Beth-Uhlenbeck Eq.	*Generalized Beth-Uhlenbeck formula:* medium modified binding energies, medium modified scattering phase shifts
(4) chemical & physical picture	
Cluster virial approach: all bound states (clusters) scattering phase shifts of all pairs	*Correlated medium:* phase space occupation by all bound states in-medium correlations, quantum condensates

For this, the chemical picture is necessary, which is realized by a cluster decomposition of the self-energy, see [10]. In particular, considering only the bound state part of the free A-particle propagator, the nuclear statistical equilibrium (NSE) is obtained. However, also the scattering part of the free A-particle propagator must be considered to obtain the correct second virial coefficient in the case $A = 2$, as given by the Beth-Uhlenbeck formula [15]. A cluster Beth-Uhlenbeck formula has been worked out [16] which considers the chemical picture, where scattering states between two arbitrary clusters A and B are taken into account.

The inclusion of density effects for these approximations, as collected in Table 1, is not simple. A generalized Beth-Uhlenbeck formula has been worked out in Ref. [12]. In this approach, the Pauli blocking and the dissolution of bound states with increasing density is shown.

A more detailed approach should also take into account that the medium is not considered as uncorrelated, but correlations in the medium have to be taken into account. The cluster-mean-field (CMF) approximation has been worked out [11] where also clustering of the medium is treated. This is of relevance in the region where correlations and bound state formation dominate.

These effects have been considered recently [17] in warm dense matter. Improving the NSE, medium modifications of the bound states are discussed. For a consistent description, in particular of the correct second virial coefficient, scattering phase shifts have been implemented. If clustering is relevant, the medium cannot be approximated by an uncorrelated, ideal Fermion gas of quasiparticles, but correlations must be taken into account.

Here we focus on the cluster virial expansion [16]. We consider density effects such as single-particle quasiparticle shifts and Pauli blocking, which are responsible for the dissolution of bound states and the appearance of a new state of matter. In the present section we treat nuclear clusters which are dissolved to form a nuclear Fermi liquid when density increases. Later on we go a step forward to investigate the dissolution of hadrons by medium effects to form the quark-gluon plasma.

As shown in [17], using the cluster decomposition of the self-energy which takes into account, in particular, cluster formation, we obtain

$$n_n^{tot}(T,\mu_n,\mu_p) = \frac{1}{V}\sum_{A,\nu,P} N f_{A,Z}[E_{A,\nu}(P;T,\mu_n,\mu_p)],$$

$$n_p^{tot}(T,\mu_n,\mu_p) = \frac{1}{V}\sum_{A,\nu,P} Z f_{A,Z}[E_{A,\nu}(P;T,\mu_n,\mu_p)], \quad (13)$$

where **P** denotes the center of mass (c.o.m.) momentum of the cluster (or, for $A=1$, the momentum of the nucleon). The internal quantum state ν contains the proton number Z and neutron number $N = A - Z$ of the cluster,

$$f_{A,Z}(\omega;T,\mu_n,\mu_p) = \frac{1}{\exp[(\omega - N\mu_n - Z\mu_p)/T] - (-1)^A} \quad (14)$$

is the Bose or Fermi distribution function for even or odd A, respectively, that is depending on $\{T,\mu_n,\mu_p\}$. The integral over ω is performed within the quasiparticle approach, the quasiparticle energies $E_{A,\nu}(P;T,\mu_n,\mu_p)$ are depending on the medium characterized by $\{T,\mu_n,\mu_p\}$. Expressions for the in-medium modifications are given in [17].

In Equation (13) the sum is to be taken over the mass number A of the cluster, the center-of-mass momentum **P**, and the intrinsic quantum number ν. The summation over ν concerns the bound states as far as they exist, as well as the continuum of scattering states. Solving the few-body problem what is behind the calculation of the A-nucleon T matrices in the Green function approach, we can introduce different channels (c) characterized, e.g., by spin and isospin quantum numbers. We assume that these channels decouple. In contrast to the angular momentum which is not conserved, e.g., for tensor forces, the contribution of different channels to Equation (13) is additive. The remaining intrinsic quantum numbers will be denoted by ν_c, it concerns the bound states as far as they exist (ground states and excited states), as well as the continuum of scattering states.

We analyze the contributions of the clusters ($A \geq 2$), suppressing the thermodynamic variables $\{T,\mu_n,\mu_p\}$. We have to perform the integral over the c.o.m. momentum **P** what, in general, must be done numerically since the dependence of the in-medium quasiparticle energies $E_{A,\nu}(P;T,\mu_n,\mu_p)$ on **P** is complicated. We have in the non-degenerate case $[\sum_P \to V/(2\pi)^3 \int d^3P]$

$$\frac{1}{V}\sum_{\nu,P} f_{A,Z}[E_{A,\nu}(P)] = \sum_c e^{(N\mu_n+Z\mu_p)/T} \int \frac{d^3P}{(2\pi)^3} \sum_{\nu_c} g_{A,\nu_c} e^{-E_{A,\nu_c}(P)/T} = \sum_c \int \frac{d^3P}{(2\pi)^3} z_{A,c}(P) \quad (15)$$

with $g_{A,c} = 2s_{A,c} + 1$ the degeneration factor in the channel c. The partial density of the channel c at **P** contains the intrinsic partition function

$$z_{A,c}(P;T,\mu_n,\mu_p) = e^{(N\mu_n+Z\mu_p)/T}\left\{\sum_{\nu_c}^{\text{bound}} g_{A,\nu_c} e^{-E_{A,\nu_c}(P)/T} \Theta[-E_{A,\nu_c}(P) + E_{A,c}^{\text{cont}}(P)]\right\} + z_{A,c}^{\text{cont}}(P). \quad (16)$$

It can be decomposed in the bound state contribution and the contribution of scattering states $z_{A,c}^{\text{cont}}(P;T,\mu_n,\mu_p)$. We emphasize that the subdivision of the intrinsic partition function into a bound and a scattering contribution is artificial and not of physical relevance.

The summations over A, c and **P** of (16) remain to be done for the EOS (13), and Z may be included in c. The region in the parameter space, in particular **P**, where bound states exist, may be restricted what is expressed by the step function $\Theta(x) = 1, x \geq 0;\ = 0$ else. The continuum edge of scattering states is denoted by $E_{A,c}^{\text{cont}}(P; T, \mu_n, \mu_p)$. The intrinsic partition function $z_{A,c}^{\text{cont}}(P)$ contains the scattering state contribution (non degenerate case)

$$z_{A,c}^{\text{cont}}(P) = \int_0^\infty \frac{dE}{2\pi} e^{\left(-E - P^2/(2M_A) + N\mu_n + Z\mu_p\right)/T} 2\sin^2\delta_c(E) \frac{d\delta_c(E)}{dE} . \tag{17}$$

Going beyond the ordinary Beth-Uhlenbeck formula [15], for nuclear matter the generalized Beth-Uhlenbeck formula has been worked out in Ref. [12]. Here, the single-particle contribution is described by quasiparticles, and to avoid double counting, the corresponding mean-field term must be subtracted from the contribution of scattering states what leads to the appearance of the term $\sin^2 \delta_c(E)$.

The NSE follows as a simple approximation where the sum is performed only over the bound states, and medium effects are neglected. We obtain the model of a mixture of non-interacting bound clusters, which can react so that chemical equilibrium is established. This approximation may be applicable for the low-density, low-temperature region where the components are nearly freely moving, and intrinsic excitations are not of relevance. However, the continuum of scattering states (the intrinsic quantum number ν_c may contain the relative momentum) are of relevance at increasing temperature, which is clearly seen in the exact Beth-Uhlenbeck expression [15] for the second virial coefficient. Another argument to take the continuum of scattering states into account is the dissolution of bound states at increasing density. Here, the thermodynamic properties behave smoothly because the contribution of the bound state to the intrinsic partition function is replaced by the corresponding contribution of the scattering states.

It turns out advantageous for analyses of the thermodynamics of the Mott transition, to avoid the separation into a bound and a scattering state part of the spectrum and to include the discrete part of the spectrum into the definition of the phase shifts, so that these are merely parameters of a polar representation of the complex A-particle cluster propagator. The partition function then takes the form

$$z_{A,c}(P; T, \mu_n, \mu_p) = \int_{-\infty}^\infty \frac{dE}{2\pi} e^{\left(-E - P^2/(2M_A) + N\mu_n + Z\mu_p\right)/T} 2\sin^2\delta_c(E) \frac{d\delta_c(E)}{dE} . \tag{18}$$

As an example let us consider the case $A = 2$. A consistent description of the medium effects should contain not only the mean-field shift of the quasiparticle energies but also the Pauli blocking, as shown in this work for conserving approximations. Within the generalized Beth-Uhlenbeck approach [12], Pauli blocking modifies the binding energy of the bound state (deuteron) in the isospin-singlet channel as well as the scattering phase shifts. The integrand of the intrinsic partition function is shown in Figure 3.

The intrinsic partition function $z_{A,c}^{\text{cont}}(P)$ cannot be divided unambiguously into a bound state contribution and a contribution of scattering states. As seen from Figure 3, the sum of both contributions behaves smoothly when with increasing density the bound state is dissolved. Therefore, we emphasize that one should consider only the total intrinsic partition function including both, bound and scattering contribution, as expressed by the generalized phase shifts as shown in Figure 3.

Whereas the two-body problem can be solved, e.g., using separable interaction potentials, and the account of medium effects has been investigated [12], the evaluation of the contribution of clusters with mass numbers $A > 2$ to the EoS is challenging. The medium modification of the cluster binding energies has been calculated, e.g., using a variational approach [17]. Problematic is the inclusion of scattering states, in particular the treatment of different channels describing the decomposition of the A-particle cluster.

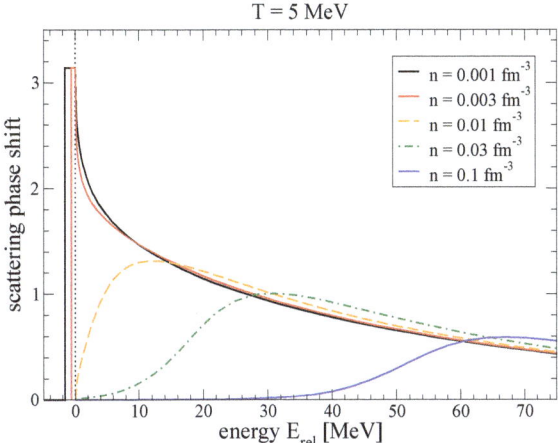

Figure 3. Integrand of the intrinsic partition function as function of the intrinsic energy in the deuteron channel. Different densities of the medium are considered, the temperature is $T = 5$ MeV. From [18].

In this work, we discuss the concept of a cluster-virial expansion and corresponding generalized cluster-Beth Uhlenbeck approaches. This concept is based on the chemical picture where bound states are considered as new components and can be treated in the same way as "elementary" particles. It is generally accepted that a second virial coefficient can be introduced for systems consisting of atoms, but also with molecules as components. The problem is the introduction of an effective interaction between the components (including the quantum symmetry postulates for fermions or bosons), and intrinsic excitations of the bound components are described in some approximations. Within the NSE, we describe a nuclear system as a mixture of "free" nucleons in single-particle states as well as of clusters (nuclei with $A > 1$). Taking into account the interaction between these components of a nuclear system, from measured scattering phase shifts (for instance $\alpha - \alpha$ scattering) a virial EoS can be derived. Results have been presented in Ref. [19].

On a more microscopic level, we consider here interacting quarks which can form bound states (hadrons), and the general approach should include both cases, the region of the quark-gluon plasma and, after the confinement transition, the region of well established hadrons. A main difficulty is the introduction of an effective interaction which can be made by fitting empirical data. However, a systematic quantum statistical approach is needed to derive such effective interaction from a fundamental Lagrangian, and to introduce the cluster states performing consistent approximations and avoiding double counting of the contributions to the EoS and other physical properties.

The Green function technique as well as the path-integral approach are such systematic quantum statistical approaches. Different contributions to the EoS are represented by Feynman diagrams, and double counting is clearly excluded. A selection of diagram classes can be performed to recover the chemical picture. As seen from Equation (11), after separation of the center-of-mass momentum $\vec{P}$, the propagator for the A-particle bound states is

$$G_{A,\nu}^{\text{bound}}(1\ldots A; 1'\ldots A'; z_A) = \langle 1 \ldots A | \psi_{A\nu P} \rangle \frac{1}{z_A - E_{A,\nu}^{(0)}(P)} \langle \psi_{A\nu P} | 1' \ldots A' \rangle \quad (19)$$

where ν covers only the bound state part of the internal quantum state of the A-particle cluster. As a new element, the bound state propagator is introduced as indicated in Figure 4. This bound state propagator has the same analytical form like the single particle propagator (12), besides the appearance of the internal wave function that determines the vertex function.

Figure 4. Splitting of the A-particle cluster propagator into a bound and scattering contribution. Note that the internal quantum number has been dropped.

As discussed above, different approximations are obtained such as the nuclear statistical equilibrium (NSE) and the cluster mean-field (CMF) approximation using the chemical picture. These approximations are based on the bound state part of the A-particle propagator. They give leading contributions in the low-density, low temperature range where bound states dominate the composition of the many-particle system.

To improve the approximation, the remaining part of continuum correlations has also been taken into account. From the point of view of the physical picture, these contributions arise in higher orders of the virial expansion of the equation of state. As example, the formation of the A-particle bound state is seen in the A-th virial coefficient, the mean-field shift due to a cluster B in the $(A+B)$-th virial coefficient. The chemical picture indicates which high-order virial coefficients of the virial expansion are essential, if the many-particle system is strongly correlated so that bound states are formed.

In an improved approximation, the scattering part of the A-particle propagator has to be considered. It contributes also to the A-th virial coefficient. The scattering processes within the A-particle system can have different channels. As an example we discuss here binary elastic scattering processes between sub-clusters A_1 and A_2 of the system of A particles, $A = A_1 + A_2$. Binary phase shifts $\delta_{A_1,A_2}(E)$ are introduced that describe the corresponding scattering experiments. They can also be calculated within few-body theory. Besides the effective interaction between the sub-clusters that are depending on the internal wave function of the sub-clusters, also virtual transitions to excited states may be taken into account. In general, the effective interaction is non-local in space and time, i.e., momentum and frequency dependent.

A generalized cluster Beth-Uhlenbeck formula, see Equation (16), is obtained when in particle loops not the free propagator, but quasiparticle Green's functions are used. If the quasiparticle shift is calculated in Hartree-Fock approximation, the first order term of the interaction must be excluded from the ladder T_2^{ladder} matrix to avoid double counting. The bound state part is not affected, it is determined by an infinite number of diagrams. The scattering part is reduced subtracting the Born contribution as shown in Equation (18) by the $2\left[\sin(\delta_c)\right]^2$ term; for the derivation see Ref. [12].

The continuum correlations that are not considered in the NSE give a contribution to the second virial coefficient in the chemical picture. We can extract from the continuum part two contributions [17]: resonances that can be treated like new particles in the law of mass action, and the quasiparticle shift of the different components contributing to the law of mass action. Both processes are expected to represent significant contributions of the continuum. After projecting out these effects, the residual contribution of the two-nucleon continuum is assumed to be reduced. One can try to parametrize the residual part, using the ambiguity in defining the bound state contribution. Eventually the residual part of the continuum correlations can be neglected.

The correct formulation of the cluster-virial expansion [16] is considered to be a main ingredient towards a unified EoS describing quark matter as well as nuclear matter. One has to introduce the interaction between the constituents in a systematic way. The account of correlations in the continuum is essential near the confinement phase transition where the hadronic bound states disappear. Conserving approximations may lead to acceptable results for the EoS. An important issue is the account of correlation in the medium, in particular when considering the Pauli blocking in the phase space of the elementary constituents.

3. Φ—Derivable Approach to the Cluster Virial Expansion for Nuclear Matter

Recently, it has been suggested [20] that the cluster virial expansion for many-particle systems [16] can be formulated within the Φ–derivable approach [3,4]. This approach is straightforwardly generalized to A–particle correlations in a many-fermion system

$$\Omega = \sum_{l=1}^{A} \Omega_l = \sum_{l=1}^{A} \left\{ c_l \left[\text{Tr} \ln \left(-G_l^{-1} \right) + \text{Tr} \left(\Sigma_l \, G_l \right) \right] + \sum_{\substack{i,j \\ i+j=l}} \Phi[G_i, G_j, G_{i+j}] \right\}, \quad (20)$$

where the full A–particle Green's function obeys the Dyson equation

$$G_A^{-1} = G_A^{(0)^{-1}} - \Sigma_A, \quad (21)$$

where $G_A^{(0)}$ is the free A–particle Green's function and the selfenergy is defined as a functional derivative of the two-cluster irreducible Φ functional

$$\Sigma_A(1 \ldots A, 1' \ldots A', z_A) = \frac{\delta \Phi}{\delta G_A(1 \ldots A, 1' \ldots A', z_A)}. \quad (22)$$

This generalization of the Φ–derivable approach fulfills by its construction the conditions of stationarity of the thermodynamical potential with respect to variations of the cluster Green's functions

$$\frac{\delta \Omega}{\delta G_A(1 \ldots A, 1' \ldots A', z_A)} = 0. \quad (23)$$

The Φ functional for our purpose of defining a cluster decomposition of the system with inclusion of residual interactions among the clusters captured by a second virial coefficient is given by a sum of all two-loop diagrams that can be drawn with cluster Green's functions G_{i+j} and their subcluster Green's functions G_i and G_j for a given bipartition with the appropriate vertex functions $\Gamma_{i+j;ij}$. This generalization of the so-called "sunset" diagram case is depicted diagrammatically in Figure 5.

Figure 5. Left panel: The Φ functional for the general case of A–particle correlations in a many-fermion system, whereby all bipartitions $A = i + j$ into lower order clusters of sizes i and j shall be considered; Right panel: Equivalent representation of the diagram in the left panel with the highest order cluster Green's function and the vertex functions replaced by the cluster T matrix, see Figure 6.

Herewith we have generalized the notion of the Φ–derivable approach to that of a system where the hierarchy of higher order Green functions is built successively from the tower of all Greens functions starting with the fundamental one G_1. The open question is how to define the vertex functions joining the cluster Greens functions. As a heuristic first step one may always introduce local coupling constants, like in the Lee model discussed in the context of a Φ–derivable approach by Weinhold et al. [6]. Introducing nonlocal formfactors at the vertices will correspond to a separable representation of the interaction. The definition of the interaction can be absorbed in the introduction of the cluster T-matrix in the corresponding channel which, in the ladder approximation, will reproduce the analytical properties encoded in the full cluster Green's function concerning bound state poles and scattering state continuum. This equivalence is depicted in Figure 6.

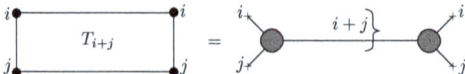

Figure 6. Diagrammatic representation for the replacement of the higher order Green's function G_{i+j} and the corresponding vertex functions in the Φ functional for the cluster virial expansion by the T_{i+j} matrix for binary collisions in the channel with the partition i, j.

The T_A matrix fulfills the Bethe-Salpeter equation in ladder approximation

$$T_{i+j}(1,2,\ldots,A;1',2',\ldots A';z) = V_{i+j} + V_{i+j}G^{(0)}_{i+j}T_{i+j}, \quad (24)$$

which in the separable approximation for the interaction potential,

$$V_{i+j} = \Gamma_{i+j}(1,2,\ldots,i;i+1,i+2,\ldots,i+j)\Gamma_{i+j}(1',2',\ldots,i';(i+1)',(i+2)',\ldots,(i+j)'), \quad (25)$$

leads to the closed expression for the T_A matrix

$$T_{i+j}(1,2,\ldots,i+j;1',2',\ldots(i+j)';z) = V_{i+j}\left\{1-\Pi_{i+j}\right\}^{-1}, \quad (26)$$

where the generalized polarization function

$$\Pi_{i+j} = \text{Tr}\left\{\Gamma_{i+j}G^{(0)}_i\Gamma_{i+j}G^{(0)}_j\right\} \quad (27)$$

has been introduced and the one-frequency free i−particle Green's function is defined by the $(i-1)$-fold Matsubara sum

$$\begin{aligned}G^{(0)}_i(1,2,\ldots,i;\Omega_i) &= \sum_{\omega_1\ldots\omega_{i-1}}\frac{1}{\omega_1-E(1)}\frac{1}{\omega_2-E(2)}\cdots\frac{1}{\Omega_i-(\omega_1+\ldots\omega_{i-1})-E(i)}\\&= \frac{(1-f_1)(1-f_2)\ldots(1-f_i)-(-)^i f_1 f_2 \ldots f_i}{\Omega_i-E(1)-E(2)-\ldots E(i)}.\end{aligned} \quad (28)$$

Note that for these Green's functions holds the relationship ($\Omega_{i+j} = \Omega_i + \Omega_j$)

$$G^{(0)}_{i+j} = G^{(0)}_{i+j}(1,2,\ldots,i+j;\Omega_{i+j}) = \sum_{\Omega_i} G^{(0)}_i(1,2,\ldots,i;\Omega_i) G^{(0)}_j(i+1,i+2,\ldots,i+j;\Omega_j). \quad (29)$$

Another set of useful relationships follows from the fact that in the ladder approximation both, the full two-cluster ($i+j$ particle) T matrix (26) and the corresponding Greens' function

$$G_{i+j} = G^{(0)}_{i+j}\left\{1-\Pi_{i+j}\right\}^{-1} \quad (30)$$

have similar analytic properties determined by the $i+j$ cluster polarization loop integral (27) and are related by the identity

$$T_{i+j}G^{(0)}_{i+j} = V_{i+j}G_{i+j}, \quad (31)$$

which is straightforwardly proven by multiplying Equation (26) with $G^{(0)}_{i+j}$ and using Equation (30). Since these two equivalent expressions in Equation (31) are at the same time equivalent to the two-cluster irreducible Φ functional introduced above in Equations (20) and Figure 5, the functional relations

$$T_{i+j} = \delta\Phi/\delta G_{i+j}^{(0)}, \tag{32}$$

$$V_{i+j} = \delta\Phi/\delta G_{i+j} \tag{33}$$

follow and may become useful in proving cancellations that are essential for the relationship to the Generalized Beth-Uhlenbeck approach as discussed below.

3.1. Generalized Beth-Uhlenbeck EoS from the Φ–Derivable Approach

Now we return to the question how the relation between the cluster Φ–derivable approach to the partition Function (20) and the generalized Beth-Uhlenbeck equation for the cluster density may be established. To this end we consider the partial density of the A–particle state defined as

$$n_A(T,\mu) = -\frac{\partial \Omega_A}{\partial \mu}. \tag{34}$$

Taking into account that any analytic complex function $F(\omega)$ has the spectral representation

$$F(iz_n) = \int_{-\infty}^{\infty} \frac{d\omega}{2\pi} \frac{\mathrm{Im} F(\omega)}{\omega - iz_n}, \tag{35}$$

we perform the Matsubara summation[3] in Equation (20)

$$\sum_{z_n} \frac{c_A}{\omega - iz_n} = f_A(\omega) = \frac{1}{\exp[(\omega - \mu)/T] - (-1)^A}. \tag{36}$$

Using the relation $\partial f_A(\omega)/\partial \mu = -\partial f_A(\omega)/\partial \omega$ we get for Equation (34) now

$$n_A(T,\mu) = -d_A \int \frac{d^3q}{(2\pi)^3} \int \frac{d\omega}{2\pi} f_A(\omega) \frac{\partial}{\partial \omega} \left[\mathrm{Im}\ln\left(-G_A^{-1}\right) + \mathrm{Im}\left(\Sigma_A G_A\right)\right] + \sum_{\substack{i,j \\ i+j=A}} \frac{\partial \Phi[G_i, G_j, G_A]}{\partial \mu}, \tag{37}$$

where a partial integration over ω has been performed and the degeneracy factor d_A for cluster state has been introduced, stemming from the trace operation in the internal spaces.

Now we use the fact that for two-loop diagrams of the sunset type a cancellation holds [21,22] which we generalize here for cluster states

$$d_A \int \frac{d^3q}{(2\pi)^3} \int \frac{d\omega}{2\pi} f_A(\omega) \frac{\partial}{\partial \omega}(\mathrm{Re}\Sigma_A\, \mathrm{Im} G_A) - \sum_{\substack{i,j \\ i+j=A}} \frac{\partial \Phi[G_i, G_j, G_A]}{\partial \mu} = 0. \tag{38}$$

Using generalized optical theorems [8,12] we can show that

$$\frac{\partial}{\partial \omega}\left[\mathrm{Im}\ln\left(-G_A^{-1}\right) + \mathrm{Im}\Sigma_A\, \mathrm{Re} G_A\right] = 2\mathrm{Im}\left[G_A\, \mathrm{Im}\Sigma_A \frac{\partial}{\partial \omega} G_A^* \,\mathrm{Im}\Sigma_A\right] = -2\sin^2\delta_A \frac{\partial \delta_A}{\partial \omega}, \tag{39}$$

where the phase shifts δ_A have been introduced via the polar representation of the complex A–particle propagator $G_A = |G_A|\exp(i\delta_A)$. With these ingredients follows from the cluster Φ–derivable approach the cluster virial expansion for the density in the form of a generalized Beth-Uhlenbeck EoS

---

[3] For odd A, $z_n = (2n+1)\pi T + \mu$ are the fermionic Matsubara frequencies and for even A, $z_n = 2n\pi T + \mu$ the bosonic ones.

$$n(T,\mu) = \sum_{i=1}^{A} n_i(T,\mu) = \sum_{i=1}^{A} d_i \int \frac{d^3q}{(2\pi)^3} \int \frac{d\omega}{2\pi} f_i(\omega) 2 \sin^2 \delta_i \frac{\partial \delta_i}{\partial \omega} \,. \tag{40}$$

In this way we have drawn the connection between the cluster virial expansion of Ref. [16] reviewed in the previous section with the Φ−derivable approach [3,4]. In the following subsection we consider the example of deuterons in nuclear matter in order to elucidate the application of the approach.

3.2. Deuterons in Nuclear Matter

Within the Φ−derivable approach [3,4] the grand canonical thermodynamic potential for a dense fermion system with two-particle correlations is given as

$$\Omega = -\text{Tr}\{\ln(-G_1)\} - \text{Tr}\{\Sigma_1 G_1\} + \text{Tr}\{\ln(-G_2)\} + \text{Tr}\{\Sigma_2 G_2\} + \Phi[G_1, G_2] \,, \tag{41}$$

where the full propagators obey the Dyson-Schwinger equations

$$G_1^{-1}(1,z) = z - E_1(p_1) - \Sigma_1(1,z); \quad G_2^{-1}(12,1'2',z) = z - E_1(p_1) - E_2(p_2) - \Sigma_2(12,1'2',z), \tag{42}$$

with selfenergies

$$\Sigma_1(1,1') = \frac{\delta \Phi}{\delta G_1(1,1')} \,; \quad \Sigma_2(12,1'2',z) = \frac{\delta \Phi}{\delta G_2(12,1'2',z)} \,, \tag{43}$$

which are defined by the choice for the Φ functional, a two-particle irreducible set of diagrams such the ones in Figure 7.

Figure 7. (**Left**) panel: Two-particle irreducible Φ functional describing two-particle correlations (double line with arrow) of elementary fermions (single arrowed lines); (**Right**) panel: Cluster Hartree approximation, following from replacing the two-cluster T matrix in the Φ functional of Figure 5 with the two-cluster potential V_{i+j}.

The functional for the thermodynamic potential (41) is constructed such that the requirement of its stationarity,

$$\frac{\partial \Omega}{\partial G_1} = \frac{\partial \Omega}{\partial G_2} = 0 \,, \tag{44}$$

in thermodynamic equilibrium is equivalent to Equation (4)

$$n = -\frac{1}{V}\frac{\partial \Omega}{\partial \mu} = \frac{1}{V}\sum_1 \int_{-\infty}^{\infty} \frac{d\omega}{\pi} f_1(\omega) S_1(1,\omega) \,, \tag{45}$$

where $S_1(1,\omega) = 2\Im G_1(1,\omega + i\eta)$ is the fermion spectral function and Equation (45) expresses particle number conservation in a system with volume V.

Having introduced the notion of a cluster expansion of the Φ functional we want to suggest a definition which eliminates the unknown vertex functions in favour of the T_{A+B} matrix which describes the nonperturbative binary collisions of $A-$ and $B-$ particle correlations in the channel $A + B$, see Figure 6. The application of this scheme to the simplest case of two-particle correlations in the deuteron channel in nuclear matter results in the selfenergy [12]

$$\Sigma(1,z) = \sum_{2} \int \frac{d\omega}{2\pi} S(2,\omega) \left\{ f(\omega) V(12,12) - \int \frac{dE}{\pi} \Im T(12,12; E + i\eta) \frac{f(\omega) + g(E)}{E - z - \omega} \right\}, \quad (46)$$

where $f(\omega) = [\exp(\omega/T) + 1]^{-1}$ is the Fermi function and $g(\omega) = [\exp(\omega/T) - 1]^{-1}$ the Bose function. The decomposition (46) corresponds to a cluster decomposition of the nucleon density

$$n(\mu, T) = n_{\mathrm{qu}}(\mu, T) + 2 n_{\mathrm{corr}}(\mu, T), \quad (47)$$

where the correlation density

$$n_{\mathrm{corr}} = \int \frac{dE}{2\pi} g(E) 2 \sin^2 \delta(E) \frac{d\delta(E)}{dE}, \quad (48)$$

contains besides the bound state a scattering state contribution as can be seen from examining the derivative of the phase shift shown in Figure 3. The one-particle density of free quasiparticle nucleons $n_q u$ is reduced in order to fulfil the baryon number conservation in the presence of deuteron correlations and contains a selfenergy contribution due to the deuteron correlations in the medium. This improvement of the quasiparticle picture due to the correlated medium accounted for by the consistent definition of the selfenergy as a derivative of the Φ Functional (20) is the reason the continuum correlations (48) are reduced by the factor $2 \sin^2 \delta$ as compared to the traditional Beth-Uhlenbeck formula [15,23]. For details, see [8,12]. With the definition of the Φ functional via the T matrix in Figure 6 we were able to show the correspondence between the generalized Beth-Uhlenbeck approach and the Φ-derivable approach for the nonrelativistic potential model approach to two-particle correlations in a warm, dense Fermion system [12,16]. Now we would like to discuss its application to a relativistic model for correlations in quark matter: mesons, diquarks and baryons.

4. Cluster Virial Expansion for Quark-Hadron Matter within the Φ Derivable Approach

Finally, we would like to sketch how the Φ derivable approach can be employed to define a cluster virial expansion for quark-hadron matter consisting of quarks (Q), mesons (M), diquarks (D) and baryons (B) that can represent a unified quark-hadron matter EoS. The thermodynamical potential for this system obtains the form very similar to the case of clustered nuclear matter, i.e.,

$$\Omega = \sum_{i=Q,M,D,B} c_i \left[\mathrm{Tr} \ln \left(-G_i^{-1} \right) + \mathrm{Tr} \left(\Sigma_i G_i \right) \right] + \Phi \left[G_Q, G_M, G_D, G_B \right], \quad (49)$$

$$= \sum_{i=Q,M,D,B} d_i \int \frac{d^3 q}{(2\pi)^3} \int \frac{d\omega}{2\pi} \left\{ \omega + 2T \ln \left[1 - e^{-\omega/T} \right] \right\} 2 \sin^2 \delta_i \frac{\partial \delta_i}{\partial \omega}. \quad (50)$$

where $c_i = 1/2$ ($c_i = -1/2$) for bosonic (fermionic) states and d_i are the degeneracy factors that stem from the trace operation in the internal spaces of the quark, meson, diquark and baryon states. We suggest that in going from Equations (49) to (50) the same cancellations apply that were used above for the density formula and that are known to apply also for the entropy [21,22] would allow to derive this generalized Beth-Uhlenbeck equation of state for the thermodynamic potential, i.e., the negative pressure, once we restrict ourselves to the minimal set of two-particle irreducible diagrams in defining the Φ functional by the class of sunset type diagrams only, as given in Figure 8.

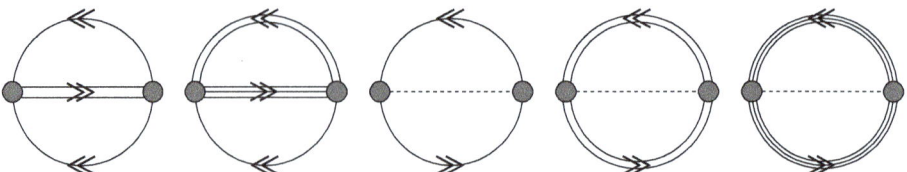

Figure 8. The contributions to the Φ functional for the quark-meson-diquark-baryon system.

From this Φ functional follow the selfenergies defining the full Greens functions of the system by functional derivation

$$\Sigma_i = \frac{\delta \Phi [G_Q, G_M, G_D, G_B]}{\delta G_i}.$$ (51)

The resulting Feynman diagrams for the selfenergy contributions are given in Figure 9.

Note that it is immediately plain from this formulation that in the situation of confinement, when the propagators belonging to colored excitations (quarks and diquarks) and thus to states that could not be populated would be cancelled, the system simplifies considerably. When all closed loop diagrams containing quarks and diquarks are neglected, this system reduces to a meson-baryon system. Out of the five closed-loop diagrams of Figure 8 remains then only the rightmost one from which the two selfenergy diagrams in Figure 9 emerge that contain only meson and baryon lines.

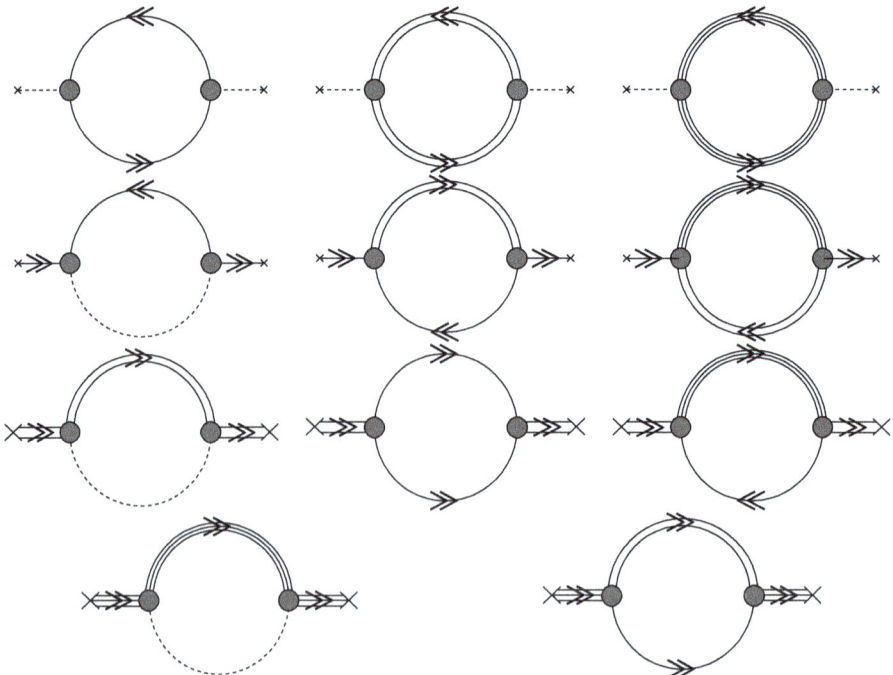

Figure 9. The selfenergy contributions for the Greens functions of the quark-meson-diquark-baryon system, defined by the Φ functional contributions shown in Figure 4. From top to down the four rows of diagrams show the selfenergies for the full propagators of mesons, quarks, diquarks and baryons, respectively.

4.1. Relativistic Density Functional Approach to Nuclear Matter

In the limit of quark (and gluon) confinement, the meson-baryon system can be further reduced when the mesonic degrees of freedom are not considered as dynamic ones but just as their meanfield values coupled to the baryon degrees of freedom with effective, possibly density- and temperature-dependent couplings, as sketched in Figure 10.

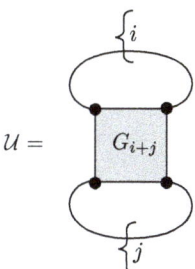

Figure 10. Effective density-functional for the interaction of species i and j given by a density-dependent coupling G_{i+j} as the local limit of a Φ functional.

In this case, the Φ−derivable approach reduces to a selfconsistent relativistic meanfield theory of nuclear matter,

$$\Omega = T \sum_{i=n,p,\Lambda,\ldots} c_i \left[\text{Tr} \ln S^{-1}_{i,qu} + \sum_{j=S,V} n_{i,j} \Sigma_{i,j} \right] + U[\{n_{i,S}, n_{i,V}\}] \, , \qquad (52)$$

where the functional $U[\{n_{i,j}\}]$ stands for the interaction in the system and is expressed in terms of the couplings of bilinears of the baryon spinors that define their scalar and vector densities $n_{i,S}$ and $n_{i,V}$, resp. The thermodynamic potential (52), where the role of the Φ functional is now played by the U functional that depends not on the propagators but on the densities and is therefore called a density functional. The thermodynamic potential (52) shares with the Φ−derivable approach the fulfillment of the conditions of stationarity

$$\frac{\partial \Omega}{\partial n_{i,S}} = \frac{\partial \Omega}{\partial n_{i,V}} = 0 \, , \ i = n, p, \Lambda, \ldots \, , \qquad (53)$$

and selfconsistency, which is provided by the fact that the (now real) selfenergies in the Dyson equations are obtained as derivatives of the U functional

$$\frac{\partial U}{\partial n_{i,S}} = \Sigma_{i,S} \, , \ \frac{\partial U}{\partial n_{i,V}} = \Sigma_{i,V} \, . \qquad (54)$$

The baryon quasiparticle propagators $S_{i,qu}$ fulfill the Dyson equations $S^{-1}_{i,qu} = S^{-1}_{i,0} - \Sigma_{i,S} - \Sigma_{i,V}$, diagrammatically given as

$$\text{(55)}$$

To this class of relativistic density functional models for baryonic matter belong the density-dependent relativistic meanfield models known as, e.g., DD2, NL3, KVOR, TM1. Their nonrelativistic relatives are the density functional models of the Skyrme type.

Recently, these models have been augmented with the inclusion of hadronic excluded volume effects. For an elaborate version of such corrections, see [24]. The origin of the excluded volume effects is the quark substructure of the baryons which entails the quark exchange effects between baryons that are a consequence of the Pauli principle on the quark level of description.

4.2. Quark Pauli Blocking in Hadronic Matter

The step to cancel all diagrams that contain propagators of colored excitations with the argument of confinement may be a too drastic step when we want to describe matter with a high density so that phase space occupation effects shall become important and the hadronic bound states "remember" their quark substructure. Formally these effects are included into the selfconsistent description of quark-hadron matter within the Φ−derivable approach outlined above in Equation (50) with the Φ functional given by the diagrams in Figure 8. In the limit of the confined phase, however, it is important to restore these quark substructure effects as they will drive the system towards deconfinement for sufficiently large densities. This can be accomplished by including selfenergy diagrams containing quark and diquark lines in Figure 9 in a perturbative manner by considering their propagators not fully selfconsistently, but in first order with respect to their selfenergies.

$$\text{(diagram)} + \mathcal{O}(\Sigma^{(2)}) \qquad (56)$$

$$\text{(diagram)} + \mathcal{O}(\Sigma^{(2)}) \qquad (57)$$

Here the quark and diquark quasiparticle propagator lines in Equations (56) and (57) are defined by a Dyson-Schwinger equation as given generically in Equation (55), but with a density functional for the effective interaction that is appropriate for the decription os quark matter and will be discussed in detail in Section 4.4.

In such a way two quark-diquark substructure contributions to the baryon selfenergy appear due to the corrections (56) and (57) beyond the quasiparticle approximation for quark and diquark propagators

$$\text{(diagram)} = \Omega(\Sigma^{(0)}) + \text{(diagram)} + \text{(diagram)} + \mathcal{O}(\Sigma^{(2)}). \qquad (58)$$

They contain one closed baryon line and are therefore of first order in the baryon density. By functional derivative w.r.t. the baryon propagator line (cutting) an effective quark and diquark exchange interaction can be obtained from those contributions to the baryon selfenergy shown in Equation (58). The diagram for the quark exchange interaction between baryons resulting from cutting the baryon line in the first of the two diagrams in Equation (58) is shown in (59) in two forms which are topologically equivalent,

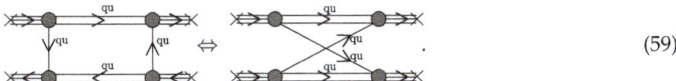 (59)

Analogously, the diquark exchange interaction is obtained by cutting the baryon line in the second of the two diagrams in Equation (58).

The quark Pauli blocking effects in nuclear matter have been evaluated in a nonrelativistic approximation with a confining potential model in Refs. [25,26] where it was found that the result for the repulsive density-dependent nucleon-nucleon interaction corresponds well to the repulsive part of the effective Skyrme interaction functional in Ref. [27]. Note that the resulting EoS has been successfully applied in predicting massive hybrid stars with quark matter cores [28]. A flaw of the nonrelativistic calculations of the quark Pauli blocking effect is that the quark mass is a fixed parameter so that partial chiral symmetry restoration in dense hadronic matter as a selfenergy effect on the quark propagator (consistent with the quark exchange) is not accounted for. This question has recently been taken up by Blaschke, Grigorian and Röpke who demonstrated that a chirally improved calculation of the quark Pauli blocking effect results in an EoS for nuclear matter that is similar to the DD2 model with excluded volume [24].

Strange quarks as well as strange hadrons belong to the system of our model, as it is formulated for any flavour. Interesting new aspects, which are expected from this approach concern, for instance, the Pauli blocking between baryons, including hyperons, in dense matter. This shall be of relevance for the discussion of the hyperon puzzle in compact stars [29,30].

The quark Pauli blocking effect applies also to mesons and corresponding expressions for selfenergy effects can be extracted from the Φ−derivable approach in a similar manner as for the baryons. This has been outlined in Ref. [31]. In a nonrelativistic potential model calculation, an effective quark exchange potential for the π-π interaction has been derived [32] which reproduces the scattering length of the pion interaction in the isospin = 2 channel, see also [33].

Let us now turn to the other limit of the Φ−derivable approach to a unified EoS for quark-hadron matter, the case of deconfined quarks. In this case, also chiral symmetry is restored, and due to the resulting lowering of the mass threshold the meson and baryon states become unbound (Mott effect). Their contribution to the thermodynamics as captured in the corresponding phase shift functions is gradually vanishing at high temperatures and chemical potentials with just chiral quark matter remaining asymptotically. As a paradigmatic example for the treatment of Mott dissociation of hadrons in hot, dense matter let us consider the case of pion dissociation in hot quark matter.

4.3. Mott Dissociation of Pions in Quark Matter

In order to describe the problem of mesons in quark matter within the Φ−derivable approach we define the Φ functional and the corresponding selfenergy in Figure 11.

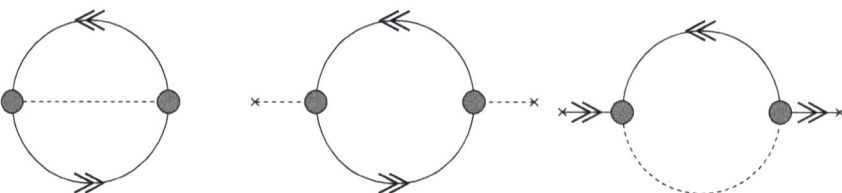

Figure 11. The Φ functional (left panel) for the case of mesons in quark matter, where the bosonic meson propagator is defined by the dashed line and the fermionic quark propagators are shown by the solid lines with arrows. The corresponding meson and quark selfenergies are shown in the middle and right panels, respectively.

The meson polarization loop $\Pi_M(q,z)$ in the middle panel of Figure 11 enters the definition of the meson T matrix (often called propagator)

$$T_M^{-1}(q,\omega+i\eta) = G_S^{-1} - \Pi_M(q,\omega+i\eta) = |T_M(q,\omega)|^{-1}e^{-i\delta_M(q,\omega)}, \quad (60)$$

which in the polar representation introduces a phase shift $\delta_M(q,\omega) = \arctan(\Im T_M/\Re T_M)$, that results in a generalized Beth-Uhlenbeck equation of state for the thermodynamics of the consistently coupled quark-meson system [34].

$$\Omega = \Omega_{MF} + \Omega_M, \quad (61)$$

where the selfconsistent quark meanfield contribution is

$$\Omega_{MF} = \frac{\sigma_{MF}^2}{4G_S} - 2N_c N_f \int \frac{d^3p}{(2\pi)^3} \left[E_p + T\ln\left(1+e^{-(E_p-\Sigma_+-\mu)/T}\right) + T\ln\left(1+e^{-(E_p+\Sigma_-+\mu)/T}\right) \right], \quad (62)$$

with the quasiparticle energy shift for quarks (antiquarks) due to mesonic correlations given by $\Sigma_\pm = \sum_{M=\pi,\sigma} \mathrm{tr}_D[\Sigma_M \Lambda_\pm \gamma_0]/2$ and the positive (negative) energy projection operators $\Lambda_\pm = (1\pm\gamma_0)/2$. The mesonic contribution to the thermodynamics is

$$\Omega_M = d_M \int \frac{d^3k}{(2\pi)^3} \int \frac{d\omega}{2\pi} \left\{\omega + 2T\ln\left[1-e^{-\omega/T}\right]\right\} 2\sin^2\delta_M(k,\omega) \frac{\delta_M(k,\omega)}{d\omega}, \quad (63)$$

where similar to the case of deuterons in nuclear matter the factor $2\sin^2\delta_M$ accounts for the fact that mesonic correlations in the continuum are partly already accounted for by the selfenergies Σ_M defining the improved selfconsistent quasiparticle picture. In the previous works of Refs. [34–38] on this topic, however, the effect of the backreaction from mesonic correlations on the quark meanfield thermodynamics had been disregarded. In Figure 12 we show the phase shift of the pion as a quark-antiquark state for different temperatures, below and above the Mott dissociation temperature. The shape of these functions and their evolution with increasing temperature over the Mott dissociation resembles the similar behaviour of the deuteron phase shift in nuclear matter at increasing density, see Figure 3.

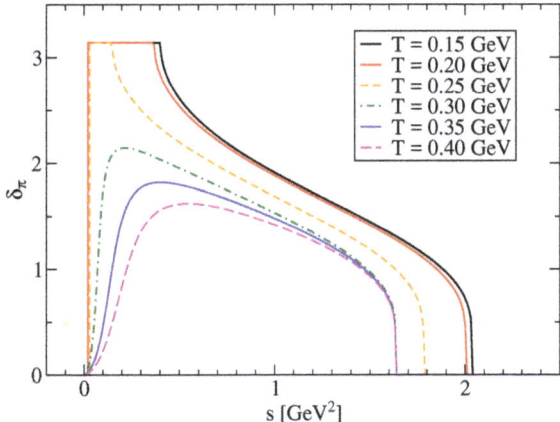

Figure 12. Phase shift of the pion as a quark-antiquark state for different temperatures, below and above the Mott dissociation temperature, from Ref. [34,39].

Here we note from the Φ−derivable approach that for consistency the quark propagator in the quark meanfield thermodynamic potential shall contain effects from the selfenergy Σ_M due to the coupling to the mesonic correlations as in the right panel of Figure 11. This total quark selfenergy is the given by $\Sigma(\mathbf{p}, p_0) = \sigma_{MF} + \Sigma_M(\mathbf{p}, p_0)$, where for a local NJL model with scalar coupling constant G_S the meanfield contribution is

$$\sigma_{MF} = 2N_f N_c G_S \int \frac{d^3 p}{(2\pi)^3} \frac{m}{E_p} [1 - f(E_p - \mu) - f(E_p + \mu)], \tag{64}$$

and the contribution due to scalar/pseudoscalar mesons (corresponding to the diagram shown in the rightmost panel of Figure 11) is given by [40]

$$\Sigma_M(0, p_0) = d_M \int \frac{d^4 q}{(2\pi)^4} \pi \varrho_M(\mathbf{q}, q_0) \left\{ \frac{(\gamma_0 + m/E_q)[1 + g(q_0) - f(E_q - \mu)]}{q_0 - p_0 + E_q - \mu - i\eta} + \frac{(\gamma_0 - m/E_q)[g(q_0) + f(E_q + \mu)]}{q_0 - p_0 - E_q - \mu - i\eta} \right\}, \tag{65}$$

where $\varrho_M = (-1/\pi)\Im T_M(\mathbf{q}, \omega + i\eta)$ is the meson spectral density and $E_q = \sqrt{q^2 + m^2}$ is the quark dispersion law with the quark mass $m = m_0 + \sigma_{MF}$. One can observe the similarity of this result (65) with that for a Dirac fermion coupled to a pointlike scalar meson, as given in [41].

Finally, let us consider the aspect of quark deconfinement in dense matter within a relativistic density functional approach which can be considered as a local limit of the Φ−derivable approach to quark-hadron matter (50) that obtains a form similar to the relativistic density functional theory for hadronic matter. The important difference lies in the density functional that in the case of quark matter shall account for confining effects, see [25,42]. In the final part of this section, we outline the recently developed relativistic version of the so-called string-flip model that proved rather successful for phenomenological applications to compact stars, supernovae and neutron star mergers.

4.4. Relativistic Density Functional Approach to Quark Matter

Hence, one obtains a contribution to the energy density functional of quark matter correspondingly [43]. In analogy to the Walecka model of nuclear matter [44], the relativistic density-functional approach to interacting quark matter can be obtained from the path integral approach based on the partition function [42],

$$\mathcal{Z} = \int D\bar{q}Dq \exp\left\{\int_0^\beta d\tau \int_V d^3x \left[\mathcal{L}_{\text{eff}} + \bar{q}\gamma_0 \hat{\mu} q\right]\right\}, \quad (66)$$

with $q = (q_u, q_d)^T$, $\hat{\mu} = \text{diag}(\mu_u, \mu_d)$ and effective Lagrangian density $\mathcal{L}_{\text{eff}} = \mathcal{L}_{\text{free}} - U(\bar{q}q, \bar{q}\gamma_0 q)$. The interaction is given by the potential $U(\bar{q}q, \bar{q}\gamma_0 q)$, which is a nonlinear functional of the scalar and vector quark field-currents. In the mean-field approximation, this potential can be expanded around the expectation values of the field currents, $n_s = \langle \bar{q}q \rangle$ and $n_v = \langle \bar{q}\gamma_0 q \rangle$ respectively,

$$U(\bar{q}q, \bar{q}\gamma_0 q) = U(n_s, n_v) + (\bar{q}q - n_s)\Sigma_s + (\bar{q}\gamma_0 q - n_v)\Sigma_v + \dots, \quad (67)$$

with scalar and vector self-energies, Σ_s and Σ_v. By appropriate rearranging of the quantities and performing the path integrals of Equation (66) one gets the thermodynamic potential

$$\Omega = -T \ln \mathcal{Z} = \Omega^{\text{quasi}} + U(n_s, n_v) - n_s \Sigma_s - n_v \Sigma_v. \quad (68)$$

The quasi-particle term (for the case of isospin symmetry and degenerate flavors)

$$\Omega^{\text{quasi}} = -2N_c N_f T \int \frac{d^3p}{(2\pi)^3} \left\{ \ln\left[1 + e^{-\beta(E^* - \mu^*)}\right] + \ln\left[1 + e^{-\beta(E^* + \mu^*)}\right] \right\} \quad (69)$$

can be calculated by using the ideal Fermi gas distribution for quarks with the quasiparticle energy $E^* = \sqrt{p^2 + M^2}$, the effective mass $M = m + \Sigma_s$ and effective chemical potential $\mu^* = \mu - \Sigma_v$. The self energies are determined by the density derivations

$$\Sigma_s = \frac{\partial U(n_s, n_v)}{\partial n_s}, \quad \text{and} \quad (70)$$

$$\Sigma_v = \frac{\partial U(n_s, n_v)}{\partial n_v}. \quad (71)$$

In this approach the stationarity of the thermodynamical potential

$$0 = \frac{\partial \Omega}{\partial n_s} = \frac{\partial \Omega}{\partial n_v} \quad (72)$$

is always fulfilled. For the case of isospin asymmetry, see Ref. [42].

In the mean-field approximation, the correlation energy can be obtained by folding the string-length distribution function for a given density with some interaction potential [25,45,46]. Moreover, the average string length between quarks in uniform matter is related to the scalar density, n_s, being proportional to $n_s^{-1/3}$. To capture this phenomenology, the following density functional of the interaction is adopted,

$$U(n_s, n_v) = D(n_v) n_s^{2/3} + a n_v^2 + \frac{b n_v^4}{1 + c n_v^2}. \quad (73)$$

The first term captures aspects of (quark) confinement through the density dependent scalar self-energy,

$$\Sigma_s = \frac{2}{3} D(n_v) n_s^{-1/3}, \quad (74)$$

defining the effective quark mass $M = m + \Sigma_s$. We also employ higher-order quark interactions [47], by inclusion of the third term in Equation (73), for the description of hybrid stars (neutron stars with a quark matter core) in order to obey the observational constraint of 2 $M_\odot$. To this end, the denominator in the last term of Equation (73) guarantees that for the appropriate choice of the parameters b and c, causality is not violated (i.e., the speed of sound $c_s = \sqrt{\partial P/\partial \varepsilon}$ does not exceed the speed of light). All terms in Equation (73) that contain the vector density contribute to the shift defining the effective chemical potentials $\mu^* = \mu - \Sigma_V$, where

$$\Sigma_V = 2an_v + \frac{4bn_v^3}{1 + cn_v^2} - \frac{2bcn_v^5}{(1 + cn_v^2)^2} + \frac{\partial D(n_v)}{\partial n_v} n_s^{2/3} . \tag{75}$$

The SFM modification takes into account the effective reduction of the in-medium string tension, $D(n_v) = D_0 \phi(n_v; \alpha)$. It is understood as a consequence of the modification of the pressure on the color field lines by the dual Meissner effect, since the reduction of the available volume corresponds to the reduction of the non-perturbative dual superconductor QCD vacuum that determines the strength of the confining potential between the quarks. The reduction of the string tension is modeled via a Gaussian function of the baryon density n_v,

$$\phi(n_v; \alpha) = \exp\left[-\alpha(n_v \cdot \text{fm}^3)^2\right] , \tag{76}$$

similar the available volume fraction in dense nuclear matter (see [24] and Section 5.3 below), as it shall be related to the available volume of nonperturbative QCD vacuum that according to the dual superconductor model is responsible for the formation of stringy color field configurations because of the dual Meissner effect. A detailed discussion of the role of the parameters a, b, c and α is given in Ref. [42].

5. Applications in Core-Collapse Supernovae and Neutron Stars

In the recent years the equation of state (EoS) has been significantly constraint, in particular at densities around and in excess of nuclear saturation density (ρ_0). At densities $\rho \leq \rho_0$, chiral effective field theory is the ab-initio approach to the nuclear many-body problem of dilute neutron matter [48–54]. Moreover, the high-precision observation of massive neutron stars of about 2 $M_\odot$ [55–57] constraints the supersaturation-density EoS, i.e., sufficient stiffness is required. The latter aspect challenges the appearance of additional particle degrees of freedom, e.g., hyperons and quarks, which tend to soften the EoS at $\rho > \rho_0$.

While neutron stars feature matter in β-equilibrium at zero temperature, the challenge lays in the development of EoS for core-collapse supernova applications, which cover a large domain of temperature, density and isospin asymmetry (cf. Figure 1a of Ref. [58]). At $T \leq 0.5$ MeV, time-dependent processes determine the nuclear composition, where heavy nuclei dominate. With increasing temperature, towards $T \simeq 0.5$ MeV, complete chemical and thermal equilibrium known as NSE (nuclear statistical equilibrium) is achieved, where, the nuclear composition is determined from the three independent variables: T, ρ (or n_B), y_C (baryonic charge fraction). Note that there is a strong density dependence, i.e., these heavy nuclei, originally belonging to the iron group with $A \gtrsim 56$, become increasingly heavier with increasing density. This phenomenon is well known also from the neutron star crust, where due to the low temperatures only a single nucleus exists for at given density, instead of a (broad) distribution as in the supernova case. At $\rho = \rho_0$ and at temperatures above $T = 5 - 10$ MeV, nuclei dissolve at the liquid-gas phase transition into homogeneous nuclear matter composed of (quasi-free) nucleons [16,59,60].

It becomes evident that first-principle calculations covering the entire domain are presently inexistent. Instead, model EOS are being developed for astrophysical applications. These combine several domains with different degrees of freedom, i.e., heavy nuclei at low temperatures, inhomogeneous nuclear matter with light and heavy nuclei together with the free nucleons (mean field), and homogeneous matter at high temperatures and densities. The latter has long been subject to investigations of a possible phase transition to the quark-gluon plasma.

The role of the EOS in simulations of core-collapse supernovae was explored within the failed scenario and consequently the formation of black holes, with focus on the dynamics and the neutrino signal [61–64]. In the multi-dimensional framework, neutrino-driven supernova explosions were the subjects of investigation [65–67], where it was found that such explosions are favored for soft EOS [68] with an earlier onset of shock revival and generally higher explosion energies, in comparison to stiff EOS [69]. Moreover, the role of the nuclear symmetry energy has been reviewed [70]. This an important nuclear matter parameter has a strong density dependence and becomes more tightly constrained by experiments, nuclear theory and observations [71,72].

5.1. Heavy Nuclear Clusters with $A \gtrsim 56$

At low temperatures ($T < 0.5$ MeV) time-dependent nuclear processes determine the evolution, which corresponds to the outer core of the stellar progenitor, with the nuclear composition of dominantly silicon, sulfur as well as carbon and oxygen. In some cases, even parts of the hydrogen-rich helium envelope are taken into account, e.g., in simulations of supernova explosions following the shock evolution for tens of seconds through parts of the stellar envelope. Therefore, small nuclear reaction networks are commonly employed in supernova studies [73,74]. They are sufficient for the nuclear energy generation.

At $T \gtrsim 0.5$ MeV, NSE is fulfilled and the relation $\mu_{(A,Z)} = Z\mu_p + (A-Z)\mu_n$ between the chemical potential of nuclei $\mu_{(A,Z)}$, with atomic mass A and charge Z, and the chemical potentials of neutron μ_n and proton μ_p holds. The NSE conditions found in the collapsing stellar core feature a broad distribution of nuclei with a pronounced peak around the iron-group, at low densities (see Figure 2 of Ref. [58]). In simulations of supernovae, this nuclear distribution is classified by the NSE average, including nuclear shell effects as discussed [75], which extends beyond the commonly used single-nucleus approximation. This is important for the consideration of weak processes with heavy nuclei. In particular, rates for electron captures on protons bound in these heavy nuclei [76] are averaged over the NSE composition and provided to the community as a table. In addition, coherent neutrino-nucleus scattering is considered [77], where even inelastic contributions are take into account [78], as well as nuclear (de)excitations [79,80]. The profound understanding of weak processes associated with nuclear transitions is also important for the understanding of the physics of the neutron star crust (cf. Ref. [81] and references therein), in particular for accreting neutron stars leading to the phenomenon of deep-crustal heating [82–84].

5.2. Light Nuclear Clusters with $A \leq 4$

In the domain corresponding to NSE, there is a rather narrow density domain where light nuclear clusters such as ^{2}H, ^{3}H, ^{3}He and ^{4}He can exist, at finite temperatures on the order of few MeV [59]. There are two aspects related to the presence of these light nuclear clusters, modification of the nuclear EOS due to these degrees of freedom, and the neutrino response due to the inclusion of a large variety of weak processes (cf. Table 1 in Ref. [85]). Common model supernova EOS with (light) clusters are based on the modified NSE [75]. However, the dissolving of the clusters towards high density is mimicked by the geometric excluded volume approach, as well as by hand with increasing temperature. This applies equally for light and heavy nuclear clusters. In comparison with the state-of-the-art quantum statistical approach [86,87] the deficits of NSE are revealed [58], while the cluster-virial EOS can provide the constraint at low densities [16]. It relates to the overestimation of the abundance of light clusters and in particular the too late dissolving of clusters into homogeneous matter.

Besides the NSE approach with 'all' (light) clusters included, in simulations of core-collapse supernovae the simplified nuclear composition $(n, p, \alpha, \langle A, Z \rangle)$, with only ^{4}He as representative light cluster, has long been employed [68,69]. It leads to an overestimate of the abundances of the unbound baryons and ^{4}He. This has important consequences for the supernova results, since such simplistic approach overestimates the neutrino response with the neutrons and protons, which in turn changes the supernova neutrino fluxes and spectra, however, with only a mild impact on the overall supernova dynamics [58].

In supernova simulations the temperatures are generally too high for any significant abundance of any light cluster. Only when the supernova explosion onset has been launched and the remnant central proto-neutron star deleptonizes and cools via the emission of neutrinos, light clusters can start to play a role. However, it has been demonstrated that neutral-current reactions are dominated by scattering on free neutrons, which is the most abundant nuclear species due to the generally large neutron excess (>95%) of supernova matter. Scattering reactions with light clusters play a negligible role. On the other hand, the neutrino response for light clusters is dominated by charged-current (break-up) reactions involving deuteron, triton and helium-3. However, the ν_e charged-current opacity is dominated by absorption on free neutrons. The situation is different for $\bar{\nu}_e$, since it was shown that protons and light clusters have similar abundances [58,85]. Taking properly into medium modifications for the cross sections and the proper phase-space of the contributing particles, the final rates are generally small. They never reach values as for the standard rates with protons, and hence the impact from charged-current weak processes with light clusters was found to be negligible [58,85]. Note that this study was based on the NSE approach which generally overestimates the abundance of light clusters. Hence, any improved treatment of the nuclear composition will most likely even reduce the impact of light clusters and the associated weak processes.

5.3. Homogeneous Matter at Supersaturation Density and Phase Transition to Quark Matter

With increasing densities, around ρ_0 (depending on the temperature), the transition to homogeneous nuclear matter proceeds where the EOS becomes less and less constrained by nuclear physics. As discussed above, the quark substructure of baryons shall become apparent at supersaturation densities and manifest itself by a nucleonic hard core repulsion due to quark Pauli blocking. This effect is likely to be enhanced by chiral symmetry restoration at high densities. To explore its role for the nuclear EOS at supersaturation densities the geometric excluded volume mechanism can be employed [24], where the available volume of the nucleons, $V_N = V\,\phi(\rho;\text{v})$, is proportional to the total volume V of the system where as proportionality factor occurs the available volume fraction

$$\phi(\rho;\text{v}) = \exp\left[-|\text{v}|\text{v}\,(\rho - \rho_0)^2\right]. \tag{77}$$

This is a density functional taken here according to [24] in a Gaussian form similar to (76). Depending on the sign of the excluded volume parameter, v, it allows to model both, stiffening and softening of the supersaturation-density EOS based on some reference model. This approach has been applied to confirm the independence of supernova simulations, e.g., the supernova shock dynamics as well as the evolution of neutrino luminosities and average energies, on the supersaturation-density EOS [88]. In studies of neutron stars, this approach results in significantly altered neutron star properties, e.g., maximum masses and radii [42,47]. The latter property is currently constraint only poorly, from the analysis of observations of low-mass x-ray binary systems [89–91].

Another uncertain aspect of the supersaturation-density EoS is the possibility of a phase transition from nuclear matter, with hadrons as degrees of freedom, to the deconfined quark gluon plasma with quarks and gluons as the new degrees of freedom. This has long been explored in the context of cold neutron stars. Unfortunately, the evaluation of the partition function of Quantum Chromodynamics (QCD)—the theory of strongly interacting matter—is possible only at vanishing baryon density by means of large-scale numerical simulations of this gauge field theory in a representation on space-time lattices [92,93]. These numerical ab-initio solutions predict a smooth cross-over transition at a temperature of $T = 154 \pm 9$ MeV at $\mu_B \simeq 0$, see Refs. [94–96]. Consequently, to study the role of quark degrees of freedom at high baryon densities, effective models for low-energy QCD have to be employed. Generally, the nuclear and quark matter phases are modeled separately and a phase transition construction is employed. This so-called *two-phase approach* results in a first-order phase transition by design. Note further that perturbative QCD, which is valid in the limit of asymptotic freedom, where the smallness of the coupling between quarks allows the usage of perturbative methods and corrections to the behaviour of an ideal gas of ultra-relativistic particles [97] are small, become applicable only at extremely high temperatures and densities exceeding by far the values attainable in compact stars or ultrarelativistic heavy-ion collisions. Instead, for studies of neutron stars and supernovae, effective quark matter models have been commonly employed, such as the thermodynamic bag model [98], models based on the Nambu-Jona-Lasinio (NJL) type [99–101], the recently developed vector-interaction enhanced chiral bag model [102,103] and in particular the density-functional-based DD2-SFM hybrid EoS approach [42] that has been outlined above. With the finite-temperature extension of the DD2F-SFM EoS it was possible to show that the deconfinement phase transition may serve as an explosion mechanism for massive ($\sim$50 $M_\odot$) blue-supergiant stars [104] long sought-for. At the same time, it explains the occurrence of a population of neutron stars born with high masses.

It has been realized that repulsive interactions are the necessary ingredient to provide a sufficient stiffness for the EOS at high densities in order to yield massive neutron stars with quark-matter core (known as hybrid stars) in agreement with the current constraint of 2 $M_\odot$. Moreover, higher-order vector repulsion terms [42,47] can lead to the 'twin' phenomenon (cf. [105–108] and references therein). It relates to the existence of compact stellar objects with similar-to-equal masses but different radii, to a strong phase transition with a large latent heat. As a consequence, the hybrid stars for such an EoS appear in the mass-radius diagram on a disconnected "third family" branch, separated from the branch of neutron stars (second family) by a sequence of unstable configurations. As has been demonstrated impressively for the DD2-SFM hybrid EoS [42], by varying only the available volume parameter α describing the screening of the confining interaction in dense matter, the twin phenomenon can be obtained at high masses of $\sim$2 $M_\odot$ as well as for typical pulsar masses of 1.3–1.4 $M_\odot$ or even below. This feature allows to discuss such hybrid stars in the context of the first multi-messenger observation of the binary compact star merger GW170817 [109], where such a scenario appears as an alternative to the conservative one of a binary neutron star merger with a relatively soft EoS [110–112]. In this interesting situation the NICER (Neutron Star Interior Composition Explorer)[4] NASA mission has the potential to rule out the soft EoS scenario, when it would measure for its primary target, the nearest millisecond pulsar PSR J0437-4715 with a mass of 1.44 ± 0.07 $M_\odot$ a large radius of, say, 14 km with the expected high precision of 0.5 km. Such a measurement would contradict the constraint on compactness of neutron stars extracted from GW170817 in the double neutron star merger scenario that constrains the radius in the mass range of 1.4 $M_\odot$ to $R < 13.4$ km, see [113]. Such a measurement would imply the discovery of the third family of compact stars [112] with an onset mass in the mass range 1.16–1.60 $M_\odot$ extracted from the gravitational wave signal of binary inspiral [109].

6. Cluster Formation and Quark Deconfinement Transition in Heavy-Ion Collisions

6.1. Light Cluster Formation and Symmetry Energy in Low-Energy Heavy-Ion Collisions

The only possibility to probe the properties of hot and dense nuclear matter in the laboratory are heavy ion collisions (HIC). The fragment distributions, their energy spectra and correlations measured in the detectors are used to infer the properties of the initial state ("fireball") produced in HIC. To reconstruct the initial state, one has to model the time evolution of the expanding hot and dense matter, including the formation of correlations and clusters.

A strict quantum statistical approach to this nonequilibrium process is not available at present. A possible method is the Zubarev nonequilibrium statistical operator which is able to describe the

---

[4] https://www.nasa.gov/nicer

formation of correlations in the expanding hot and dense matter [13,114]. First steps to describe cluster formation ($A < 4$) in expanding matter have been performed [115], but have to be worked out further, in particular to include larger clusters ($A \geq 4$). Kinetic codes based on a single-particle description, but also QMD and AMD codes which simulate cluster formation using the coalescence model and simpler concepts, have to be improved to obtain a microscopic description of cluster formation. Work in this direction is in progress [14,116,117]. Alternatively, the freeze-out concept has been used to model the expansion process of the "fireball". Within the chemical freeze-out approach, it is assumed that at a certain instant of the expanding and down-cooling fireball, the composition is frozen because the "chemical" reactions become slow so that the composition remains unchanged. This concept of a sudden freeze-out has been applied to HIC not only at moderate energies, but also at very high energies [14]. We give a short summary of some results obtained from HIC experiments at moderate energies ($\approx$35 MeV/A), where the production of light clusters was measured and interpreted within a freeze-out model. The main issue of these investigation was to show the relevance of clustering in nuclear systems and the necessity to describe medium effects. In particular, the equation of state of nuclear matter is of interest, at moderate temperatures ($T \leq 20$ MeV) and at subsaturation densities.

A first series of experiments was related to the symmetry energy [118–122].

In contrast to the standard treatment of symmetry energy within mean-field approaches, see, for instance, [123], it does not vanish in the zero density limit, but is significantly determined by cluster formation in the low-density region. This is a very obvious result, and more or less trivial in the density region where medium effects can be neglected. However, going to higher densities, medium effects, in particular the dissolution of bound states, must be included so that a smooth transition to the near-saturation density region is expected, where mean-field approaches can be applied. It has been shown in [118–122] that general expressions for the symmetry energy can be obtained which reproduce the results of experiments at low densities, determined by cluster formation, but agree with mean-field approaches at high densities. A cluster-virial approach may improve the description in the intermediate region.

The direct observation of medium effects in the nuclear matter EoS from HIC experiments is more involved. The composition, in particular the yield of light clusters ($A \leq 4$), is obtained in the low-density limit from a mass-action law using the binding energies of free nuclei. We discussed this approach above, Section 2, as nuclear statistical equilibrium (NSE). A special ratio of cluster yields Y_i, the so-called chemical constant $K_A = Y_A/(Y_p^Z Y_n^N)$, can be considered which in this low-density limit is solely a function of the temperature and the volume, but not the chemical potentials. Because in chemical equilibrium the chemical potentials of the cluster A, consisting of Z protons and $N = A - Z$ neutrons, are related as $\mu_A = Z\mu_p + N\mu_n$, the chemical potentials cancel in the low-density limit. This simple approach, neglecting any density effects, has been disproved by experiments with HIC [124]. The reason is the modification of the binding energies if the density is increasing. In particular, self-energy shifts and Pauli blocking lead to the reduction of the binding energy and the dissolution at the so called Mott density. The measured chemical constants can be used for the experimental determination of in-medium cluster binding energies and Mott points in nuclear matter [125].

We can discus these experimental results as clear indication for the need to consider medium effects. Within a QS approach including quasiparticle shifts and correlations in the continuum [17], it was possible to reproduce the data for the chemical constants of the light elements d, t, h, α obtained from the cluster yields. More simple models for medium corrections such as the semiempirical excluded volume concept [75] can be adapted to reproduce the data [126].

Also for these experiments, a nuclear matter EoS is needed which describes cluster formation and the medium modification, as well as the treatment of continuum correlations. A cluster-virial approach may improve the calculation of the EoS in a wide region of the phase diagram.

In conclusion, medium effects, in particular self-energy shifts and Pauli blocking for light clusters, are verified by recent HIC experiments. An improved cluster viral approach should be worked out to describe adequately the contributions of correlations in the continuum, expressed by in-medium scattering phase shifts between the different constituents. The freeze-out model to describe the expanding fireball may be considered as an approximation to treat the nonequilibrium process. Kinetic approaches, for instance transport codes allowing for cluster production (such as QMD or AMD), may be developed further to include in-medium correlation effects. This would allow for a systematic and consistent treatment of HIC experiments. Nevertheless, the correct description of the thermodynamic equilibrium, in particular the cluster viral approach, is a benchmark for all nonequilibrium approaches, and should be advanced in the future.

6.2. Deconfinement Transition in Relativistic Heavy-Ion Collisions

The main goal of the theoretical developments towards a unified EoS for quark-hadron matter is to achieve a most reliable prediction for the behavior of warm, dense strongly interacting matter including the deconfinement transition, which is described here as a Mott dissociation of baryonic and mesonic bound states, triggered by the restoration of the dynamically broken chiral restoration. Dynamical chiral symmetry breaking is a rather robust phenomenon that can be described in a broad variety of chiral quark models of the Nambu-Jona-Lasinio type, i.e., with a four-fermion interaction of the current-current form with a sufficiently strong coupling to allow for a nontrivial solution of the gap equation for the quark mass. This is a nonperturbative effect that cannot be obtained in any finite order of perturbation theory and is nowadays an obligatory element of modern descriptions of quark matter. At finite temperatures, however, such models often fail to provide a sensible description of the EoS of QCD matter since the quark matter pressure dominates over the hadronic one already at unphysically low temperatures, due to the lack of a quark confinement mechanism in those models. A simple way out is provided by adopting a bag pressure for mimicking confinement. This is a too simple concept and spoils the beauty of a dynamical description. While the details of the confinement mechanism in QCD are still debated, a viable compromise is provided by the concept of a confining density functional that is based on string-type interactions between color charges and, together with the concept of saturation of color interactions within nearest neighbors (string-flip model) allows for the treatment of such confining interactions in quark matter within a relativistic, selfconsistent quasiparticle model.

The success of the DD2(DD2F)-SFM hybrid EoS for astrophysical applications has been summarized in the previous section. The applications for heavy-ion collisions, in the isospin-symmetric case, are still under way. A necessary requirement for a sensible description of the complex system of an ultrarelativistic heavy-ion collision calls for a numerical simulation code like THESEUS, the three-fluid hydrodynamics-based event simulator extended by UrQMD simulations of final-state interactions [127]. Such a description is most appropriate for the investigation of possible effects of a phase transition in the baryon stopping regime, i.e., when the projectile and target fluids collide and form a highly compressed baryonic matter system, i.e., in the collision energy range $\sqrt{s_{NN}} = 2 - 20$ which is covered by the THESEUS program and by the high-statistics collison experiments like NA61, RHIC BES or RHIC FXT and the upcoming NICA and FAIR experiments. Therefore, we have chosen THESEUS as the tool for identifying the QCD phase transition and for investigating the effects of a first-order phase transition on heavy-ion collision observables. Previous studies of this question have been performed with THESEUS [128] and with the three-fluid hydrodynamics code in it [129–131] using model EoS of three kinds: purely hadronic, crossover and with a strong first-order transition [5]. The upgrade of the EoS with the DD2(DD2F)-SFM hybrid EoS as described in this work is under way.

---

[5] Recently, a thermodynamically consistent generalization of the excluded-volume improved RDF approach to the hadronic EoS has been suggested which employs a density- and temperature-dependent excluded volume parameter. Within this setting, a second first-order phase transition with a critical endpoint in the QCD phase diagram has been obtained [132]. Such a formulation may be most convenient, e.g., for Bayesian studies of the structure of the QCD phase diagram to be extracted from data of heavy-ion collision experiments.

In particular, one expects modifications from the lower density region of the phase transition and its temperature dependence (for a direct comparison, see the right panel of Figure 2 in [133]) and a better description of flow observables due to the increased stiffening of the high-density part of the new hybrid EoS when compared to the one in Ref. [127].

A main goal of the unified approach to the quark-hadron EoS as outlined in this work is the possibility to obtain an EoS with a second critical endpoint in the QCD phase diagram associated with the chiral/deconfinement transition. The first one, corresponding to the liquid-gas phase transition in nuclear matter is already an integral part of the RDF description of nuclear matter within the DD2(DD2F) part of the present approach. We want to point out that with this approach one may have achieved a systematic formulation of a theory for quark-hadron matter that allows to address also the presently puzzling questions of chemical freezeout of hadrons and nuclei like:

1. Can the success of the thermal statistical model in describing the production of nuclear clusters as measured by the ALICE experiment at LHC [134] be interpreted so that they freeze out directly when hadronizing the QGP so that they may be viewed as preformed multiquark systems already in the QGP?
2. What are the necessary ingredients to understand chemical freezeout of hadrons and clusters kinetically [135]?

With these prospects for the development of a unified quark-hadron matter EoS we want to conclude the present work.

7. Conclusions

We have outlined an approach to a unified equation of state for quark-hadron matter on the basis of a Φ−derivable approach to the generalized Beth-Uhlenbeck equation of state for a cluster decomposition of thermodynamic quantities like the density. To this end we have summarized the cluster virial expansion for nuclear matter and demonstrated the equivalence of the Green's function approach to the Φ−derivable formulation. As an example, the formation and dissociation of deuterons in nuclear matter was discussed. We have formulated the cluster Φ−derivable approach to quark-hadron matter which allows to take into account the specifics of chiral symmetry restoration and deconfinement in triggering the Mott-dissociation of hadrons. Applications to the phenomenology of nuclear clusters and quark deconfinement in the astrophysics of supernovae and compact stars as well as in heavy-ion collisions are outlined.

This approach unifies the description of a strongly coupled quark-gluon plasma with that of a medium-modified hadron resonance gas description which are shown to be its limiting cases. The developed formalism shall replace the common two-phase approach to the description of the deconfinement and chiral phase transition, where separately developed equations of state for hadronic and quark matter are matched with Gibbs conditions of phase equilibrium. Roughly speaking, one would develop a Ginzburg-Landau-type density functional which allows first, second and higher-order transitions, including crossovers. Examples are the van-der-Waals EoS which has a first-order transition with critical endpoint or the RMF models of nuclear matter for the liquid-gas transition. The cluster virial expansion shall allow a formulation of the quark-hadron transition in a similar way.

Author Contributions: N.U.F.B. as first author is responsible for the final form and content of the whole paper, integrating the contributions of the coauthors, who provided Sections 2, 6 (G.R.), 3, 4 (D.B.) and 5 (T.F.).

Acknowledgments: We acknowledge support by the Polish National Science Center (NCN) under grant number UMO-2016/23/B/ST2/00720 (TF) and grant number UMO-2014/13/B/ST9/02621 (NUFB) and by the Bogoliubov-Infeld program (NUFB, TF) for collaboration between JINR Dubna and Polish Institutes, as well as the Heisenberg-Landau program for collaboration between JINR Dubna and German Institutes (GR). GR and DB are grateful for support within the MEPhI Academic Excellence programme under contract no. 02.a03.21.0005 for their work in Sections 2 and 3, respectively. The work of DB was supported in part by a grant from the Russian Science Foundation under contract number 17-12-01427. This work was supported by the COST Actions CA15213 "THOR", CA16117 "ChETEC" and CA16214 "PHAROS".

Conflicts of Interest: The authors declare no conflict of interest.

References

1. Natowitz, J.B.; Hagel, K.; Ma, Y.; Murray, M.; Qin, L.; Wada, R.; Wang, J. Limiting temperatures and the equation of state of nuclear matter. *Phys. Rev. Lett.* **2002**, *89*, 212701. [CrossRef] [PubMed]
2. Bazavov, A.; Bhattacharya, T.; Cheng, M.; DeTar, C.; Ding, H.T.; Gottlieb, S.; Gupta, R.; Hegde, P.; Heller, U.M.; Karsch, F.; et al. The chiral and deconfinement aspects of the QCD transition. *Phys. Rev. D* **2012**, *85*, 054503. [CrossRef]
3. Baym, G.; Kadanoff, L.P. Conservation Laws and Correlation Functions. *Phys. Rev.* **1961**, *124*, 287–299. [CrossRef]
4. Baym, G. Selfconsistent approximation in many body systems. *Phys. Rev.* **1962**, *127*, 1391–1401. [CrossRef]
5. Kraeft, W.D.; Kremp, D.; Ebeling, W.; Röpke, G. *Quantum Statistics of Charged Particle Systems*; Springer: Berlin, Germany, 1986.
6. Weinhold, W.; Friman, B.; Nörenberg, W. Thermodynamics of Delta resonances. *Phys. Lett. B* **1998**, *433*, 236–242. [CrossRef]
7. Weinhold, W. Thermodynamik mit Resonanzzuständen. Ph.D. Thesis, Technischen Universität Darmstadt, Darmstadt, Germany, 1998.
8. Zimmermann, R.; Stolz, H. The Mass Action Law in Two-Component Fermi Systems Revisited Excitons and Electron-Hole Pairs. *Phys. Status Solidi* **1985**, *131*, 151–164. [CrossRef]
9. Röpke, G.; Münchow, L.; Schulz, H. On the phase stability of hot nuclear matter and the applicability of detailed balance equations. *Phys. Lett. B* **1982**, *110*, 21–24. [CrossRef]
10. Röpke, G.; Münchow, L.; Schulz, H. Particle clustering and Mott transitions in nuclear matter at finite temperature. *Nucl. Phys. A* **1982**, *379*, 536–552. [CrossRef]
11. Röpke, G.; Schmidt, M.; Münchow, L.; Schulz, H. Particle clustering and Mott transition in nuclear matter at finite temperature (II). *Nucl. Phys. A* **1983**, *399*, 587–602. [CrossRef]
12. Schmidt, M.; Röpke, G.; Schulz, H. Generalized Beth-Uhlenbeck approach for hot nuclear matter. *Ann. Phys.* **1990**, *202*, 57–99. [CrossRef]
13. Röpke, G. Correlations and Clustering in Dilute Matter. In *Nuclear Particle Correlations and Cluster Physics*; Schröder, W., Ed.; World Scientific: Singapore, 2017; pp. 31–69.
14. Röpke, G.; Blaschke, D.; Ivanov, Y.B.; Karpenko, I.; Rogachevsky, O.V.; Wolter, H.H. Medium effects on freeze-out of light clusters at NICA energies. *Phys. Part. Nucl. Lett.* **2018**, *15*, 225–229. [CrossRef]
15. Beth, E.; Uhlenbeck, G. The quantum theory of the non-ideal gas. II. Behaviour at low temperatures. *Physica* **1937**, *4*, 915–924. [CrossRef]
16. Röpke, G.; Bastian, N.U.; Blaschke, D.; Klähn, T.; Typel, S.; Wolter, H.H. Cluster virial expansion for nuclear matter within a quasiparticle statistical approach. *Nucl. Phys. A* **2013**, *897*, 70–92. [CrossRef]
17. Röpke, G. Nuclear matter equation of state including two-, three-, and four-nucleon correlations. *Phys. Rev. C* **2015**, *92*, 054001. [CrossRef]
18. Röpke, G. Clustering in nuclear environment. *J. Phys.* **2014**, *569*, 012031.
19. Horowitz, C.J.; Schwenk, A. Cluster formation and the virial equation of state of low-density nuclear matter. *Nucl. Phys. A* **2006**, *776*, 55–79. [CrossRef]
20. Blaschke, D. Cluster virial expansion for quark and nuclear matter. In Proceedings of the XXII International Baldin Seminar on High Energy Physics Problems, JINR Dubna, Russia, 15–20 September 2014.
21. Vanderheyden, B.; Baym, G. Selfconsistent approximations in relativistic plasmas: Quasiparticle analysis of the thermodynamic properties. *J. Stat. Phys.* **1998**, *93*, 843–861. [CrossRef]
22. Blaizot, J.P.; Iancu, E.; Rebhan, A. Approximately selfconsistent resummations for the thermodynamics of the quark gluon plasma. 1. Entropy and density. *Phys. Rev. D* **2001**, *63*, 065003. [CrossRef]
23. Uhlenbeck, G.E.; Beth, E. The quantum theory of the non-ideal gas I. Deviations from the classical theory. *Physica* **1936**, *3*, 729–745. [CrossRef]
24. Typel, S. Variations on the excluded-volume mechanism. *Eur. Phys. J. A* **2016**, *52*, 16.
25. Röpke, G.; Blaschke, D.; Schulz, H. Pauli Quenching Effects in a Simple String Model of Quark/Nuclear Matter. *Phys. Rev. D* **1986**, *34*, 3499–3513. [CrossRef]
26. Blaschke, D.; Röpke, G. Pauli Quenching for Hadrons in Nuclear Matter: A Quark Substructure Effect. *Dubna Preprint* **1988**, unpublished.

27. Vautherin, D.; Brink, D.M. Hartree-Fock calculations with Skyrme's interaction. 1. Spherical nuclei. *Phys. Rev. C* **1972**, *5*, 626–647. [CrossRef]
28. Blaschke, D.; Tovmasian, T.; Kämpfer, B. Predicting Stable Quark Cores in Neutron Stars From a Unified Description of Quark—Hadron Matter. *Sov. J. Nucl. Phys.* **1990**, *52*, 675–678.
29. Klähn, T.; Blaschke, D. Strange Matter in Compact Stars. *EPJ Web Conf.* **2018**, *171*, 08001 [CrossRef]
30. Bastian, N.U.F.; Blaschke, D.B.; Cierniak, M.; Fischer, T.; Kaltenborn, M.A.R.; Marczenko, M.; Typel, S. Strange matter prospects within the string-flip model. *EPJ Web Conf.* **2018**, *171*, 20002 [CrossRef]
31. Dubinin, A.; Blaschke, D.; Friesen, A.; Turko, L. Pauli Blocking Effect Within the Relativistic Pion Gas. *Acta Phys. Pol. Suppl.* **2017**, *10*, 903. [CrossRef]
32. Blaschke, D.; Röpke, G. Quark exchange contribution to the effective meson meson interaction potential. *Phys. Lett. B* **1993**, *299*, 332–337. [CrossRef]
33. Barnes, T.; Swanson, E.S. A Diagrammatic approach to meson meson scattering in the nonrelativistic quark potential model. *Phys. Rev. D* **1992**, *46*, 131–159. [CrossRef]
34. Blaschke, D.; Buballa, M.; Dubinin, A.; Röpke, G.; Zablocki, D. Generalized Beth—Uhlenbeck approach to mesons and diquarks in hot, dense quark matter. *Ann. Phys.* **2014**, *348*, 228–255. [CrossRef]
35. Hüfner, J.; Klevansky, S.P.; Zhuang, P.; Voss, H. Thermodynamics of a quark plasma beyond the mean field: A generalized Beth-Uhlenbeck approach. *Ann. Phys.* **1994**, *234*, 225–244. [CrossRef]
36. Zhuang, P.; Hüfner, J.; Klevansky, S.P. Thermodynamics of a quark—Meson plasma in the Nambu-Jona-Lasinio model. *Nucl. Phys. A* **1994**, *576*, 525–552. [CrossRef]
37. Yamazaki, K.; Matsui, T. Quark-Hadron Phase Transition in the PNJL model for interacting quarks. *Nucl. Phys. A* **2013**, *913*, 19–50. [CrossRef]
38. Wergieluk, A.; Blaschke, D.; Kalinovsky, Y.L.; Friesen, A. Pion dissociation and Levinson's theorem in hot PNJL quark matter. *Phys. Part. Nucl. Lett.* **2013**, *10*, 660–668. [CrossRef]
39. Blaschke, D.; Dubinin, A.; Buballa, M. Polyakov-loop suppression of colored states in a quark-meson-diquark plasma. *Phys. Rev. D* **2015**, *91*, 125040. [CrossRef]
40. Kitazawa, M.; Kunihiro, T.; Nemoto, Y. Emergence of soft quark excitations by the coupling with a soft mode of the QCD critical point. *Phys. Rev. D* **2014**, *90*, 116008. [CrossRef]
41. Blaizot, J.P. Quantum fields at finite temperature and density. *J. Korean Phys. Soc.* **1992**, *25*, S65–S98.
42. Kaltenborn, M.A.R.; Bastian, N.U.F.; Blaschke, D.B. Quark-Nuclear Hybrid Equation of State with Excluded Volume Effects. *Phys. Rev. D* **2017**, *96*, 056024. [CrossRef]
43. Khvorostukin, A.S.; Skokov, V.V.; Toneev, V.D.; Redlich, K. Lattice QCD constraints on the nuclear equation of state. *Eur. Phys. J. C* **2006**, *48*, 531–543. [CrossRef]
44. Kapusta, J.I. Finite Temperature Field Theory. In *Cambridge Monographs on Mathematical Physics*; Cambridge University Press: Cambridge, UK, 1989.
45. Horowitz, C.J.; Piekarewicz, J. Quark models of nuclear matter: 1. Basic models and ground state properties. *Nucl. Phys. A* **1992**, *536*, 669–696. [CrossRef]
46. Horowitz, C.J.; Piekarewicz, J. Nuclear to quark matter transition in the string flip model. *Phys. Rev. C* **1991**, *44*, 2753–2764. [CrossRef]
47. Benic, S.; Blaschke, D.; Alvarez-Castillo, D.E.; Fischer, T.; Typel, S. A new quark-hadron hybrid equation of state for astrophysics—I. High-mass twin compact stars. *Astron. Astrophys.* **2015**, *577*, A40. [CrossRef]
48. Hebeler, K.; Schwenk, A. Chiral three-nucleon forces and neutron matter. *Phys. Rev. C* **2010**, *82*, 014314. [CrossRef]
49. Hebeler, K.; Lattimer, J.M.; Pethick, C.J.; Schwenk, A. Constraints on Neutron Star Radii Based on Chiral Effective Field Theory Interactions. *Phys. Rev. L* **2010**, *105*, 161102. [CrossRef] [PubMed]
50. Holt, J.W.; Kaiser, N.; Weise, W. Chiral nuclear dynamics with three-body forces. *Prog. Part. Nucl. Phys.* **2012**, *67*, 353–358. [CrossRef]
51. Sammarruca, F.; Chen, B.; Coraggio, L.; Itaco, N.; Machleidt, R. Dirac-Brueckner-Hartree-Fock versus chiral effective field theory. *Phys. Rev. C* **2012**, *86*, 054317. [CrossRef]
52. Tews, I.; Krüger, T.; Hebeler, K.; Schwenk, A. Neutron Matter at Next-to-Next-to-Next-to-Leading Order in Chiral Effective Field Theory. *Phys. Rev. Lett.* **2013**, *110*, 032504. [CrossRef] [PubMed]
53. Krüger, T.; Tews, I.; Hebeler, K.; Schwenk, A. Neutron matter from chiral effective field theory interactions. *Phys. Rev. C* **2013**, *88*, 025802. [CrossRef]

54. Coraggio, L.; Holt, J.W.; Itaco, N.; Machleidt, R.; Sammarruca, F. Reduced regulator dependence of neutron-matter predictions with perturbative chiral interactions. *Phys. Rev. C* **2013**, *87*, 014322. [CrossRef]
55. Demorest, P.B.; Pennucci, T.; Ransom, S.M.; Roberts, M.S.E.; Hessels, J.W.T. A two-solar-mass neutron star measured using Shapiro delay. *Nature* **2010**, *467*, 1081–1083. [CrossRef] [PubMed]
56. Antoniadis, J.; Freire, P.C.C.; Wex, N.; Tauris, T.M.; Lynch, R.S.; van Kerkwijk, M.H.; Kramer, M.; Bassa, C.; Dhillon, V.S.; Driebe, T.; et al. A Massive Pulsar in a Compact Relativistic Binary. *Science* **2013**, *340*, 448. [CrossRef] [PubMed]
57. Fonseca, E.; Pennucci, T.T.; Ellis, J.A.; Stairs, I.H.; Nice, D.J.; Ransom, S.M.; Demorest, P.B.; Arzoumanian, Z.; Crowter, K.; Dolch, T.; et al. The NANOGrav Nine-year Data Set: Mass and Geometric Measurements of Binary Millisecond Pulsars. *Astrophys. J.* **2016**, *832*, 167. [CrossRef]
58. Fischer, T.; Bastian, N.U.; Blaschke, D.; Cerniak, M.; Hempel, M.; Klähn, T.; Martínez-Pinedo, G.; Newton, W.G.; Röpke, G.; Typel, S. The state of matter in simulations of core-collapse supernovae—Reflections and recent developments. *Publ. Astron. Soc. Austral.* **2017**, *34*, 67 [CrossRef]
59. Typel, S.; Röpke, G.; Klähn, T.; Blaschke, D.; Wolter, H.H. Composition and thermodynamics of nuclear matter with light clusters. *Phys. Rev. C* **2010**, *81*, 015803. [CrossRef]
60. Hempel, M.; Schaffner-Bielich, J.; Typel, S.; Röpke, G. Light clusters in nuclear matter: Excluded volume versus quantum many-body approaches. *Phys. Rev. C* **2011**, *84*, 055804. [CrossRef]
61. Sumiyoshi, K.; Yamada, S.; Suzuki, H.; Chiba, S. Neutrino signals from the formation of black hole: A probe of equation of state of dense matter. *Phys. Rev. Lett.* **2006**, *97*, 091101. [CrossRef] [PubMed]
62. Fischer, T.; Whitehouse, S.C.; Mezzacappa, A.; Thielemann, F.K.; Liebendörfer, M. The neutrino signal from protoneutron star accretion and black hole formation. *Astron. Astrophys.* **2009**, *499*, 1–15. [CrossRef]
63. O'Connor, E.; Ott, C.D. Black Hole Formation in Failing Core-Collapse Supernovae. *Astrophys. J.* **2011**, *730*, 70. [CrossRef]
64. Steiner, A.W.; Hempel, M.; Fischer, T. Core-collapse Supernova Equations of State Based on Neutron Star Observations. *Astrophys. J.* **2013**, *774*, 17. [CrossRef]
65. Marek, A.; Janka, H.T.; Müller, E. Equation-of-state dependent features in shock-oscillation modulated neutrino and gravitational-wave signals from supernovae. *Astron. Astrophys.* **2009**, *496*, 475–494. [CrossRef]
66. Suwa, Y.; Takiwaki, T.; Kotake, K.; Fischer, T.; Liebendörfer, M.; Sato, K. On the Importance of the Equation of State for the Neutrino-driven Supernova Explosion Mechanism. *Astrophys. J.* **2013**, *764*, 99. [CrossRef]
67. Nagakura, H.; Iwakami, W.; Furusawa, S.; Okawa, H.; Harada, A.; Sumiyoshi, K.; Yamada, S.; Matsufuru, H.; Imakura, A. Simulations of Core-Collapse Supernovae in Spatial Axisymmetry with Full Boltzmann Neutrino Transport. *Astrophys. J.* **2018**, *854*, 136 [CrossRef]
68. Lattimer, J.M.; Swesty, F. A Generalized equation of state for hot, dense matter. *Nucl. Phys. A* **1991**, *535*, 331–376. [CrossRef]
69. Shen, H.; Toki, H.; Oyamatsu, K.; Sumiyoshi, K. Relativistic equation of state of nuclear matter for supernova and neutron star. *Nucl. Phys. A* **1998**, *637*, 435–450. [CrossRef]
70. Fischer, T.; Hempel, M.; Sagert, I.; Suwa, Y.; Schaffner-Bielich, J. Symmetry energy impact in simulations of core-collapse supernovae. *Eur. Phys. J. A* **2014**, *50*, 46. [CrossRef]
71. Lattimer, J.M.; Lim, Y. Constraining the Symmetry Parameters of the Nuclear Interaction. *Astrophys. J.* **2013**, *771*, 51. [CrossRef]
72. Tews, I.; Lattimer, J.M.; Ohnishi, A.; Kolomeitsev, E.E. Symmetry Parameter Constraints from A Lower Bound on the Neutron-Matter Energy. *Astrophys. J.* **2017** *848*, 105 [CrossRef]
73. Thielemann, F.K.; Brachwitz, F.; Höflich, P.; Martinez-Pinedo, G.; Nomoto, K. The physics of type Ia supernovae. *New Astron. Rev.* **2004**, *48*, 605–610. [CrossRef]
74. Fischer, T.; Whitehouse, S.; Mezzacappa, A.; Thielemann, F.K.; Liebendörfer, M. Protoneutron star evolution and the neutrino driven wind in general relativistic neutrino radiation hydrodynamics simulations. *Astron. Astrophys.* **2010**, *517*, A80. [CrossRef]
75. Hempel, M.; Schaffner-Bielich, J. Statistical Model for a Complete Supernova Equation of State. *Nucl. Phys. A* **2010**, *837*, 210–254. [CrossRef]
76. Juodagalvis, A.; Langanke, K.; Hix, W.R.; Martínez-Pinedo, G.; Sampaio, J.M. Improved estimate of electron capture rates on nuclei during stellar core collapse. *Nucl. Phys. A* **2010**, *848*, 454–478. [CrossRef]
77. Bruenn, S.W. Stellar core collapse: Numerical model and infall epoch. *Astrophys. J. Suppl.* **1985**, *58*, 771–841. [CrossRef]

78. Langanke, K.; Martinez-Pinedo, G.; Müller, B.; Janka, H.T.; Marek, A. Effects of Inelastic Neutrino-Nucleus Scattering on Supernova Dynamics and Radiated Neutrino Spectra. *Phys. Rev. Lett.* **2008**, *100*, 011101. [CrossRef] [PubMed]
79. Fuller, G.M.; Meyer, B.S. High-temperature neutrino-nucleus processes in stellar collapse. *Astrophys. J.* **1991**, *376*, 701–716. [CrossRef]
80. Fischer, T.; Langanke, K.; Martínez-Pinedo, G. Neutrino-pair emission from nuclear de-excitation in core-collapse supernova simulations. *Phys. Rev. C* **2013**, *88*, 065804. [CrossRef]
81. Schatz, H.; Gupta, S.; Möller, P.; Beard, M.; Brown, E.F.; Deibel, A.T.; Gasques, L.R.; Hix, W.R.; Keek, L.; Lau, R.; et al. Strong neutrino cooling by cycles of electron capture and β^- decay in neutron star crusts. *Nature* **2014**, *505*, 62–65. [CrossRef] [PubMed]
82. Haensel, P.; Zdunik, J.L. Non-equilibrium processes in the crust of an accreting neutron star. *Astron. Astrophys.* **1990**, *227*, 431–436.
83. Brown, E.F.; Bildsten, L.; Rutledge, R.E. Crustal Heating and Quiescent Emission from Transiently Accreting Neutron Stars. *Astrophys. J.* **1998**, *504*, L95–L98. [CrossRef]
84. Haensel, P.; Zdunik, J.L. Models of crustal heating in accreting neutron stars. *Astron. Astrophys.* **2008**, *480*, 459–464. [CrossRef]
85. Fischer, T.; Martínez-Pinedo, G.; Hempel, M.; Huther, L.; Röpke, G.; Typel, S.; Lohs, A. Expected impact from weak reactions with light nuclei in corecollapse supernova simulations. In Proceedings of the 13th International Symposium on Origin of Matter and Evolution of Galaxies, Beijing, China, 24–27 June 2015.
86. Röpke, G. Light nuclei quasiparticle energy shifts in hot and dense nuclear matter. *Phys. Rev. C* **2009**, *79*, 014002. [CrossRef]
87. Röpke, G. Parametrization of light nuclei quasiparticle energy shifts and composition of warm and dense nuclear matter. *Nucl. Phys. A* **2011**, *867*, 66–80. [CrossRef]
88. Fischer, T. Constraining the supersaturation density equation of state from core-collapse supernova simulations—Excluded volume extension of the baryons . *Eur. Phys. J. A* **2016**, *52*, 54. [CrossRef]
89. Steiner, A.W.; Lattimer, J.M.; Brown, E.F. The Equation of State from Observed Masses and Radii of Neutron Stars. *Astrophys. J.* **2010**, *722*, 33–54. [CrossRef]
90. Suleimanov, V.; Poutanen, J.; Revnivtsev, M.; Werner, K. Neutron star stiff equation of state derived from cooling phases of the X-ray burster 4U 1724-307. *Astrophys. J.* **2011**, *742*, 122. [CrossRef]
91. Steiner, A.W.; Lattimer, J.M.; Brown, E.F. The Neutron Star Mass-Radius Relation and the Equation of State of Dense Matter. *Astrophys. J.* **2013**, *765*, L5. [CrossRef]
92. Fodor, Z.; Katz, S. Critical point of QCD at finite T and mu, lattice results for physical quark masses. *J. High Energy Phys.* **2004**, *0404*, 050. [CrossRef]
93. Ratti, C.; Thaler, M.A.; Weise, W. Phases of QCD: Lattice thermodynamics and a field theoretical model. *Phys. Rev. D* **2006**, *73*, 014019. [CrossRef]
94. Borsányi, S.; Fodor, Z.; Katz, S.D.; Krieg, S.; Ratti, C.; Szabó, K. Fluctuations of conserved charges at finite temperature from lattice QCD. *J. High Energy Phys.* **2012**, *1*, 138. [CrossRef]
95. Bazavov, A.; Ding, H.T.; Hegde, P.; Kaczmarek, O.; Karsch, F.; Laermann, E.; Mukherjee, S.; Petreczky, P.; Schmidt, C.; Smith, D.; et al. Freeze-Out Conditions in Heavy Ion Collisions from QCD Thermodynamics. *Phys. Rev. Lett.* **2012**, *109*, 192302. [CrossRef] [PubMed]
96. Borsányi, S.; Fodor, Z.; Hoelbling, C.; Katz, S.D.; Krieg, S.; Szabó, K.K. Full result for the QCD equation of state with 2 + 1 flavors. *Phys. Lett. B* **2014**, *730*, 99–104. [CrossRef]
97. Kurkela, A.; Fraga, E.S.; Schaffner-Bielich, J.; Vuorinen, A. Constraining neutron star matter with Quantum Chromodynamics. *Astrophys. J.* **2014**, *789*, 27. [CrossRef]
98. Farhi, E.; Jaffe, R. Strange Matter. *Phys. Rev. D* **1984**, *30*, 2379. [CrossRef]
99. Nambu, Y.; Jona-Lasinio, G. Dynamical Model of Elementary Particles Based on an Analogy with Superconductivity. 1. *Phys. Rev.* **1961**, *122*, 345–358. [CrossRef]
100. Klevansky, S. The Nambu-Jona-Lasinio model of quantum chromodynamics. *Rev. Mod. Phys.* **1992**, *64*, 649–708. [CrossRef]
101. Buballa, M. NJL model analysis of quark matter at large density. *Phys. Rept.* **2005**, *407*, 205–376. [CrossRef]
102. Klähn, T.; Fischer, T. Vector Interaction Enhanced Bag Model for Astrophysical Applications. *Astrophys. J.* **2015**, *810*, 134. [CrossRef]

103. Klähn, T.; Fischer, T.; Hempel, M. Simultaneous chiral symmetry restoration and deconfinement - Consequences for the QCD phase diagram. *Astrophys. J.* **2017**, *836*, 89. [CrossRef]
104. Fischer, T.; Bastian, N.U.F.; Wu, M.R.; Typel, S.; Klähn, T.; Blaschke, D.B. High-density phase transition paves the way for supernova explosions of massive blue-supergiant stars. *arXiv* **2017**, arXiv:1712.08788. [CrossRef]
105. Haensel, P.; Potekhin, A.Y.; Yakovlev, D.G. Neutron Stars 1: Equation of State and Structure. In *Astrophysics and Space Science Library*; Springer: Berlin, Germany, 2007.
106. Read, J.S.; Lackey, B.D.; Owen, B.J.; Friedman, J.L. Constraints on a phenomenologically parametrized neutron-star equation of state. *Phys. Rev. D* **2009**, *79*, 124032. [CrossRef]
107. Zdunik, J.L.; Haensel, P. Maximum mass of neutron stars and strange neutron-star cores. *Astron. Astrophys.* **2013**, *551*, A61. [CrossRef]
108. Alford, M.G.; Han, S.; Prakash, M. Generic conditions for stable hybrid stars. *Phys. Rev. D* **2013**, *88*, 083013. [CrossRef]
109. Abbott, B.; Abbott, R.; Abbott, T.D.; Acernese, F.; Ackley, K.; Adams, C.; Adams, T.; Addesso, P.; Adhikari, R.X.; Adya, V.B.; et al. GW170817: Observation of Gravitational Waves from a Binary Neutron Star Inspiral. *Phys. Rev. Lett.* **2017**, *119*, 161101. [CrossRef] [PubMed]
110. Ayriyan, A.; Bastian, N.U.; Blaschke, D.; Grigorian, H.; Maslov, K.; Voskresensky, D.N. How robust is a third family of compact stars against pasta phase effects? *Phys. Rev. C* **2018**, *97*, 045802. [CrossRef]
111. Paschalidis, V.; Yagi, K.; Alvarez-Castillo, D.; Blaschke, D.B.; Sedrakian, A. Implications from GW170817 and I-Love-Q relations for relativistic hybrid stars. *Phys. Rev. D* **2018**, *97*, 084038. [CrossRef]
112. Blaschke, D.; Chamel, N. Phases of dense matter in compact stars. *arXiv* **2018**, arXiv:1803.01836. [CrossRef]
113. Annala, E.; Gorda, T.; Kurkela, A.; Vuorinen, A. Gravitational-wave constraints on the neutron-star-matter Equation of State. *Phys. Rev. Lett.* **2017**, *120*, 17270. [CrossRef] [PubMed]
114. Röpke, G. Nuclear matter EoS including few-nucleon correlations. *Nuovo Cim. C* **2017**, *39*, 392.
115. Kuhrts, C.; Beyer, M.; Danielewicz, P.; Röpke, G. Medium corrections in the formation of light charged particles in heavy ion reactions. *Phys. Rev. C* **2001**, *63*, 034605. [CrossRef]
116. Bastian, N.U.; Batyuk, P.; Blaschke, D.; Danielewicz, P.; Ivanov, Y.B.; Karpenko, I.; Röpke, G.; Rogachevsky, O.; Wolter, H.H. Light cluster production at NICA. *Eur. Phys. J. A* **2016**, *52*, 244. [CrossRef]
117. Bastian, N.U.; Blaschke, D.; Röpke, G. Light cluster production at NICA. *Acta Phys. Pol. Suppl.* **2017**, *10*, 899. [CrossRef]
118. Kowalski, S.; Natowitz, J.B.; Shlomo, S.; Wada, R.; Hagel, K.; Wang, J.; Materna, T.; Chen, Z.; Ma, Y.G.; Qin, L.; et al. Experimental determination of the symmetry energy of a low density nuclear gas. *Phys. Rev. C* **2007**, *75*, 014601. [CrossRef]
119. Kowalski, S.; Natowitz, J.B.; Shlomo, S.; Wada, R.; Hagel, K.; Wang, J.; Materna, T.; Chen, Z.; Ma, Y.G.; Qin, L.; et al. Symmetry energy of dilute warm nuclear matter. *Phys. Rev. Lett.* **2010**, *104*, 202501.
120. Wada, R. The Nuclear Matter Symmetry Energy at $0.03 \leq \rho/\rho_0 \leq 0.2$. *Phys. Rev. C* **2012**, *85*, 064618. [CrossRef]
121. Hagel, K.; Natowitz, J.B.; Röpke, G. The equation of state and symmetry energy of low density nuclear matter. *Eur. Phys. J. A* **2014**, *50*, 39. [CrossRef]
122. Typel, S.; Wolter, H.H.; Röpke, G.; Blaschke, D. Effects of the liquid-gas phase transition and cluster formation on the symmetry energy. *Eur. Phys. J. A* **2014**, *50*, 17. [CrossRef]
123. Li, B.A.; Chen, L.W.; Ko, C.M. Recent Progress and New Challenges in Isospin Physics with Heavy-Ion Reactions. *Phys. Rept.* **2008**, *464*, 113–281. [CrossRef]
124. Qin, L. Laboratory Tests of Low Density Astrophysical Equations of State. *Phys. Rev. Lett.* **2012**, *108*, 172701. [CrossRef] [PubMed]
125. Hagel, K.; Wada, R.; Qin, L.; Natowitz, J.B.; Shlomo, S.; Bonasera, A.; Röpke, G.; Typel, S.; Chen, Z.; Huang, M.; et al. Experimental Determination of In-Medium Cluster Binding Energies and Mott Points in Nuclear Matter. *Phys. Rev. Lett.* **2012**, *108*, 062702. [CrossRef] [PubMed]
126. Hempel, M.; Hagel, K.; Natowitz, J.; Röpke, G.; Typel, S. Constraining supernova equations of state with equilibrium constants from heavy-ion collisions. *Phys. Rev. C* **2015**, *91*, 045805. [CrossRef]
127. Batyuk, P.; Blaschke, D.; Bleicher, M.; Ivanov, Y.B.; Karpenko, I.; Merts, S.; Nahrgang, M.; Petersen, H.; Rogachevsky, O. Event simulation based on three-fluid hydrodynamics for collisions at energies available at the Dubna Nuclotron-based Ion Collider Facility and at the Facility for Antiproton and Ion Research in Darmstadt. *Phys. Rev. C* **2016**, *94*, 044917. [CrossRef]

128. Batyuk, P.; Blaschke, D.; Bleicher, M.; Ivanov, Y.B.; Karpenko, I.; Malinina, L.; Merts, S.; Nahrgang, M.; Petersen, H.; Rogachevsky, O. Three-fluid Hydrodynamics-based Event Simulator Extended by UrQMD final State interactions (THESEUS) for FAIR-NICA-SPS-BES/RHIC energies. In Proceedings of the 6th International Conference on New Frontiers in Physics (ICNFP 2017), Kolymbari, Greece, 17–26 August 2017.
129. Ivanov, Y.B. Alternative Scenarios of Relativistic Heavy-Ion Collisions: I. Baryon Stopping. *Phys. Rev. C* **2013**, *87*, 064904. [CrossRef]
130. Ivanov, Y.B. Alternative Scenarios of Relativistic Heavy-Ion Collisions: II. Particle Production. *Phys. Rev. C* **2013**, *87*, 064905. [CrossRef]
131. Ivanov, Y.B. Alternative Scenarios of Relativistic Heavy-Ion Collisions: III. Transverse Momentum Spectra. *Phys. Rev. C* **2014**, *89*, 024903. [CrossRef]
132. Typel, S.; Blaschke, D. A Phenomenological Equation of State of Strongly Interacting Matter with First-Order Phase Transitions and Critical Points. *Universe* **2018**, *4*, 32. [CrossRef]
133. Bastian, N.U.; Blaschke, D. Towards a new quark-nuclear matter EoS for applications in astrophysics and heavy-ion collisions. *J. Phys.* **2016**, *668*, 012042. [CrossRef]
134. Andronic, A.; Braun-Munzinger, P.; Redlich, K.; Stachel, J. Decoding the phase structure of QCD via particle production at high energy. *arXiv* **2017**, arXiv:1710.09425. [CrossRef]
135. Blaschke, D.; Jankowski, J.; Naskret, M. Formation of hadrons at chemical freeze-out. *arXiv* **2017**, arXiv:1705.00169. [CrossRef]

© 2018 by the authors. Licensee MDPI, Basel, Switzerland. This article is an open access article distributed under the terms and conditions of the Creative Commons Attribution (CC BY) license (http://creativecommons.org/licenses/by/4.0/).

Conference Report

The High-Density Symmetry Energy in Heavy-Ion Collisions and Compact Stars

Hermann Wolter

Faculty of Physics, University of Munich, 85748 Garching, Germany; hermann.wolter@lmu.de;
Tel.: +49-179-917-8954

Received: 5 April 2018; Accepted: 16 May 2018; Published: 14 June 2018

Abstract: High-density nuclear symmetry energy is of crucial importance in astrophysics. Information on such energy has been obtained from mass–radius determinations of neutron stars (NSs), and in the future NS mergers will increasingly contribute. In the laboratory, the symmetry energy can be studied in heavy-ion collisions (HICs) at different incident energies over a large range, from very low to several times higher saturation density. Transport theory is necessary to extract the symmetry energy from the typically non-equilibrated nuclear collisions. In this contribution, we first review the transport approaches, their differences, and recent studies of their reliability. We then discuss several prominent observables, which have been used to determine the symmetry energy at high density: collective flow, light cluster emission, and particle production. It is finally argued that the results of the symmetry energy from microscopic many-body calculations, nuclear structure, nuclear reactions, and astrophysics begin to converge but still need considerable improvements in terms of accuracy.

Keywords: nuclear symmetry energy; heavy-ion collisions; transport theory; collective flow; light cluster emission; meson production

1. Introduction

The nuclear equation of state (EoS) specifies the energy density of nuclear matter without Coulomb energy as a function of density, temperature, and asymmetry. For zero temperature, it is usually written in the lowest order in asymmetry as $E(\rho,\delta) = E_0(\rho) + E_{sym}(\rho)\delta^2$, where $\delta = (\rho_n - \rho_p)/\rho$ and ρ_n, ρ_p, and ρ are the neutron, proton, and total densities, respectively. The energy density of symmetric nuclear matter E_0 has been extensively investigated in heavy-ion collisions (HICs) in the past, and a consensus has been reached: that it is rather soft, that is, rises less than linearly with density, and is also momentum dependent [1,2]. The nuclear symmetry energy $E_{sym}(\rho)$ is less well understood. Studies with HICs at energies below about 400 MeV have constrained it fairly well around and below saturation density $\rho_0 \approx 0.16$ fm/c [3]. However, the high-density behavior is still a matter of debate. Microscopic many-body calculations still diverge considerably at higher densities. The reasons are seen in the uncertainty of three-body forces and of short-range isovector correlations and in the question of the strangeness content. The symmetry energy is often represented as an expansion around saturation density as $E_{sym}(\rho) = S_0 + \frac{L}{3}\frac{\rho-\rho_0}{\rho_0} + \frac{K_{sym}}{18}(\frac{\rho-\rho_0}{\rho_0})^2$, where S_0 is the value and L and K_{sym} proportional to the slope and curvature, respectively, of the symmetry energy at saturation. L has been determined to be in the range of 50 to 100 MeV [3] and is a measure of the stiffness of the symmetry energy.

On the other hand, the nuclear symmetry energy is a crucial input for the understanding of astrophysical objects. The structure of neutron stars (NSs) and the explodability of core-collapse supernovae (CCSNe) depend critically on the properties of neutron-rich nuclear matter and thus on the symmetry energy. Astrophysical observations have yielded important constraints on the nuclear symmetry energy. The Tolman–Openheimer–Volkov equation relates the symmetry energy directly to the mass–radius relation of NSs. The observation of NSs with masses of around 2 solar masses already excludes symmetry energies that are too soft [4]. Radii are difficult to measure, as they depend on the interpretation of γ- or X-ray emission and are even more difficult to obtain simultaneously with masses. However, Bayesian analysis methods already also set limits here [5,6]. The satellite experiment Neutron star Interior Composition Explorer (NICER) should provide more and better data.

The very recent discovery of the gravitational wave event GW170817 from a NS merger [7], together with the observation of a short gamma-ray burst (GRB) [8,9] and the optical afterglow, promises further important insights into high-density symmetry energy. In addition to determining the masses of the merging NSs, it was possible to obtain limits on their tidal deformabilities, which are also directly given by the EoS, similarly to the mass–radius relation of NSs [10]. Figure 1 shows an example of the results of interpretations of the NS merger [11,12]. The mass–radius relation for a number of models for the EoS is shown together with the constraints originating from the observation of 2 solar mass NSs, as well as from the limits of the tidal deformabilities of the GW170817 event [13]. It is seen that these constraints together already exclude several models for the EoS. If the possibility of a quark phase in the interior of NSs is considered, the constraints become more complex [14]. Further observations of NS mergers will provide more information in the future.

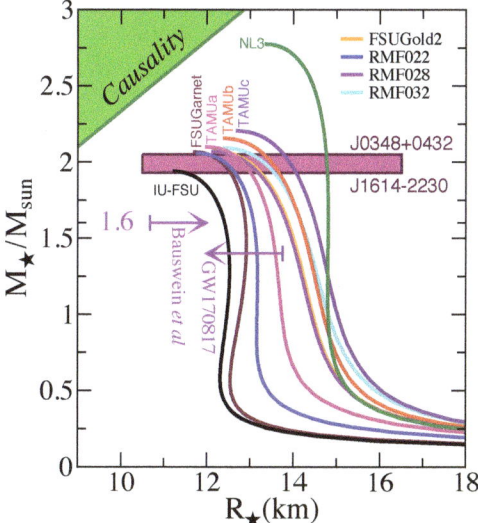

Figure 1. (Color on-line) Mass–radius relation of neutron stars (NSs) predicted for different relativistic mean-field (RMF) models identified in the legend. Shown also are the masses of two heavy NSs and the constraints following from the tidal deformability of the GW event. Figure taken from [11] with permission.

The EoS can also be investigated in terrestrial laboratories in HICs. In energetic collisions, densities of up to several times the saturation density can be reached, similar to those that are thought to exist in NSs. In HICs, one has the possibility to scan the density in certain ranges via the incident energy and colliding masses and the asymmetry by the choice of the collision system. The latter is limited by the asymmetries of available projectiles and targets but is being extended by the use of exotic beams. However, the high-density phase only exists for the short time span of the collision in the order of 1 to 100 fm/c, depending on the energy. In order to draw conclusions from the final asymptotic observables to the properties of the dense medium, the evolution of the collision has to be described in detail. This is achieved by various transport descriptions. As the system in these collisions is out of equilibrium for most of the time, transport also takes into account the non-equilibrium nature. Hydrodynamical treatments, which assume a local equilibrium, and statistical treatments, which assume a global equilibrium, are often not adequate, at least not for the whole process. On the other hand, transport descriptions are complex, and the necessary approximations, the implementations, and the accuracy are issues that also have to be discussed.

This contribution aims to review the issues, the problems, and some of the results of the study of high-density symmetry energy with emphasis on HICs. Thus we first discuss the transport theories and then give some examples of observables that have been used to constrain the symmetry energies. It is beyond the scope here to discuss in more detail than the above the constraints on the symmetry energy from NS observations. A good collection of articles regarding this can be found in the recent special volume on symmetry energy, published by B.A. Li et al. in [2]. Finally, we summarize the present status of these studies on symmetry energy.

2. Theoretical Considerations

2.1. Overview of Transport Theories

Transport theory is necessary to draw conclusions on the EoS from HIC experiments to account for the non-equilibrium nature of the process. Practical transport approaches are derived from non-equilibrium many-body physics by a chain of approximations [15,16]. One usually starts from the Martin–Schwinger real-time Green function formalism, which by a folded time path with forward and backward branches takes into account the time-reversal asymmetry. The hierarchy of many-body densities is truncated by the factorization of the two-body density and by introducing single-particle self-energies. One arrives at the Kadanoff–Baym theory for the non-local densities and self-energies. With a Wigner transform, this is cast into an equation in phase space $\{\vec{r}, \vec{p}\}$. A gradient approximation in the first order in $\hbar$ leads to a semiclassical description, where a quantity $f(\vec{r}, \vec{p}; t)$ can be interpreted as the phase-space probability. At this stage, there are still two independent Green functions, which can be transformed into the above phase-space probability and a spectral function for the off-shell particles. In the quasi-particle approximation, the spectral function is taken to be on-shell in terms of effective momenta and masses. With these approximations, one arrives at an equation of the Boltzmann–Vlasov type for the evolution of the phase-space probability $f(\vec{r}, \vec{p}; t)$ under the influence of a self-consistent mean field $U[f]$ and of two-body collisions as the dissipative mechanism, which non-relativistically reads

$$\frac{\partial f_1}{\partial t} + \frac{\vec{p}}{m}\nabla_{\vec{r}} f_1 - \nabla_{\vec{r}} U \nabla_{\vec{p}} f_1 = \qquad (1)$$
$$(\frac{2\pi}{m})^3 \int d\vec{p}_2 d\vec{p}_3 d\vec{p}_4 |\vec{v}_1 - \vec{v}_2| \sigma_{NN}(\Omega_{12}) \delta(\vec{p}_1 + \vec{p}_2 - \vec{p}_3 - \vec{p}_3)(f_3 f_4 \tilde{f}_1 \tilde{f}_2 - f_1 f_2 \tilde{f}_3 \tilde{f}_4).$$

Here $f_i \equiv f(\vec{r}, \vec{p}_i; t)$ and $\tilde{f}_i = (1 - f_i)$ are the blocking factors, which assure the Pauli principle for the final state of a collision, and which are the essential quantum ingredient in this equation apart from the initialization. The v_i are velocities, and $\sigma_{NN}(\Omega)$ is the in-medium nucleon–nucleon (NN) cross-section. The potential $U[f]$ and the cross-section are either derived from a nuclear energy density functional $E[f]$ or are parametrized. The Coulomb interaction is included separately. If the energy functional explicitly depends on f, then the potential can be momentum dependent, which adds another term to the left-hand side of the transport equation. The physical components of the equation, namely, the potential and in-medium cross-sections, are not independent, but are connected by an approximation scheme for the self-energies. An obvious choice is the Brückner scheme for the effective in-medium interaction, used in a local-density approximation. This has either been used directly [17] or in parametrized Yukawa-like form with meson–nucleon couplings, which depend on the density and may in addition depend on the energy [18,19]. If particle production, for example, of pions and Δ resonances, is to be considered, additional physics input is needed: inelastic cross-sections, potentials of the new particles, their cross-sections, and, possibly, mass distributions of unstable particles, simulating the off-shell effects.

The temperature T does not enter explicitly into the transport equation but rather into the distribution function, which may or may not be well represented by an equilibrium distribution, which could be characterized by a temperature. If the interaction is specified by an energy density functional, then these effects are implicitly included. The same functional is then evaluated for a ground-state distribution function to obtain the EoS and the symmetry energy at zero temperature.

Equation (1) is often referred to as the Boltzmann–Uehling–Uhlenbeck (BUU) equation but is also known by other names. It is a complex non-linear integro-differential equation, which is usually solved by the test particle (TP) method, where the distribution function is represented in terms of finite elements, TPs, as

$$f(\vec{r},\vec{p};t) = \frac{1}{N_{TP}} \sum_{i=1}^{AN_{TP}} g(\vec{r}-\vec{r}_i(t))\,\tilde{g}(\vec{p}-\vec{p}_i(t)). \qquad (2)$$

Here N_{TP} is the number of TPs per nucleon, $\vec{r}_i$ and $\vec{p}_i$ are the time-dependent coordinates and momenta of the TPs, and g and $\tilde{g}$ are the shape functions in coordinate and momentum space (e.g., δ-functions or Gaussians), respectively. Upon inserting this ansatz into the left-hand side of Equation (2), Hamiltonian equations of motion for the TPs are obtained, $\frac{d\vec{r}_i}{dt} = \vec{p}_i$ and $\frac{d\vec{p}_i}{dt} = -\vec{\nabla}_{r_i} U$. The collision term is simulated stochastically, by performing TP collisions with a probability depending on the cross-section and obeying the Pauli principle for the final state according to blocking factors $\bar{f}_i = (1-f_i)$.

A second family of transport approaches is the quantum molecular dynamics (QMD) model, in which the evolution of the collision is formulated in terms of the evolution of the coordinates $R_i(t)$ and momenta $P_i(t)$ of the individual nucleons, similarly to as in classical molecular dynamics, but with particles of finite width representing minimum nucleon wave packets, with the width usually assumed to be constant. These move under the influence of NN forces. The method can also be viewed as being derived from the time-dependent Hartree (TDH) method with a product trial wave-function of single-particle states in Gaussian form. One obtains equations of motion of the same form as in BUU in this case for the coordinates of the wave packets. There is also a version of anti-symmetrized molecular dynamics (AMD) [20], which takes into account the anti-symmetrization of the wave packets. The equations of motion become non-local but are of similar type. Additionally, in QMD and AMD, a stochastic two-body collision term is introduced and treated in very much the same way as in BUU, but now for nucleons and the full NN cross-section. There are also relativistic formulations for both approaches using relativistic density functionals. A review of the BUU method is given in [21], while the QMD method is reviewed in [22].

2.2. Fluctuations

The main difference between the two approaches lies in the number of fluctuations and correlations in the representation of the phase-space distribution. In the BUU approach, the phase-space distribution function is seen as a smooth function of coordinates and momenta and can be increasingly better approximated by increasing the number of TPs. In the limit of $N_{TP} \to \infty$, the TP method provides an exact solution of the BUU equation, which is strictly deterministic and has no fluctuations. However, fluctuations are a necessary companion of dissipative dynamics, as expressed by the dissipation–fluctuation theorem. In practice, they are important in the expansion phase of a HIC, which often proceeds through mechanically unstable conditions and may lead to fragmentation of the residual nuclei. If such phenomena are to be described, one has to add a fluctuation term to the equation, which leads to the Boltzmann–Langevin equation. Approximate treatments of fluctuations in HICs have been implemented in [23–26]. In practice, N_{TP} is finite, in the order of 50 to 100 depending on the assumed shape of the TP, which leads to numerical fluctuations. In early treatments, these have been gauged to reproduce the unstable properties of the medium. In QMD, fluctuations are present because of the intrinsically finite number of wave packets in the representation of the phase space. The fluctuations are smoothed and regulated by the choice of the width parameter of the wave packets. In addition, classical correlations are present if explicit two-body interactions are used. QMD can be seen as an event generator solving the time evolution of different events independently. Event-by-event fluctuations are not suppressed by increasing the number of events.

The results of simulations with the two methods are thus expected to be similar, although not necessarily identical, as far as one-body observables are concerned. Larger differences are expected for observables depending on fluctuations and correlations, such as the production of clusters and intermediate-mass fragments. Generally, the description of observables going beyond the mean-field level is a question under active discussion in transport theory. In the experiment, copious numbers of light clusters and fragments were observed in HICs, particularly at lower energies.

2.3. Code Comparison

In addition to these more fundamental differences between transport approaches, there are also differences that are caused by different implementations of the highly complex transport theories. Analyses of experimental data with seemingly similar physics input have led to rather different conclusions. The analyses of the FOPI π^-/π^+ ratios, as discussed below, represent an example. In order to reach a better understanding of possible reasons, a code-comparison project was undertaken. In a first comparison, results for standard Au+Au collisions at 100 and 400 AMeV incident energies with identical physics input were compared [27]. Eighteen commonly used transport codes, of both BUU and QMD type, participated. Comparisons of the stability of the initialized configuration; of the collision rates and the effectiveness of Pauli blocking; and of observables, such as the longitudinal and transverse flow, were discussed. The results for the flow parameter, that is, for the slope of the transverse flow at midrapidity, are shown in Figure 2 for the different codes. We are more interested here in the qualitative behavior of the different codes. It would go beyond the scope to identify the different codes and their properties in detail; one should refer rather to [27]. While there is a general agreement, quantitatively the differences were found to depend on the incident energy and amounted to approximately 30% at 100 MeV/nucleon and 15% at 400 MeV/nucleon, respectively.

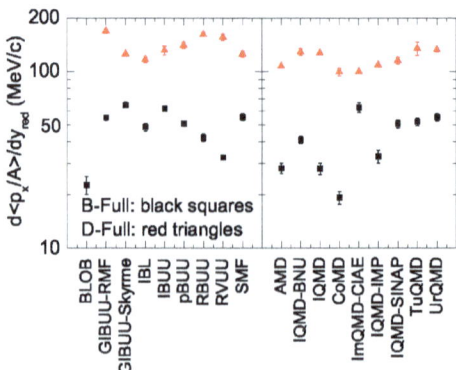

Figure 2. (Color on-line) Comparison of flow parameters (transverse momentum at midrapidity) in Au+Au collisions at 100 AMeV (square (black) symbols) and 400 AMeV (triangular (red) symbols) for codes of Boltzmann–Uehling–Uhlenbeck (BUU) type (left) and quantum molecular dynamics (QMD) type (right). For the identification of the different codes, see [27], from where this figure is taken.

In order to better understand these still appreciable differences, this was followed up by a comparison of calculations in a box with periodic boundary conditions, which approximates a calculation in infinite nuclear matter. The advantages of box calculations are that common initial conditions of a given density and temperature are easier to achieve, that the different ingredients of the transport codes can be tested separately, and that the results can be compared in many cases to exact limits. A first comparison of this kind investigated the treatment of the collision integral by switching off the mean field, without and with including the Pauli blocking, that is, by comparing Cascade calculations [28]. Without the Pauli blocking, the codes agreed to within a few percent among each other and with the exact limits, and the remaining small differences could be understood in most cases. Including the Pauli blocking, there were considerable differences. The collision rates for initializations of normal density and temperature $T = 5$ MeV for the different codes are shown in Figure 3 and compared to the exact limit, represented there by the line labeled CBOP2T5. Again, we do not discuss the different codes in detai; this is done in [28]. The results are shown for the first time step, for which the momentum distributions were still identical, and were time-averaged over the evolution, during which the momentum distribution changed in different ways for the different codes. These differences are seen not to be very important relative to the considerable differences seen regarding the exact limit. There are systematic differences between BUU and QMD and also between codes of the same kind. The main reason for the differences was found to be from fluctuations when calculating the occupation probability in the final phase-space cell of the collision partners, which led to differences in the Pauli blocking. These fluctuations, and their differences in BUU and QMD approaches, are discussed above. The consequence of the larger fluctuations in QMD codes was seen in systematically higher collision rates. On the other hand, the proper treatment of fluctuations in transport codes is still debated. The box comparisons are presently continued for the mean-field propagation and for pion production. It is anticipated that these comparisons can provide benchmark calculations against which existing and new transport codes can be compared.

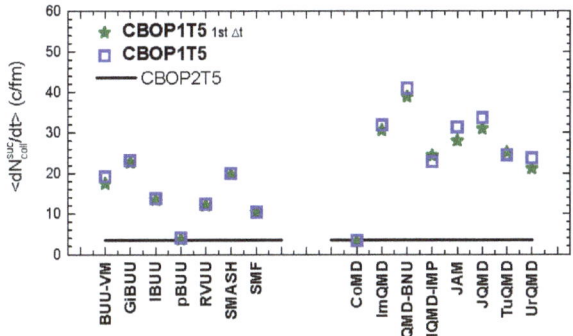

Figure 3. (Color on-line) Comparison of collision rates in box calculations at normal density and temperature $T = 5$ MeV (denoted as CBOP1T5) for different Boltzmann–Uehling–Uhlenbeck (BUU)-type (**left**) and quantum molecular dynamics (QMD)-type (**right**) codes. Star symbols give the result for the first time step; square symbols give time-averaged rates. They are compared to a numerical, essentially exact, result for this case (solid line, denoted as CBOP2T5). For the identification of different codes, see [28], from where this figure is taken.

3. Symmetry Energy in Heavy-Ion Collisions

3.1. Overview

In intermediate-energy HICs, dense and hot matter is formed for very short time periods in the order of 10^{-22} fm/c. The connection of these conditions to the asymptotic observables is provided by transport approaches as discussed above. Nuclear matter in HICs is only moderately asymmetric with asymmetries $\delta < 0.2$. Thus in HICs, both the EoS of symmetric nuclear matter E_0 and the symmetry energy E_{sym} are involved, with the symmetric matter giving the greatest contribution. In the past, the emphasis for HICs was put on the investigation of the symmetric EoS. It is thought that this has been fairly well determined by a variety of probes, mainly connected to collective phenomena [1] and to K-meson ratios [29]. The sensitivity to the symmetry energy part of the EoS can be increased by focusing on ratios or differences of observables for isospin partners, such as, for example, protons and neutrons or positively and negatively charged pions, hoping that the still-existent uncertainties in the isoscalar part of the EoS cancel out to a large extent.

The densities reached in HICs depend on the incident energy, on the system masses, and on the impact parameter of the collision. For peripheral collisions, the spectator–participant model has been useful, as it separates the warmly heated spectators of initially normal density from the strongly excited fireball, which on the other hand is observed in central collisions of similar-size ions. For energies of up to a few hundred MeV per nucleon in central collisions, densities about 20–30% in excess of saturation density are reached. In this regime, the interest is particularly on the decay and subsequent fragmentation, which contain rich information on the symmetry energy at densities below saturation in the isospin sharing of the fragments. Particularly, the value and the slope of the symmetry energy at saturation energy are extracted in these studies [3,30].

In relativistic collisions, up to several GeV per nucleon densities of up to 2–3 ρ_0 are reached. The collective motion of the final particles represents direct evidence of the symmetry energy, because the neutron–proton differences are directly driven by the isospin-dependent part of the mean field and thus by the symmetry energy. Additionally, nucleons and light clusters are emitted early in the collision, and the ratios of this pre-equilibrium emission gives complementary information on the symmetry energy at high density. Finally, in energetic collisions, mesons are produced in increasing multiplicities. The production of isospin partners depends on the asymmetry of the matter in which they are produced, and thus they are indirect probes of the effects on the symmetry energy. In the following, we discuss examples of such observables to demonstrate the possibilities and challenges to determine the density dependence of the symmetry energy in HICs.

3.2. Collective Flow

The flow of nuclear matter out of the interaction region indicates compression. The flow can be characterized by a kinetic momentum tensor. Non-sphericity indicates dynamic effects originating from the EoS, whose orientation with respect to the beam direction represents the collective sideward or directed flow. A difference in the two minor axes indicates the existence of elliptic flow. It has become customary to express both directed and elliptic flows, as well as possible higher-flow components, by means of a Fourier decomposition of the azimuthal distributions measured with respect to the orientation of the reaction plane ϕ_R [31]:

$$\frac{dN}{d(\phi - \phi_R)} = \frac{N_0}{2\pi}\left(1 + 2\sum_{n \geq 1} v_n \cos n(\phi - \phi_R)\right), \qquad (3)$$

where N_0 is the azimuthally integrated yield. The coefficients $v_n \equiv \langle \cos n(\phi - \phi_R) \rangle$ are functions of particle type, impact parameter, rapidity y, and the transverse momentum p_t; v_1 and v_2 are the directed and elliptic flows, respectively.

Directed flow indicates the repulsion or attraction of the colliding nuclei and thus the deflection in the reaction plane. It originates from the compressional properties of the EoS, but also from the momentum dependence of the potential, particularly at higher incident energies. Thus it may also reverse its sign at higher energies, such as is shown, for example, by Ivanov [32]. The elliptic flow describes the squeeze-out of the participant matter perpendicular to the reaction plane and is thus very directly connected to compression. It is a promising probe of the stiffness of the EoS and has been investigated in detail in symmetric and asymmetric nuclear matter.

An excitation function of the elliptic flow of $Z = 1$ particles in ^{197}Au+^{197}Au collisions compiled from various experiments [33] is shown in Figure 4. At lower energies, the collective angular momentum in the mean-field-dominated dynamics causes the observed in-plane enhancement of emitted reaction products; that is, $v_2 > 0$. Squeeze-out perpendicular to the reaction plane, that is, $v_2 < 0$, as a result of shadowing by the spectator remnants is observed at incident energies between about 150 and 4 GeV/nucleon with a maximum near 400 MeV/nucleon. Thus elliptic flow at these energies is particularly sensitive to the compression energy and thus to the EoS. The figure also shows that elliptic flow can be measured quite precisely, as demonstrated by the good agreement of data sets from different experiments in the overlap regions of the studied intervals in collision energy [33–35].

To determine the high-density symmetry energy, the neutron–proton differential measurement of elliptic flow is particularly promising [36]. It involves the more difficult measurement of neutron flow. Such measurements have been performed recently by the ASY-EOS collaboration at the Gesellschaft für Schwerionenforschung (GSI) in Darmstadt, Germany [37], and a result of this experiment and the analysis using a QMD code is shown in Figure 5. Here the flow ratio of neutrons over all charged particles is shown as a function of the transverse momentum per nucleon. The data are compared to two predictions using the same momentum-dependent isoscalar field and two versions of the symmetry energy, characterized by an exponent γ of a polynomial parametrization of the symmetry energy $E_{sym}^{pot} = C(\rho/\rho_0)^\gamma$, $C \approx 12$ MeV. A best fit yielded a γ value of around 0.75, which represents a somewhat soft symmetry energy. It was also shown that the sensitivity of the experiment to density was in the range of 1–2 ρ_0; that is, it tests densities higher than saturation density. This, together with a previous experiment [38] with lower sensitivity, was a first direct determination of the symmetry energy above saturation density. Experiments at the facility NICA at the Joint Institute for Nuclear Reactions at Dubna, Russia or the Facility for Anti-proton and Ion Research (FAIR) at Darmstadt, Germany, should be able to probe even higher densities and explore the region of the deconfinement phase transition.

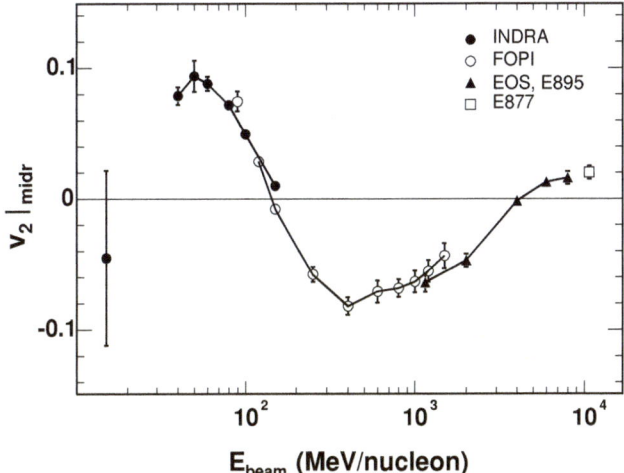

Figure 4. Elliptic flow parameter v_2 at midrapidity for ^{197}Au+^{197}Au collisions at intermediate impact parameters ($\approx$5.5–7.5 fm) as a function of incident energy. The filled and open circles represent the INDRA and FOPI data for $Z = 1$ particles, the triangles represent the equation-of-state (EoS) and E895 data for protons, and the squares represent the E877 data for all charged particles. Figure taken from [33] with permission, where the references to the data are also given.

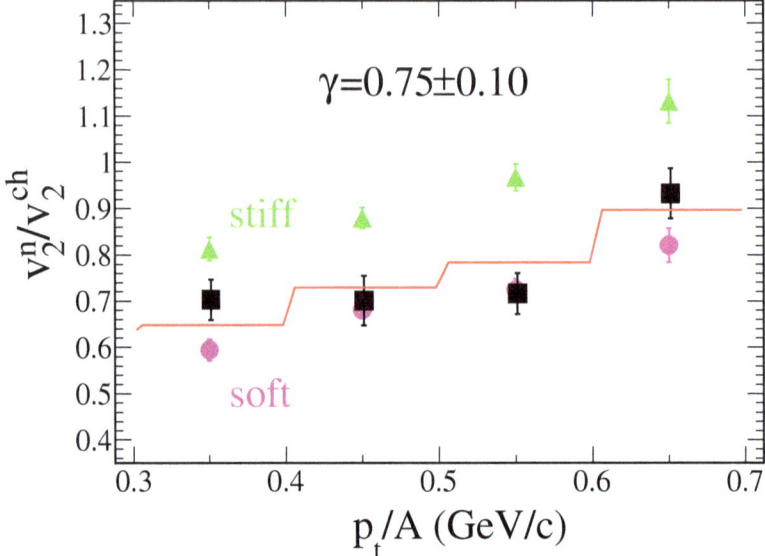

Figure 5. Elliptic flow ratio of neutrons over all charged particles for central ($b < 7.5$ fm) collisions of ^{197}Au+^{197}Au at 400 MeV/nucleon as a function of the transverse momentum per nucleon p_t/A. The black squares represent the experimental data; the green triangles and purple circles represent the QMD predictions for stiff ($\gamma = 1.5$) and soft ($\gamma = 0.5$) power-law exponents of the potential term, respectively. The solid line is the result of a linear interpolation between the predictions, weighted according to the experimental errors of the included four bins in p_t/A and leading to the indicated $\gamma = 0.75 \pm 0.10$. Figure taken from [37].

3.3. Light Cluster Emission

Pre-equilibrium nucleons and light clusters are emitted early in energetic collisions and are distinguished from equilibrium evaporation particles by their higher energies. Ratios of isospin partners should be sensitive to the symmetry potential, to both the value of the potential at the relevant density and to the momentum dependence. The latter can be characterized by an effective mass as $m^*/m = (1 + (m/\hbar^2 k)\partial U/\partial k)^{-1}$. The momentum dependence of the isospin-dependent potential U_τ, $\tau = \{n, p\}$, then leads to an isospin splitting of the effective masses for neutrons and protons. The effect of these properties of the isospin potential can be seen in Figure 6, where for a collision ^{136}Xe +^{124}Sn at 150 AMeV, the (single) yield ratio of neutrons over protons, $R(n/p; ^{136}$Xe +^{124}Sn$)$, is shown as a function of the transverse energy [39]. The calculations are shown for four combinations of the density dependence (soft vs. stiff) and the effective masses ($m_n^* > m_p^*$ vs. $m_n^* < m_p^*$). For lower transverse energies of the emitted particles, the density dependence of the symmetry energy is dominant, and the soft case has the larger repulsion for neutrons below ρ_0, while for higher transverse energies, the momentum dependence dominates and the lighter effective masses are emitted more readily. This behavior opens the possibility to separate the density and momentum dependence of the symmetry energy. At higher incident energies, the momentum dependence dominates increasingly more [40]. Similar behavior is seen for isospin partners of light clusters, such as $t/^3$He in the right panel of Figure 6.

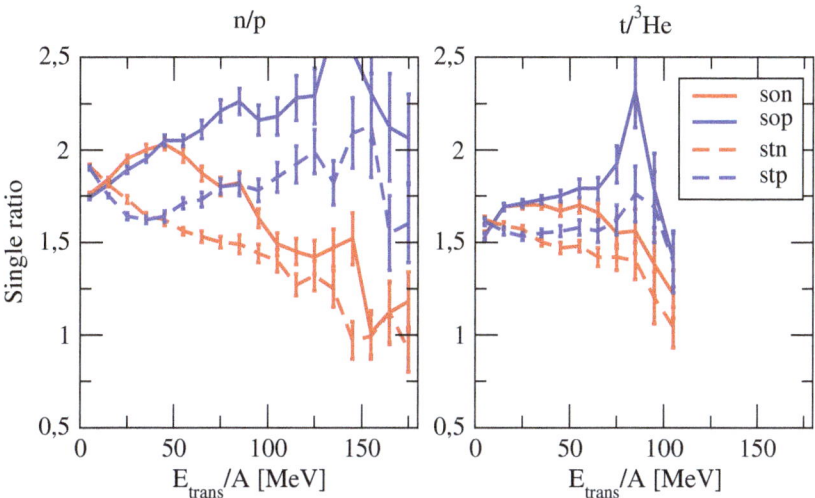

Figure 6. (**Left**) The neutron–proton ratio in $^{136}Xe + ^{124}Sn$ collisions at 150 AMeV for different choices of the symmetry energy and ordering of the effective masses, as indicated in the legend and discussed in the text: solid and dashed lines for stiff and soft symmetry energies, respectively, and red and blue lines for $m_n^* > m_p^*$ and $m_n^* < m_p^*$, respectively. Thus the labels son and sop indicate soft symmetry energy with $m_n^* > m_p^*$ and $m_n^* < m_p^*$, respectively, and the labels stn and stp are analogous for stiff symmetry energy. (**Right**) The corresponding tritium over ^3He ratio. Figure taken from [41].

In Figure 7, the n/p emission is compared to data from Michigan State University (MSU) for Sn+Sn collisions [42]. Here, not the single ratio of neutrons over protons is compared, as in Figure 6, but the ratios of the ratios for two reactions, the double ratios (DRs):

$$\text{DR}(n/p; ^{124}Sn; ^{112}Sn) = \frac{R(n/p; ^{124}Sn + ^{124}Sn)}{R(n/p; ^{112}Sn + ^{112}Sn)}, \quad (4)$$

where experimental differences in the efficiency of neutron and proton detection are expected to cancel out. Shown are the "coalescence-invariant" DRs, for which neutrons and protons of all experimentally measured clusters with $A \leq 4$ are included into the coalescence-invariant cross-sections. On the other hand, the DRs for only the free protons and neutrons show little similarity to the DRs in BUU calculations [42]. This is caused by the difficulty to describe light cluster formation in transport calculations, which depends sensitively on dynamical few-body correlations that are not accounted for in the usual BUU and only classically accounted for in QMD calculations. The experimental coalescence-invariant DRs are compared to calculations in Figure 7 with two Skyrme-type density functionals, which have a very similar stiffness of the symmetry energy of $L \approx 60$ MeV but different orderings of the effective masses. The SLy4 EoS with $m_n^* < m_p^*$ seems to fit the data somewhat better, but better data are needed for the higher-energy part of the spectrum.

Light cluster emission is of interest at both lower energies and densities in the Fermi energy regime and at higher energies and densities in the range of the NICa experiments. In the first case, it was used to investigate the symmetry energy of nuclear matter at very low densities, which exists in the expansion phase of a central low-energy reaction [43]. The ratio of cluster yields between two reactions of different isospin gave information about the change in the chemical potentials and thus about the symmetry energy via isoscaling [44,45]. At the same time, the cluster ratios gave information on the density of about 0.1–0.001 ρ_0 and temperatures of a few MeV [46]. The formation of light clusters in low-density warm matter was also confirmed by quantum statistical calculations, which take into account the medium dependence of the clusters and their eventual dissolution with increasing density [47]. Because the formation of clusters is favored by their binding energy relative to a homogeneous medium of free nucleons, the symmetry energy remains finite as the density approaches zero. The range of densities and temperatures in such investigations is also in the range of conditions in the neutrino-sphere of core-collapse supernovae and thus gives a connection between HICs and astrophysics.

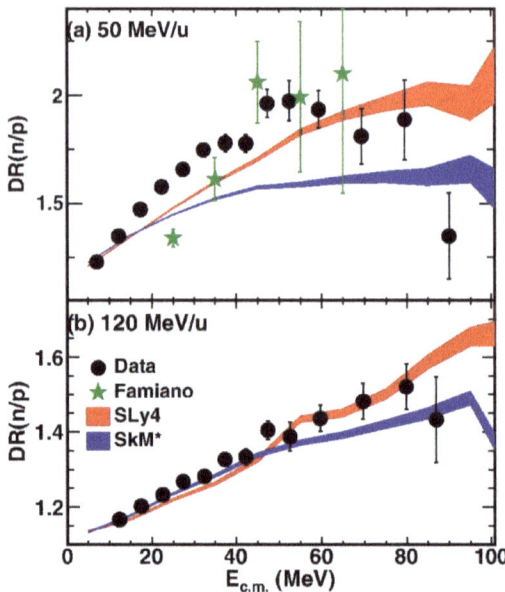

Figure 7. (Color on-line) Double ratios (DRs) of neutron over proton yields in reactions of $^{124}Sn+^{124}Sn$ over $^{112}Sn+^{112}Sn$ at energies of 50 and 120 AMeV as a function of the center-of-mass energy of the emitted particles. The experimental data represent coalescence-invariant cross-sections, for which neutrons and protons of all experimentally measured clusters with $A \leq 4$ are included. The calculations are Boltzmann–Uehling–Uhlenbeck (BUU) calculations with the two Skyrme-like functionals SKM and SLy4, which have a different ordering of the neutron and proton effective masses but are otherwise very similar; see text. The figure is taken from [42] with permission.

Cluster emission could also be an interesting probe for the EoS at the high densities reached in ultrarelativistic HICs, such as in NICa experiments. As discussed in [48,49], clusters may exist and survive in HICs near the deconfinement phase transition, and their flow can be used as a probe of the phase transition. Thus cluster formation in HICs is both an important probe and at the same time a challenge to transport descriptions.

3.4. Particle Production

The n/p asymmetry of the compressed system also determines the ratio of newly produced particles, which thus can serve as indicators of the symmetry energy in the high-density phase. Pions are produced predominantly via the Δ resonances $NN \rightarrow N\Delta$ and the subsequent decay $\Delta \rightarrow N\pi$. The ratio of the isospin partners π^-/π^+ can thus serve as a probe of the high-density symmetry energy. As analyzed in [50], there are competing effects on the Δ and pion production from the isospin-dependent mean fields and the Δ production threshold conditions.

The results from recent theoretical analyses of the π^-/π^+ ratio using different models of symmetry energies and different program codes are collected in the right panel of Figure 8, while the corresponding symmetry energy density dependencies are shown in the left panel [51–54].

These are compared to the data of the Four-PI collaboration (FOPI) [55]. For each model, the results for two parameter sets of different stiffness are shown (stiffer: blue; softer: red). As is seen, the results of the different models are very different not only quantitatively; even the trend with the stiffness of the symmetry energy is not consistent. A reason may lie in different modeling of the $\Delta - \pi$ dynamics, as well as in the competing mean-field and threshold effects, for which slightly different treatments might lead to large differences [50]. The code comparison of pion production, mentioned above, should further serve to aid the understanding of these discrepancies. This issue needs clarification in view of the sensitivity of the pion observables and the data situation. More information should be gained by discussing not the energy-integrated yield ratios but the spectral behavior of the ratio, as different energy pions are expected to probe different stages of the evolution of the reaction [56].

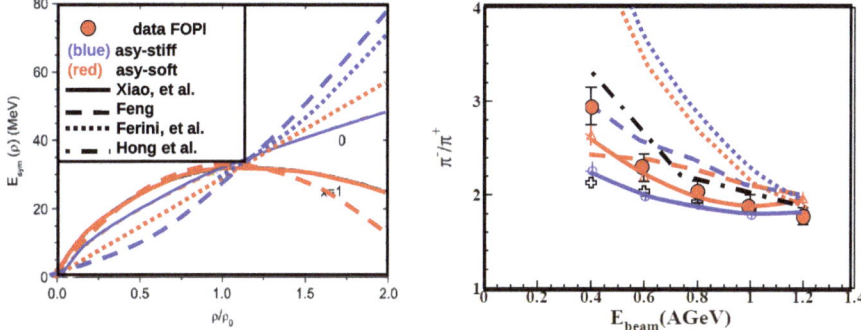

Figure 8. (**Right**) The π^-/π^+ ratio in $Au + Au$ collisions as a function of incident energy as measured by the FOPI collaboration and calculated by different groups, as indicated in the legend of the left panel and discussed in the text. The blue and red lines refer to the softer and stiffer symmetry energies, respectively, used in the different models, which are identified by the signature of the line. (**Left**) This panel shows the density dependence of the symmetry energy used in the models. Figure taken from [41].

It has also been suggested that the ratio of the anti-strange kaon isospin partners, K^0/K^+, could be a useful observable for the symmetry energy [57]. Indeed, kaon production has been one of the most useful observables to determine the EoS of symmetric nuclear matter [29]. The anti-strange kaons weakly interact with nuclear matter and are thus a direct probe of the dense matter in which they are produced. Theoretical analyses show a similar, if not greater, sensitivity to the symmetry energy compared to pion ratios.

4. Discussion and Summary

Constraints for the density dependence of the nuclear symmetry energy determined from HICs are shown in the left panel of Figure 9 for the range of densities presently explored, which also includes some results from nuclear-structure studies. We briefly discuss the various constraints shown in the figure. At very low densities, the symmetry energy was determined from the light cluster yields of the decay of the participant in low-energy HICs (solid triangles) [46], as discussed in Section 3.3. The very precise description of masses of nuclei by energy density functionals constrained the symmetry energy around saturation density—in fact about 40% below—because of the surface contribution, as is shown in the figure (solid circle and square) [3]. The analysis of the shift of the isobaric analog resonances (IAS) yielded rather stringent constraints on the asymmetry density in nuclei (dotted contour) [58]. Transport analysis of HICs, mainly of Sn+Sn but also other systems, at energies from 50 to about 200 AMeV probed the symmetry energy below saturation density from observations of isospin transport between the residual fragments (gray contour) [30]. The constraints from n/p flow ratios at the relativistic energies discussed in Section 3.2 [37] are shown by the red area (as well as the yellow area from an earlier measurement with smaller precision [38]) in the density region of 1.5–2 ρ_0, where they were the most sensitive. One result of the analysis of the π^-/π^+ ratio is shown by the blue line [51], which is also shown in Figure 8. As discussed there, this result is controversial and in any case was in conflict with the result of the flow measurements.

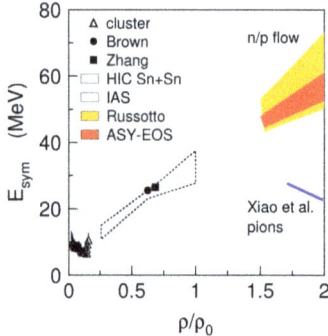 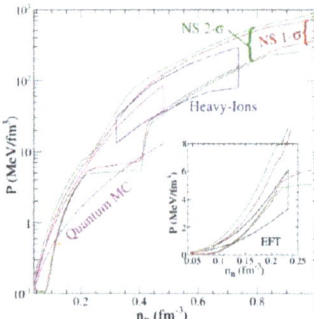

Figure 9. (**Left panel**) Constraints deduced for the density dependence of the zero-temperature symmetry energy from the ASY-EOS data (red area, [37]) in comparison with the FOPI-LAND result (yellow area, [38]) as a function of the reduced density ρ/ρ_0. The low-density results of [30,58–60] as reported in [3] are given by the symbols, the gray area (heavy-ion collision—HIC), and the dashed contour (isobaric analog resonances—IAS); see also text. Figure taken from [37]. (**Right panel**) Synopsis of constraints on the equation of state (EoS) of neutron star (NS) β-stable matter from microscopic calculations (quantum Monte Carlo and effective field theory (EFT)), heavy-ion collisions, and NS observations in a pressure–density diagram. Figure taken from A. Steiner et al. [6] with permission.

The results of the density dependence on the symmetry energy from the nuclear structure and reactions were seen to converge reasonably well; the disagreements from the pion ratio observations will hopefully be clarified via the code comparison investigations discussed in Section 2.3. In the right panel of Figure 9, these results are contrasted with results from microscopic calculations; constraints from HICs, in this case from [1]; and constraints from NS observations. The microscopic calculations were performed with chiral effective field theory (EFT) using various many-body techniques [61,62]. The NS constraints originate from mass–radius studies and from the maximum mass of observed NSs [5,6], but constraints from NS mergers are not yet given. It is seen that overall, the results tended to converge, but a reduction in the uncertainties from all sources, many-body calculations, structure, reactions, and astrophysics is expected in the future.

Acknowledgments: Hermann Wolter acknowledges support by the Universe Cluster of Excellence and by the Heisenberg-Landau program, both of the German Research Foundation (DFG). The author wishes to thank his collaborators and colleagues, whose work has been used in this overview.

Conflicts of Interest: The author declares no conflict of interest.

References

1. Danielewicz, P.; Lacey, R.; Lynch, W.G. Determination of the equation of state of dense matter. *Science* **2002**, *298*, 1592.
2. Li, B.A.; Ramos, À.; Verde, G.; Vidaña, I. Topical Issue on the Nuclear Symmetry Energy. *Eur. Phys. J. A* **2014**, *50*, 1–3.
3. Horowitz, C.J.; Brown, E.F.; Kim, Y.; Lynch, W.G.; Michaels, R.; Ono, A.; Piekarewicz, J.; Tsang, M.B.; Wolter, H.H. A way forward in the study of the symmetry energy: experiment, theory, and observation. *J. Phys. G* **2014**, *41*, 093001.

4. Demorest, P.D.; Pennucci, T.; Ransom, S.M.; Roberts, M.S.E.; Hessels, J.W.T. A two-solar-mass neutron star measured using Shapiro delay. *Nature* **2010**, *467*, 1081–1083.
5. Steiner, A.W.; Lattimer, J.M.; Brown, E.F. The equation of etate from observed masses and radii of neutron stars. *Astrophys. J.* **2010**, *722*, 33.
6. Steiner, A.W.; Lattimer, J.M.; Brown, E.F. The neutron star mass-radius relation and the equation of state of dense matter. *Astrophys. J. Lett.* **2013**, *765*, L5.
7. Abott, P.B.; Abbott, R.; Abbott, T.D.; Acernese, F.; Ackley, K.; Adams, C.; Adams, T.; Addesso, P.; Adhikari, R.X.; Adya, B.; et al. GW170817: Implications for the stochastic gravitational-wave background from compact binary coalescences. *Phys. Rev. Lett.* **2017**, *119*, 161101.
8. Goldstein, A.; Veres, P.; Burns, E.; Briggs, M.S.; Hamburg, R.; Kocevski, D.K.; Wilson-Hodge, C.A.; Preece, R.D.; Poolakkil, S.; Roberts, O.J.; et al. An ordinary short gamma-ray burst with extraordinary implications: Fermi-GBM detection of GRB 170817A. *Astrophys. J. Lett.* **2017**, *848*, L14.
9. Savchenko, V.; Ferrigno, C.; Kuulkers, E. INTEGRAL detection of the first prompt gamma-ray signal coincident with the gravitational-wave event GW170817. *Astrophys. J. Lett.* **2017**, *848*, L15.
10. Hinderer, T. Tidal LOVE numbers of neutron stars. *Astrophys. J.* **2008**, *677*, 1216.
11. Fattoyev, F.J.; Piekarewicz, J.; Horowitz, C.J. Neutron skins and neutron stars in the multimessenger era. *Phys. Rev. Lett.* **2018**, *120*, 172702.
12. Krastev, P.G.; Li, B.A. Imprints of the nuclear symmetry energy on the tidal deformability of neutron stars. *arXiv* **2018**, arXiv:1801.04620.
13. Bauswein, A.; Just, O.; Janka, H.-T.; Stergioulas, N. Neutron-star radius constraints from GW170817 and future detections. *Astrophys. J. Lett.* **2017**, *850*, L34.
14. Paschalides, V.; Yagi, K.; Alvarez-Castillo, D.; Blaschke, D.; Sedrakian, A. Implications from GW170817 and I-Love-Q relations for relativistic hybrid stars. *Phys. Rev. D* **2018**, *97*, 084038.
15. Danielewicz, P. Quantum theory of nonequilibrium processes I. *Ann. of Phys.* **1984**, *152*, 239–304.
16. Botermans, W.; Malfliet, R. Quantum transport theory of nuclear matter. *Phys. Rep.* **1990**, *198*, 115–194.
17. Hofmann, F.; Keil, C.M.; Lenske, H. Density dependent hadron field theory for asymmetric nuclear matter and exotic nuclei. *Phys. Rev. C* **2001**, *64*, 034314.
18. Fuchs, C.; Lenske, H. Rearrangement in the density-dependent field theory of relativistic nuclei. *Phys. Lett. B* **1995**, *345*, 355.
19. Fuchs, C.; Lenske, H.; Wolter, H.H. Density dependent hadron field theory. *Phys. Rev. C.* **1995**, *52*, 3043.
20. Ono, A.; Horiuchi, H.; Maruyama, T.; Ohnishi, A. Antisymmetrized version of molecular dynamics with two-nucleon collisions and its application to heavy ion reactions. *Prog. Theor. Phys.* **1992**, *87*, 1185.
21. Bertsch, G.F.; Gupta, S.D. A guide to microscopic models for intermediate energy heavy ion collisions. *Phys. Rep.* **1988**, *160*, 189.
22. Aichelin, J. Quantum molecular dynamics: A dynamical microscopic n-body approach to investigate fragment formation and the nuclear equation of state in heavy ion collisions. *Phys. Rep.* **1991**, *202*, 233–360.
23. Napolitani, P.; Colonna, M. Bifurcations in Boltzmann—Langevin one body dynamics for fermionic systems. *Phys. Lett. B* **2013**, *726*, 382–386.
24. Napolitani, P.; Colonna, M. Frustrated fragmentation and re-aggregation in nuclei: A non-equilibrium description in spallation. *Phys. Rev. C* **2015**, *92*, 034607.
25. Colonna, M.; di Toro, M.; Guarnera, A.; Maccarone, S.; Zielinska-Pfabé, M.; Wolter, H.H. Fluctuations and dynamical instabilities in heavy-ion reactions. *Nucl. Phys. A* **1998**, *642*, 449–460.
26. Colonna, M. Fluctuations and Symmetry Energy in Nuclear Fragmentation Dynamics. *Phys. Rev. Lett.* **2013**, *110*, 042701.
27. Xu, J.; Chen, L.-W.; Tsang, M.B.; Wolter, H.H.; Zhang, Y.-X.; Aichelin, J.; Colonna, M.; Cozma, D.; Danielewicz, P.; Feng, Z.-Q.; et al. Understanding transport simulations of heavy-ion collisions at 100 A and 400 A MeV: Comparison of heavy-ion transport codes under controlled conditions. *Phys. Rev. C* **2016**, *93*, 044609.
28. Zhang, Y.-X.; Wang, Y.-J.; Colonna, M.; Danielewicz, P.; Ono, A.; Tsang, M.B.; Wolter, H.H.; Xu, J.; Chen, L.-W.; Cozma, D.; et al. Comparison of heavy-ion transport simulations: Collision integral in a box. *Phys. Rev. C.* **2018**, *97*, 034625.
29. Fuchs, C.; Faessler, A.; Zabrodin, E.; Zheng, Y.-M. Probing the nuclear equation of state by K^+ production in heavy-ion collisions. *Phys. Rev. Lett.* **2001**, *86*, 1974.

30. Tsang, M.B.; Stone, J.R.; Camera, F.; Danielewicz, P.; Gandolfi, S.; Hebeler, K.; Horowitz, C.J.; Lee, J.; Lynch, W.G.; Kohley, Z. et al. Constraints on the symmetry energy and neutron skins from experiments and theory. *Phys. Rev. C* **2012**, *86*, 015803.
31. Poskanzer, A.M.; Voloshin, S.A. Methods for analyzing anisotropic flow in relativistic nuclear collisions. *Phys. Rev. C* **1998**, *58*, 1671.
32. Ivanov, Y.B. Directed flow in heavy-ion collisions and its implications for astrophysics. *Universe* **2017**, *3*, 79.
33. Andronic, A.; Łukasik, J.; Reisdorf, W.; Trautmann, W. Systematics of stopping and flow in Au + Au collisions. *Eur. Phys. J. A* **2006**, *30*, 31–46.
34. Reisdorf, W.; Leifels, Y.; Andronic, A.; Averbeck, R. Barret, V. Systematics of azimuthal asymmetries in heavy ion collisions in the 1A GeV regime. *Nucl. Phys. A* **2012**, *876*, 1–60.
35. Łukasik, J.; Auger, G.; Bellaize, M.L.; Be, N.; Bittiger, R.; Bocage, F.; Borderie, B.; Bougault, R.; Bouriquet, F.; Charvet, J.L.; Chbihi, A.; et al. Directed and elliptic flow in ^{197}Au+^{197}Au at intermediate energies. *Phys. Lett. B* **2005**, *608*, 223–230.
36. Trautmann, W.; Wolter, H.H. Elliptic flow and the symmetry energy at supra-saturation density. *Int. J. Mod. Phys. E* **2012**, *21*, 1230003.
37. Russotto, P.; Gannon, S.; Kupny, S.; Lasko, P.; Acosta, L.; Adamczyk, M.; Al-Ajlan, A.; Amorini, F. Results of the ASY-EOS experiment at GSI: The symmetry energy at suprasaturation density. *Phys. Rev. C* **2016**, *94*, 034608.
38. Russotto, P.; Wu, P.Z.; Zoric, M.; Chartier, M.; Leifels, Y.; Lemmon, R.C.; Li, Q.; Łukasik, J.; Pagano, A.; Pawłowski, P.; et al. Symmetry energy from elliptic flow in ^{197}Au + ^{197}Au. *Phys. Lett. B* **2011**, *697*, 471–476.
39. Wolter, H.H.; Zielinska-Pfabe, M.; Decowski, P.; Colonna, M.; Bougault, R.; Chbihi, A. Symmetry energy dependence of light fragment production in heavy ion collisions. *EPJ Web Conf.* **2014**, *66*, 03097.
40. Giordano, V.; Colonna, M.; di Toro, M.; Greco, V.; Rizzo, J. Isospin emission and flow at high baryon density: A test of the symmetry potential. *Phys. Rev. C* **2010**, *81*, 064611.
41. Wolter, H.H. The nuclear symmetry energy in heavy ion collisions. *Phys. Part. Nucl.* **2015**, *46*, 781.
42. Coupland, D.D.S.; Youngs, M.; Chajecki, Z.; Lynch, W.G.; Tsang, M.B.; Zhang, Y.X.; Famiano, A.; Ghosh, T.K.; Giacherio, B.; Kilburn, M.A.; et al. Probing effective nucleon masses with heavy-ion collisions. *Phys. Rev. C* **2016**, *94*, 011601.
43. Kowalski, S.; Natowitz, J.B.; Shlomo, S.; Wada, R.; Hagel, K.; Wang, J.; Materna, T.; Chen, Z.; Ma, Y.G.; Qin, L.; et al. Experimental determination of the symmetry energy of a-low density nuclear gas. *Phys. Rev. C* **2007**, *75*, 014601.
44. Tsang, M.B.; Friedman, W.A.; Gelbke, C.K.; Lynch, W.G.; Verde, G.; Xu, H.S. Isotopic scaling in nuclear reactions. *Phys. Rev. Lett.* **2001**, *86*, 5023.
45. Tsang, M.B.; Friedman, W.A.; Gelbke, C.K.; Lynch, W.G.; Verde, G.; Xu, H.S. Conditions for isoscaling in nuclear reactions. *Phys. Rev. C* **2001**, *64*, 041603.
46. Natowitz, J.B.; Röpke, G.; Typel, S.; Blaschke, D.; Bonasera, A.; Hagel, K.; Klähn, T.; Kowalski, S.; Qin, L.; Shlomo, S.; Wada, R.; Wolter, H.H. Symmetry energy of dilute warm nuclear matter. *Phys. Rev. Lett.* **2010**, *104*, 202501.
47. Typel, S.; Röpke, G.; Klähn, T.; Blaschke, D.; Wolter, H.H. Composition and thermodynamics of nuclear matter with light clusters. *Phys. Rev. C* **2010**, *81*, 015803.
48. Bastian, N.-U.; Batyuk, P.; Blaschke, D.; Danielewicz, P.; Yu, B. Light cluster production at NICA. *Eur. Phys. J. A* **2016**, *52*, 244.
49. Röpke, G.; Blaschke, D.; Ivanov, Y.B.; Karpenko, I.; Rogachevsky, O.V.; Wolter, H.H. Medium effects on freeze-out of light clusters at NICA energies. *Phys. Part. Nucl. Lett.* **2018**, *15*, 225.
50. Ferini, G.; Colonna, M.; Gaitanos, T.; di Toro, M. Aspects of particle production in isospin-asymmetric matter. *Nucl. Phys. A* **2005**, *762*, 147–166.
51. Xiao, Z.; Li, B.A.; Chen, L.W.; Yong, G.C.; Zhang, M. Circumstantial evidence for a soft nuclear symmetry energy at suprasaturation densities. *Phys. Rev. Lett.* **2009**, *102*, 062502.
52. Prassa, V.; Ferini, G.; Gaitanos, T.; Wolter, H.H.; Lalazissis, G.; di Toro, M. In-medium effects on particle production in heavy ion collisions. *Nucl. Phys. A* **2007**, *789*, 311–333.
53. Feng, Z.Q.; Jin, G.M. Probing high-density behavior of symmetry energy from pion emission in heavy-ion collisions. *Phys. Lett. B* **2010**, *683*, 140–144.

54. Hong, J.; Danielewicz, P. Subthreshold pion production within a transport description of central Au + Au collisions. *Phys. Rev. C* **2014**, *90*, 024605.
55. Reisdorf, W. The FOPI Collaboration. Systematics of pion emission in heavy ion collisions in the 1 A GeV regime. *Nucl. Phys. A* **2007**, *781*, 459–508.
56. Tsang, M.B. *Pion Production in Rare-Isotope Collisions*; Nuclear Symmetry Energy 2017 (NuSYM2017); Grand Accélérateur National D'Ions Lourds: Caen, France, 2017.
57. Ferini, G.; Gaitanos, T.; Colonna, M.; di Toro, M.; Wolter, H.H. Isospin effects on subthreshold kaon production at intermediate energies. *Phys. Rev. Lett.* **2006**, *97*, 202301.
58. Danielewicz, P.; Lee, J. Symmetry energy II: Isobaric analog states. *Nucl. Phys. A* **2014**, *922*, 1–70.
59. Brown, B.A. Constraints on the Skyrme Equations of State from Properties of Doubly Magic Nuclei. *Phys. Rev. Lett.* **2013**, *111*, 232502.
60. Zhang, Z.; Chen, L.-W. Constraining the symmetry energy at subsaturation densities using isotope binding energy difference and neutron skin thickness. *Phys. Lett. B* **2013**, *726*, 234–238.
61. Zhang, Z.; Chen, L.-W.; Pethick, C.J.; Schwenk, A. Constraints on neutron star radii based on chiral effective field theory interactions. *Phys. Rev. Lett.* **2010**, *105*, 161102.
62. Gandolfi, S.; Carlson, J.; Reddy, S. Maximum mass and radius of neutron stars, and the nuclear symmetry energy. *Phys. Rev. C* **2012**, *85*, 032801.

© 2018 by the author. Licensee MDPI, Basel, Switzerland. This article is an open access article distributed under the terms and conditions of the Creative Commons Attribution (CC BY) license (http://creativecommons.org/licenses/by/4.0/).

Conference Report

Equation of State for Dense Matter with a QCD Phase Transition

Sanjin Benić [†]

Physics Department, Faculty of Science, University of Zagreb, Zagreb 10000, Croatia; sanjinb@phy.hr
† Current address: Yukawa Institute for Theoretical Physics, Kyoto University, Kyoto 606-8502, Japan.

Received: 14 January 2018; Accepted: 21 February 2018; Published: 1 March 2018

Abstract: We construct a dense matter equation of state (EoS) starting from a hadronic density dependent relativistic mean-field model with a DD2 parametrization including the excluded volume corrections at low densities. The high density part is given by a Nambu–Jona–Lasinio (NJL) model with multi-quark interactions. This EoS is characterized by increasing speed of sound below and above the phase transition region. The first order transition region has a large latent heat leaving a distinctive signature in the mass-radii relations in terms of twin stars.

Keywords: neutron star; equation of state; phase transition; quark matter

1. Introduction

We know from finite temperature lattice Quantum Chromodynamics (QCD) that the transition from hadrons to quarks and gluons is a crossover [1]. At finite chemical potential and especially at small temperatures and large chemical potentials, appropriate for neutron stars, the situation is completely different—first principle lattice calculations have a sign problem and so it is an open question whether the transition remains a crossover or if it becomes a first order with the corresponding critical point somewhere in the temperature-chemical potential plane.

This question could hopefully be resolved by colliding heavy ions, but one should be aware of various uncertainties related to the system size and lifetime as well as the possibility that it just may not be possible to reach sufficiently high densities where we would find deconfined quark matter. By contrast, in the case of compact stars, we have a huge system which is long lived and potentially very dense.

As on the theoretical side we still cannot get a first principle information on finite density QCD, in this work, we will assume some model equation of state (EoS). Additionally, we assume the EoS has a first order transition from hadron to quarks at finite density. Our purpose is then to provide a systematic model study of the EoS and the mass-radii (M–R) relations of compact stars which can be used to verify this scenario with simultaneous observations of masses and radii.

2. Twin Stars

Compact stars can be divided into families, such as white dwarfs, neutron stars and sometimes hybrid stars with a quark core and a nuclear mantle as a third family is considered. Purely quark stars is another example of a third family, but we do not consider such possibility as it involves more assumptions. In the case of third families, it is possible to get the so-called twin stars phenomena ([2,3]), where the neutron and the hybrid stars would have same mass but different radii. The main interest here is that, to get twin stars, we need strong first order transition and so measuring twins would provide evidence for first order transitions in QCD.

Recently, astrophysical observations pointed out the existence of several compact stars with masses at $2M_\odot$ [4–6], pressing a clear understanding of the interior structure in terms of its equation of state (EoS) at very high density. The purpose of our work in [7] was to revisit the twin stars scenario with the goal of understanding qualitatively and quantitatively (within a model) systematics of twin stars at $2M_\odot$. Some more recent works include Bayesian analysis [8,9], relation to the CEP [10], impact of rotation [11], twins in protoneutron stars [12] and a triplet of stars with same mass and different radii [13] (see Alford et al. [14] and Zacchi et al. [15] for classification studies).

3. Equation of state

How should a strong first order transition be engineered? We first calculate the hadron and the quark EoS separately and we use Maxwell construction—this is always first order by construction. To get strong first order, the two EoS must have different slopes on the $p - \mu_B$ plot.

For hadrons, we use the density dependent relativistic mean field model with the DD2 parameterization [16]. We can use this EoS (or any kind of hadron EoS, for that matter) until some density where the quarks from different baryons start to overlap. Increasing the number of baryons further is disfavored because of Pauli blocking effects between quarks. We mimic this effect by taking into account the finite size of hadrons via the excluded volume approach (see, e.g., [17,18] and references therein). We introduce a quantity Φ

$$\Phi = \frac{V_{av}}{V} = 1 - v \sum_{i=n,p} n_i, \qquad (1)$$

which is the ratio of the available volume V_{av} for hadrons and the total volume of the system. This can also be written in terms of the excluded volume parameter denoted by small v and the number densities, as in the second line of Equation (1). By looking at single particle energies,

$$E_i = \mu_i - V_i - \frac{v}{\Phi} \sum_{j=p,n} p_j, \qquad (2)$$

where μ_i is the chemical potential and V_i the vector mean field, we see that v acts in a similar way as the vector mean field and so an EoS should become more stiff with excluded volume. In this work, we consider two flavor case so that the hyperon problem is avoided by a phase transition to quark matter.

For quarks, we use the Nambu–Jona–Lasinio (NJL) model with scalar and vector interactions and we add the higher order interactions in the scalar

$$\mathcal{L}_{\text{scal}} = \frac{g_{20}}{\Lambda^2}(\bar{q}q)^2 + \frac{g_{40}}{\Lambda^8}(\bar{q}q)^4, \qquad (3)$$

and in the vector channel

$$\mathcal{L}_{\text{vec}} = \frac{g_{02}}{\Lambda^2}(\bar{q}\gamma^\mu q)^2 + \frac{g_{04}}{\Lambda^8}(\bar{q}\gamma^\mu q)^4. \qquad (4)$$

Refer to Benić [7] and Benić et al. [19] for more details. It turns out that higher order scalar interactions do not change the chiral transition much. However, the higher order vector interactions should be more and more important as we increase the density in terms of stiffness of the EoS [7,19].

4. Impact of the Choice of the Equation of State on the Mass-Radii Relations

Now, we discuss some EoS systematics where we control the repulsion in the hadron and the quark EoS. Characteristic EoS are shown on Figure 1. If there is no repulsion (Figure 1a) in both phases, then this particular model cannot pass the $2M_\odot$ constraint. Typically, one introduces some quark repulsions as in Figure 1b. This delays the onset of quarks and at the same time reduces the latent heat. Because quark EoS becomes stiffer we can get to $2M_\odot$ and typically at first we have hybrid stars. Increasing the quark repulsions more and more it becomes hard to get the hadron and the quark EoS to cross. This is the familiar problem of vector interactions in the quark phase [20].

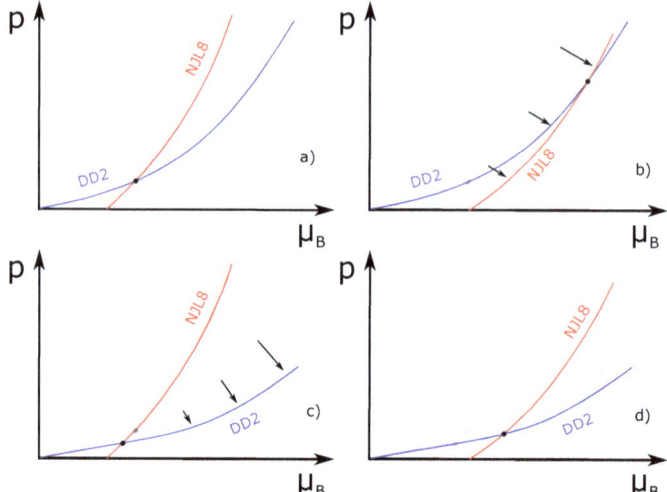

Figure 1. Generic systematics of the different possibilities of the hybrid Equation of State (EoS) within the density dependent relativistic mean field hadronic model with the DD2 parametrization and Nambu-Jona-Lasinio model with 8-quark interactions (DD2-NJL8) hybrid EoS model in terms of the absence (soft EoS) or presence (stiff EoS) of the repulsive interactions: (**a**) soft-to-soft; (**b**) soft-to-stiff; (**c**) stiff-to-soft; and (**d**) stiff-to-stiff EoS.

On the other hand, if we have some repulsions in hadron phase and no repulsion in quark phase (see Figure 1c), then we get an interesting situation that the onset of quarks is lowered and the latent heat becomes increased. Because we put more repulsions, we can pass the $2M_\odot$ limit and typically we get neutron stars. In other words, with the simultaneous combination of the increase of the repulsions in the hadron phase, large latent heat and a soft quark phase, hybrid stars quickly become unstable.

The most interesting situation is when we turn on repulsions in the hadron and the quark phase, as shown in Figure 1d. Then, we can explore also hybrid stars with very stiff quark matter EoS which was not possible previously. Because both the quark and the hadron phase are now stiff, we can easily pass the $2M_\odot$ constraint. In particular, in this window of parameters, we can get twin stars because there is also considerable latent heat.

Figure 2 summarizes the previous discussion. Without repulsions in either of the phases. we can only have the conventional hybrid stars within this particular model. Adding repulsions in either phase hybrid stars turn to neutron stars. If there are some finite quark and hadron repulsions, we can get twin stars.

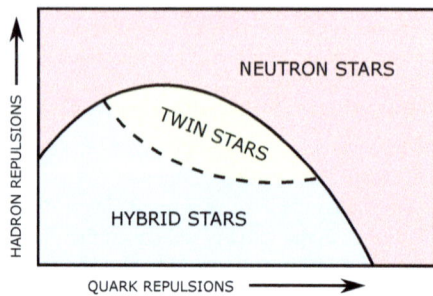

Figure 2. Summary of impact of repulsions in the hadron and in the quark phase on the M–R characteristics.

5. Results

Now, we show some selected results of model calculations performed in [7]. For a density functional approach to such class of EoS, see [21]. Further astrophysical implications are discussed in [22–24]. In the left panel of Figure 3, we show the pressure and the speed of sound as a function of the energy density. First, let us appreciate that the DD2 EoS with the excluded volume is significantly stiffer than the standard DD2 EoS. After the phase transition, we change the stiffness of quark matter by the 8-quark vector coupling. We see that its effect on the stiffness becomes more and more significant by increasing the density. It is essentially the reason why the transition can achieve considerable latent heat. In other words, changing the higher order vector coupling controls the high density part of the EoS while it does not influence the phase transition much. The latter is completely controlled by the excluded volume of the hadron EoS.

In the right panel, we show the M–R relations using this hybrid EoS model. First, because of the very stiff hadron EoS, we get stars with large radii: of the order of 14.5 km or even 15 km. Second, because the latent heat is also quite large, we can get the twin stars: the radii difference is around a 0.5 km or 1 km depending on the model details.

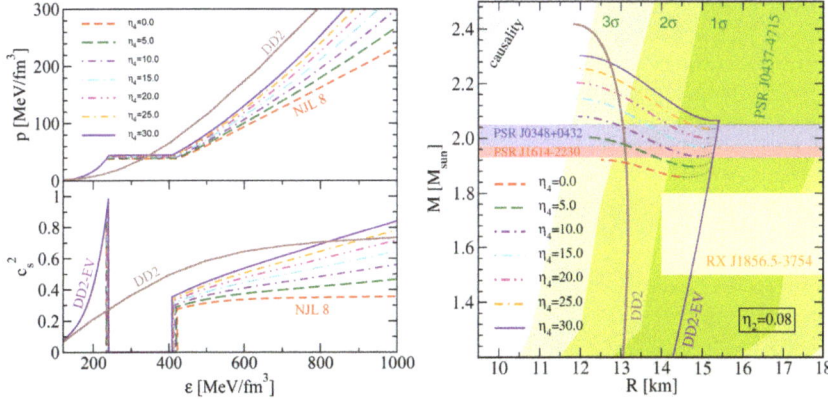

Figure 3. (**Left**) Equation of state (EoS) and speed of sound as a function of the energy density; and (**Right**) M–R relation for the corresponding EoS. Different curves correspond to a variation of the high density quark EoS in terms of a NJL 8-vector coupling parameter η_4. Figure from [7].

6. Conclusions

To conclude, we showed that, to get twin stars at $2M_\odot$, we need strong repulsions in nuclear and in quark matter. Measurements of twin stars has strong potential, as it could exclude the approaches with stiffening EoS in the transition regions [25–27]. On the other hand, the two EoS scenario (quark stars and neutron stars) cannot be excluded by radii measurements provided that the quark star has a large radii as well. In our case, twin stars at $2M_\odot$ require very stiff EoS already in the hadron phase, that is, below the hadron–quark transition. The EoS at these densities controls the radius of the star, and so in our calculation we get big, dilute stars. Measurements of very small radii may disfavor our scenario.

Finally, we make a couple of remarks on the possible caveats. Provided we measure twin stars we can say transition in beta equilibrium is strong first order but this does not mean transition in symmetric matter is also first order. Some effective models suggest chiral transition should become more strong in symmetric matter [28] but in QCD this question does not have a definite answer yet. Additionally, to get twins, we need strong first order but this is not completely correct. It is in fact sufficient for the EoS to just be very soft in a wide region of densities [29,30].

Acknowledgments: S. B. would like to thank A. Ayriyan, D. E. Alvarez-Castillo, D. Blaschke, T. Fischer and S. Typel for the collaborations on the work presented in this material. S. B. would like to thank the organizers for providing a pleasant environment for discussions. S. B. is partially supported by the Croatian Science Foundation under Project No. 8799. S. B. acknowledges partial support by the COST Action CA15213 ITC Conference Grant.

Conflicts of Interest: The author declares no conflict of interest.

References

1. Bhattacharya, T.; Buchoff, M.I.; Christ, N.H.; Ding, H.T.; Gupta, R.; Jung, C.; Karsch, F.; Lin, Z.; Mawhinney, R.D.; McGlynn, G.; et al. QCD Phase Transition with Chiral Quarks and Physical Quark Masses. *Phys. Rev. Lett.* **2014**, *113*, 082001.
2. Glendenning, N.K.; Kettner, C. Nonidentical neutron star twins. *Astron. Astrophys.* **2000**, *353*, L9.
3. Schertler, K.; Greiner, C.; Schaffner-Bielich, J.; Thoma, M.H. Quark phases in neutron stars and a 'third family' of compact stars as a signature for phase transitions. *Nucl. Phys. A* **2000**, *677*, 463–490.
4. Demorest, P.; Pennucci, T.; Ransom, S.; Roberts, M.; Hessels, J. Shapiro Delay Measurement of A Two Solar Mass Neutron Star. *Nature* **2010**, *467*, 1081–1083.
5. Antoniadis, J.; Freire, P.C.; Wex, N.; Tauris, T.M.; Lynch, R.S.; van Kerkwijk, M.H.; Kramer, M.; Bassa, C.; Dhillon, V.S.; Driebe, T.; et al. A Massive Pulsar in a Compact Relativistic Binary. *Science* **2013**, *340*, 1233232.
6. Fonseca, E.; Pennucci, T.T.; Ellis, J.A.; Stairs, I.H.; Nice, D.J.; Ransom, S.M.; Demorest, P.B.; Arzoumanian, Z.; Crowter, K.; Dolch, T.; et al. The NANOGrav Nine-year Data Set: Mass and Geometric Measurements of Binary Millisecond Pulsars. *Astrophys. J.* **2016**, *832*, 167.
7. Benić, S.; Blaschke, D.; Alvarez-Castillo, D.E.; Fischer, T.; Typel, S. A new quark-hadron hybrid equation of state for astrophysics—I. High-mass twin compact stars. *Astron. Astrophys.* **2015**, *577*, A40.
8. Alvarez-Castillo, D.; Ayriyan, A.; Benic, S.; Blaschke, D.; Grigorian, H.; Typel, S. New class of hybrid EoS and Bayesian M-R data analysis. *Eur. Phys. J. A* **2016**, *52*, 69.
9. Alvarez-Castillo, D.E.; Blaschke, D. Proving the CEP with compact stars? *arXiv* **2013**, arXiv:1304.7758.
10. Alvarez-Castillo, D.; Benic, S.; Blaschke, D.; Han, S.; Typel, S. Neutron star mass limit at $2M_\odot$ supports the existence of a CEP. *Eur. Phys. J. A* **2016**, *52*, 232.
11. Bejger, M.; Blaschke, D.; Haensel, P.; Zdunik, J.L.; Fortin, M. Consequences of a strong phase transition in the dense matter equation of state for the rotational evolution of neutron stars. *Astron. Astrophys.* **2017**, *600*, A39.
12. Hempel, M.; Heinimann, O.; Yudin, A.; Iosilevskiy, I.; Liebendörfer, M.; Thielemann, F.K. Hot third family of compact stars and the possibility of core-collapse supernova explosions. *Phys. Rev. D* **2016**, *94*, 103001.
13. Alford, M.; Sedrakian, A. Compact stars with sequential QCD phase transitions. *Phys. Rev. Lett.* **2017**, *119*, 161104.
14. Alford, M.G.; Han, S.; Prakash, M. Generic conditions for stable hybrid stars. *Phys. Rev. D* **2013**, *88*, 083013.
15. Zacchi, A.; Tolos, L.; Schaffner-Bielich, J. Twin Stars within the SU(3) Chiral Quark-Meson Model. *Phys. Rev. D* **2017**, *95*, 103008.
16. Typel, S.; Wolter, H.H. Relativistic mean field calculations with density dependent meson nucleon coupling. *Nucl. Phys. A* **1999**, *656*, 331–364.
17. Typel, S. Variations on the excluded-volume mechanism. *Eur. Phys. J. A* **2016**, *52*, 16.
18. Typel, S.; Blaschke, D. Phenomenological Equation of State of Strongly Interacting Matter with First-Order Phase Transitions and Critical Points. *Universe* **2018**, *4*, 32.
19. Benic, S. Heavy hybrid stars from multi-quark interactions. *Eur. Phys. J. A* **2014**, *50*, 111.
20. Klähn, T.; Łastowiecki, R.; Blaschke, D. Implications of the measurement of pulsars with two solar masses for quark matter in compact stars and heavy-ion collisions: A Nambu-Jona-Lasinio model case study. *Phys. Rev. D* **2013**, *88*, 085001.
21. Kaltenborn, M.A.R.; Bastian, N.U.F.; Blaschke, D.B. Quark-nuclear hybrid star equation of state with excluded volume effects. *Phys. Rev. D* **2017**, *96*, 056024.
22. Alvarez-Castillo, D.E.; Bejger, M.; Blaschke, D.; Haensel, P.; Zdunik, L. Energy bursts from deconfinement in high-mass twin stars. *arXiv* **2015**, arXiv:1506.08645.
23. Paschalidis, V.; Yagi, K.; Alvarez-Castillo, D.; Blaschke, D.B.; Sedrakian, A. Implications from GW170817 and I-Love-Q relations for relativistic hybrid stars. *arXiv* **2017**, arXiv:1712.00451.
24. Fischer, T.; Bastian, N.U.F.; Wu, M.R.; Typel, S.; Klähn, T.; Blaschke, D.B. High-density phase transition paves the way for supernova explosions of massive blue-supergiant stars. *arXiv* **2017**, arXiv:1712.08788.

25. Masuda, K.; Hatsuda, T.; Takatsuka, T. Hadron-Quark Crossover and Massive Hybrid Stars with Strangeness. *Astrophys. J.* **2013**, *764*, 12.
26. Alvarez-Castillo, D.E.; Benic, S.; Blaschke, D.; Lastowiecki, R. Crossover transition to quark matter in heavy hybrid stars. *Acta Phys. Polon. Supp.* **2014**, *7*, 203–208.
27. Kojo, T.; Powell, P.D.; Song, Y.; Baym, G. Phenomenological QCD equation of state for massive neutron stars. *Phys. Rev. D* **2015**, *91*, 045003.
28. Ueda, H.; Nakano, T.Z.; Ohnishi, A.; Ruggieri, M.; Sumiyoshi, K. QCD phase diagram at finite baryon and isospin chemical potentials in Polyakov loop extended quark meson model with vector interaction. *Phys. Rev. D* **2013**, *88*, 074006.
29. Alvarez-Castillo, D.E.; Blaschke, D. Mixed phase effects on high-mass twin stars. *Phys. Part. Nucl.* **2015**, *46*, 846–848.
30. Ayriyan, A.; Bastian, N.U.; Blaschke, D.; Grigorian, H.; Maslov, K.; Voskresensky, D.N. How robust is a third family of compact stars against pasta phase effects? *arXiv* **2017**, arXiv:1711.03926.

© 2018 by the author. Licensee MDPI, Basel, Switzerland. This article is an open access article distributed under the terms and conditions of the Creative Commons Attribution (CC BY) license (http://creativecommons.org/licenses/by/4.0/).

Conference Report

Charged ρ Meson Condensate in Neutron Stars within RMF Models

Konstantin A. Maslov [1,2,*], Evgeni E. Kolomeitsev [2,3] and Dmitry N. Voskresensky [1,2]

1 Department of Theoretical Nuclear Physics, National Research Nuclear University (MEPhI), Kashirskoe sh. 31, Moscow 115409, Russia; d.voskresen@gmail.com
2 Joint Institute for Nuclear Research, Joliot-Curie 6, Dubna 141980, Russia; E.Kolomeitsev@gsi.de
3 Department of Physics, Matej Bel University, Tajovského 40, 97401 Banská Bystrica, Slovakia
* Correspondence: maslov@theor.mephi.ru

Received: 30 November 2017; Accepted: 12 December 2017; Published: 26 December 2017

Abstract: Knowledge of the equation of state (EoS) of cold and dense baryonic matter is essential for the description of properties of neutron stars (NSs). With an increase of the density, new baryon species can appear in NS matter, as well as various meson condensates. In previous works, we developed relativistic mean-field (RMF) models with hyperons and Δ-isobars, which passed the majority of known experimental constraints, including the existence of a $2\,M_\odot$ neutron star. In this contribution, we present results of the inclusion of ρ^--meson condensation into these models. We have shown that, in one class of the models (so-called KVOR-based models, in which the additional stiffening procedure is introduced in the isoscalar sector), the condensation gives only a small contribution to the EoS. In another class of the models (MKVOR-based models with additional stiffening in isovector sector), the condensation can lead to a first-order phase transition and a substantial decrease of the NS mass. Nevertheless, in all resulting models, the condensation does not spoil the description of the experimental constraints.

Keywords: neutron stars; equation of state; ρ meson condensation; maximum mass; Δ resonances

1. Introduction

The equation of state (EoS) of strongly interacting hadronic matter is an essential input for describing the properties of neutron stars (NSs). A viable EoS has to fulfill various constraints following from both astrophysical observations and nuclear experimental data [1]. The discovery of the most massive pulsar with the mass $M = 2.01 \pm 0.04\,M_\odot$ [2] put a severe constraint, ruling out many soft EoSs. Currently, one of the most challenging tasks for phenomenological models of the EoS is to pass simultaneously the maximum NS mass constraint, requiring the EoS to be stiff, and the so-called flow constraint [3] coming from the analysis of flows in heavy ion collisions, which requires a soft EoS. This is hard to achieve within traditional models.

Relativistic mean-field (RMF) framework is a convenient and successful tool for constructing the nuclear equation of state. With an increase of the density hyperons and Δ isobars can appear in the NS matter [4]. In most of known model, this leads to a decrease of the maximum NS mass to unrealistic values. Another reason of the EoS softening is the possible appearance of the charged ρ-meson condensate in NS matter [5]. An RMF EoS can be made more flexible by introducing the dependence on the scalar field of the effective couplings and masses of all hadrons. In [6], we constructed new models within this approach, which fulfill the maximum NS mass constraint and the flow constraint simultaneously together with many other constraints, even with hyperons and Δ isobars included. In the current contribution, we demonstrate results of the inclusion of the charged ρ-meson condensate into our models. More details on the calculations can be found in [7].

2. Description of the Model

Our model was initially formulated in [8] and extended in [6,7,9]. Without the ρ^- meson condensation, the energy density of the model reads:

$$E[\{n_b\},\{n_l\},f] = \sum_b E_{\text{kin}}(p_{F,b}, m_b \Phi_b(f), s_b) + \sum_{l=e,\mu} E_{\text{kin}}(p_{F,l}, m_l, s_l) + \frac{m_N^4 f^2}{2C_\sigma^2}\eta_\sigma(f)$$

$$+ \frac{1}{2m_N^2}\left[\frac{C_\omega^2 n_V^2}{\eta_\omega(f)} + \frac{C_\rho^2 n_I^2}{\eta_\rho(f)} + \frac{C_\phi^2 n_S^2}{\eta_\phi(f)}\right], \qquad (1)$$

$$E_{\text{kin}}(p_F,m,s) = (2s+1)\int_0^{p_F}\frac{p^2 dp}{2\pi^2}\sqrt{p^2+m^2},$$

$$n_V = \sum_b x_{\omega b} n_b, \quad n_I = \sum_b x_{\rho b} t_{3b} n_b, \quad n_S = \sum_H x_{\phi H} n_H.$$

Here, we introduced the dimensionless scalar field $f = g_{\sigma N}\chi_{\sigma N}(\sigma)\sigma/m_N$. The isospin projection of baryon b is t_{3b}, and $p_{F,j} = (6\pi^2 n_j/(2s_j+1))^{1/3}$ denotes the Fermi momentum of a fermion j, with s_j and n_j being the spin and density of a species j, respectively, $j = (b,l)$, l labels leptons. In the infinite hadronic matter without meson condensates, the energy density depends only on the ratios of the meson coupling constants, masses and their corresponding scaling functions, namely

$$C_M = \frac{g_{MN} m_N}{m_M}, \quad M=\sigma,\omega,\rho, \quad C_\phi = \frac{g_{\omega N} m_N}{m_\phi},$$

$$\eta_\omega(f) = \frac{\Phi_\omega^2(f)}{\chi_{\omega N}^2(f)}, \eta_\rho(f) = \frac{\Phi_\rho^2(f)}{\chi_{\rho N}^2(f)}, \eta_\phi(f) = \frac{\Phi_\phi^2(f)}{\chi_{\phi H}^2(f)}, \eta_\sigma(f) = \frac{\Phi_\sigma^2[\sigma(f)]}{\chi_{\sigma N}^2[\sigma(f)]} + \frac{2C_\sigma^2}{m_N^4 f^2}U[\sigma(f)].$$

Here, the the self-interaction potential $U(\sigma)$ entering the Lagrangian of the model is hidden into the scaling function $\eta_\sigma(f)$. The coupling constant ratios for various baryons are defined as $x_{Mb} = g_{Mb}/g_{MN}$, $x_{\phi H} = g_{\phi H}/g_{\omega N}$. We refer the reader to [6,9] for explicit expressions for the scaling functions $\eta_m(f)$ and values of the parameters for all our models. Below, we use $\chi_{Mb} = \chi_{MN}$, $\chi_{\phi H} = 1$.

The baryon coupling ratios with vector mesons $x_{\omega B}$ and $x_{\rho B}$ are determined by the quark SU(6) symmetry. The baryon coupling ratios with the scalar field $x_{\sigma B}$ follow from the potentials

$$U_B(n_0) = C_\omega^2 m_N^{-2} x_{\omega B} n_0/\eta_\omega(f(n_0)) - x_{\sigma B}(m_N + m_N^*(n_0))$$

in the isospin-symmetric matter (ISM) at the saturation density $n = n_0$. The Δ potential $U_\Delta(n_0) \equiv U_\Delta$ is a subject of large uncertainties. Here we assume $U_\Delta = -50$ MeV, following from the most realistic estimate [10]. The values of the parameters for all included baryon species are given in [6].

3. Charged ρ Condensate

The ρ meson field is described by the following Lagrangian [5,8]

$$\mathcal{L}_\rho = -\frac{1}{4}\vec{R}_{\mu\nu}\vec{R}^{\mu\nu} + \frac{1}{2}m_\rho^2 \Phi_\rho^2 \vec{\rho}_\mu \vec{\rho}^\mu - \sum_b g_{\rho b}\chi_{\rho b}\vec{\rho}_\mu \vec{j}_{I,b}^\mu, \quad \vec{j}_{I,b}^\mu = \bar{\psi}_b \gamma^\mu \vec{t}_b \psi_b, \qquad (2)$$

$$\vec{R}_{\mu\nu} = \partial_\mu \vec{\rho}_\nu - \partial_\nu \vec{\rho}_\mu + g'_\rho \chi'_\rho(f)[\vec{\rho}_\mu \times \vec{\rho}_\nu] + \mu_{\text{ch},\rho}\delta_{\nu 0}[\vec{n}_3 \times \vec{\rho}_\mu] - \mu_{\text{ch},\rho}\delta_{\mu 0}[\vec{n}_3 \times \vec{\rho}_\nu],$$

where the chemical potential $\mu_{\text{ch},\rho}$ is introduced for charged mesons and $(\vec{n}_3)^a = \delta^{a3}$ is the unit vector in the isospin space. We treat ρ meson as a gauge boson of a hidden local symmetry and introduce the non-Abelian coupling constant g'_ρ and its scaling function $\chi'_\rho(f)$. Hidden local symmetry requires that $g'_\rho = g_{\rho N}$, which we use here, and for simplicity we consider $\chi'_\rho(f) = 1$. In the standard ansatz for the ρ-meson mean fields only the $\rho_0^{(3)}$ component is non-zero. This ansatz was used in obtaining

Equation (3). The charged ρ^- meson condensation can be introduced by using the new ansatz for the ρ meson field:

$$\rho_0^{(3)} \neq 0, \quad \rho_i^{\pm} = (\rho_i^{(1)} + \rho_i^{(2)})/\sqrt{2} \neq 0, \quad i = 1, 2, 3. \tag{3}$$

It can be shown that within this ansatz the minimum of energy is realized if the condition

$$\rho_i^{(+)}\rho_j^{(-)} - \rho_i^{(-)}\rho_j^{(+)} = 0$$

is fulfilled. This is equivalent to the ratio $\rho_i^{(+)}/\rho_i^{(-)}$ being constant and independent of the spatial index i. Thus we can assume that $\rho_i^{(-)} = a_i \rho_c$ and $\rho_i^{(+)} = a_i \rho_c^{\dagger}$, where we defined the complex amplitude of the charged ρ meson field ρ_c, and $\vec{a} = \{a_i\}$ is the spatial unit vector. In such terms the thermodynamic potential can be minimized by two distinct solutions for the $\rho_0^{(3)}$ and ρ_c fields. The first one is the traditional solution with only $\rho_0^{(3)}$ being non-zero. The second solution is

$$\rho_0^{(3)} = \frac{\mu_{ch,\rho} - m_\rho \Phi_\rho}{g_\rho \chi_\rho'}, \quad |\rho_c|^2 = \frac{(-n_I - n_\rho)\theta(-n_I - n_\rho)}{2 m_\rho \eta_\rho^{1/2} \chi_\rho'}, \tag{4}$$

$$n_\rho = \frac{m_N^2 \eta_\rho^{1/2} \Phi_\rho}{C_\rho^2 \chi_\rho'}(m_\rho \Phi_\rho - \mu_{ch,\rho}). \tag{5}$$

The electric charge density of ρ^- is $n_{ch,\rho} = -2m_\rho \Phi_\rho(f)|\rho_c|^2 < 0$. The ρ^- meson condensate gives the following contribution to the energy density:

$$\Delta E_{ch,\rho}[\{n_b\}; f] = -\frac{C_\rho^2}{2 m_N^2 \eta_\rho}(n_I + n_\rho)^2 \theta(-n_I - n_\rho) - \mu_{ch,\rho} n_{ch,\rho}, \tag{6}$$

where $\theta(-n_I - n_\rho) = 1$ for $n_I + n_\rho < 0$ and zero otherwise. In the presence of the condensate, the charge neutrality condition is modified to be $\sum_b Q_b n_b - n_e - n_\mu + n_{ch,\rho} = 0$. In the beta-equilibrium matter (BEM) of a NS the chemical potentials are related through conditions $\mu_e = \mu_\mu = \mu_{ch,\rho}$, $\mu_b = \mu_n - Q_b \mu_l$. All equations are solved self-consistently with the equation of motion for the scalar field $\partial(E + \mu_{ch,\rho} n_{ch,\rho})/\partial f = 0$. Once the equilibrium concentrations are obtained, the pressure can be evaluated as $P = \sum_{j=b,l,\{ch,\rho\}} \mu_j n_j - E$.

4. Numerical Results

In [7], we considered the MKVOR* and KVORcut03 model. We have shown that, in the KVORcut03 model, the ρ^- condensate does not appear in the most realistic case with the hyperons and/or Δs taken into account. Thus, in this contribution, we focus on results for the MKVOR* model. Below, we present the results of the inclusion of the ρ^- condensate into the MKVOR* model with the universal mass scaling and check the sensitivity of the results to varying the scaling functions $\Phi_\rho(f)$ and $\eta_\rho(f)$. In the following, we denote the MKVOR* model with hyperons and Δs included as MKVOR*H$\Delta\phi$, and the inclusion of ρ^- condensate is denoted by the ρ suffix.

4.1. Inclusion of the Condensate

In this section, we present the results for the MKVOR* model with the universal meson mass scaling $\Phi_m(f) = 1 - f$, $m = \{\sigma, \omega, \rho, \phi\}$, in accordance with [11–13]. In the left panel of Figure 1, we show the particle concentrations and the scalar field f as functions of the total baryon densities. There exists a region of densities where several solutions for the particle fractions and the scalar field exist. This means that the system must prefer the branch of solutions with a lower energy. The density, for which such a transition from one branch of solutions to another happens, in our model, equals $n_c^{(I)} = 2.81\, n_0$. It is shown in Figure 1 by dotted vertical lines in the left and middle panels.

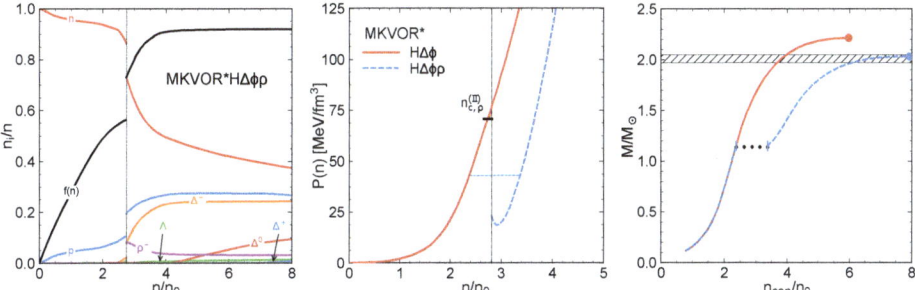

Figure 1. Left panel: Particle fractions together with the scalar field f as functions of the total baryon density n in the BEM for the MKVOR*H$\Delta\phi\rho$ model. Only the energetically favorable regions of the solutions' branches are shown. **Middle panel**: The pressure as a function of the density for MKVOR*H$\Delta\phi$ and MKVOR*H$\Delta\phi\rho$ models. The short vertical dash shows the critical density for the 2nd order phase transition (PT). The vertical line in the left and middle panels denotes the density, at which the ρ-condensed phase becomes energetically more favorable than the normal one. **Right panel**: NS mass as a function of the central density for the same models as in the middle panel. Vertical dashes denote boundaries of the Maxwell construction region (shown by the dotted line), where no stable NS configurations exist. The horizontal stripe denotes the observational constraint $M_{max} \geq (2.01 \pm 0.04)\, M_\odot$ [2] and large dots denote the maximum masses for the models.

The 1st order PT results in a van der Waals-like shape of the pressure, which, neglecting the possible pasta formation, should be replaced by a Maxwell construction, spanning over the densities $2.37\, n_0 \lesssim n \lesssim 3.37\, n_0$. The resulting pressure as a function of the density is shown in the middle panel of Figure 1, where we compare the pressure with and without condensate for the MKVOR*H$\Delta\phi$ model. The condensate appears not only on the new branch of solutions, but also at the old one with the critical density $n_c^{(II)} \simeq 2.74\, n_0$, which is marked by a horizontal dash.

In the right panel of Figure 1, we show the NS mass as a function of the central density for MKVOR*H$\Delta\phi$ with and without the inclusion of the condensate. We see that the ρ^- condensation in this model results in a substantial decrease of the maximum NS mass from 2.21 $M_\odot$ to 2.03 $M_\odot$. Nevertheless, even after such a reduction, the NS maximum mass still passes the observational constraint.

4.2. Variation of the ρ-Meson Effective Mass

A strong phase transition to the ρ^--condensed state relies on the strong decrease of its effective mass. In this subsection, we study the effect of limiting the decrease of the ρ-meson effective mass using the following mass scaling function:

$$\Phi_\rho(f) = \begin{cases} 1 - f & , f \leq f_s \\ (1 - f_s)\left[1 + \frac{\xi}{1+b_\rho\xi}\left(\frac{\xi}{2+b_\rho} - 1\right)\right] & , f > f_s \end{cases} , \quad \xi = \frac{f - f_s}{1 - f_s}. \tag{7}$$

This expression defines a one-parametric family of scaling functions, with a minimum value of the function $\Phi_{\rho,min}$ as the parameter. For a given $\Phi_{\rho,min}$, the value of f_s is

$$f_s = 1 - \Phi_{\rho,min} - \delta\Phi_\rho,$$

where we introduce a constant offset $\delta\Phi_\rho = 0.1$ allowing for a smooth transition at $f = f_s$. The parameter $b_\rho = \frac{\Phi_{\rho,\min}}{\delta\Phi_\rho} - 1$ assures that $\Phi'_\rho(f = 1) = 0$. Under the choice $\Phi_{\rho,\min} = 0$, we will understand also $\delta\Phi = 0$, which leads to $\Phi_\rho(f) = 1 - f$. In the left panel of Figure 2, we show the scaling function $\Phi_\rho(f)$ given by Equation (7) as a function of the scalar field f for $\Phi_{\rho,\min} = 0, 0.3, 0.5, 0.7$, which we examine below. We see that for $\Phi_{\rho,\min} > 0$ this function monotonously decreases, but asymptotically reaches $\Phi_\rho(1) = \Phi_{\rho,\min}$. Thus this function $\Phi_\rho(f)$ is suitable for studying the effect of varying the decrease rate of the effective mass.

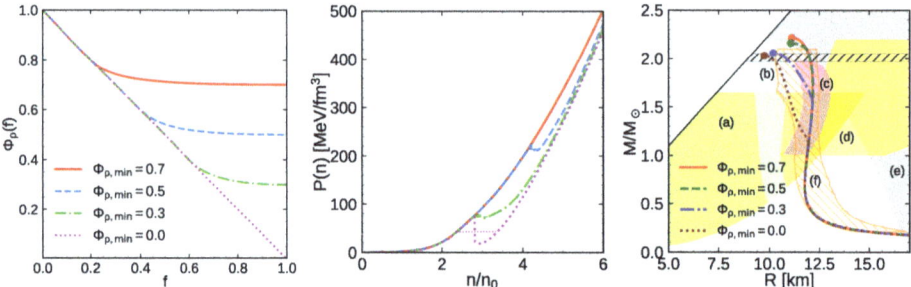

Figure 2. **Left panel**: the scaling function $\Phi_\rho(f)$ as a function of the scalar field f defined by Equation (7) for $\Phi_{\rho,\min} = 0, 0.3, 0.5, 0.7$. **Middle panel**: the pressure as a function of the baryon density for the same values of $\Phi_{\rho,\min}$. **Right panel**: mass-radius curves for the same values of the parameters. For comparison, we show the emprical constraints: (a) [14]; (b) [15]; (c) [16]; (d) [17]; (e) [18]; and (f) [19–21]. The horizontal band shows the maximum NS mass constraint within the uncertainty range [2].

The resulting pressure as a function of the density is shown in the middle panel of Figure 2. The 1st order PT proves to be present for $\Phi_{\rho,\min} < 0.7$. However, the corresponding critical density grows and the pressure loss decreases as we increase $\Phi_{\rho,\min}$. For $\Phi_{\rho,\min} = 0.7$ no condensate appears in the model. The mass-radius curves for this models shown in the right panel of Figure 2. The limiting of the decrease of $\Phi_\rho(f)$ leads to an increase of the maximum NS mass. The maximum NS mass $\Phi_{\rho,\min} = 0.3, 0.5, 0.7$ are 2.06, 2.16, and 2.21 $M_\odot$, respectively, proven to be larger than 2.03 $M_\odot$ in the case of $\Phi_\rho(f) = 1 - f$. Thus taming the ρ-meson effective mass decrease is an efficient way to control the effect of the ρ^- condensate.

4.3. Variation of the $\eta_\rho(f)$

Another reason for the large ρ^- condensate fraction is the particular shape of the $\eta_\rho(f)$ function in the MKVOR* model. The sharp decrease in the $\eta_\rho(f)$ for $f \gtrsim 0.5$ is needed to quench the scalar field growth in the isospin-asymmetric matter, implementing the cut-mechanism of stiffening of the EoS [22]. However, as we shall see in this section, this choice of $\eta_\rho(f)$ corresponds to the maximum ρ^- condensate in the NS matter. To show this, we investigate a set of scaling functions $\eta_\rho(f)$, smoothly changing their behavior from a sharp decrease at $f \gtrsim 0.5$ to a monotonous growth for all f.

The family of $\eta_\rho(f)$ we use here consists of functions $\eta_\rho^{(i)}(f)$, $i = 1\ldots 17$ within three different analytic parameterizations labeled by an integer number. The details on the parameterizations can be found in [7]. The choice of $\eta_\rho^{(1)}$ corresponds to the original $\eta_\rho(f)$ of the MKVOR* model (tail 2). Dependence of $\eta_\rho^{(i)}(f)$, $i = 1\ldots 17$ on the scalar field f is shown in the left panel of Figure 3. We see that with an increase of the model number gradually changes the $\eta_\rho^{(i)}$ behavior from sharply decreasing for $f \gtrsim 0.5$ to a monotonously growing function, thus switching off the cut-mechanism [22], which limits the growth of the scalar field.

In the right panel of Figure 3, we show the maximum NS mass as a function of the model index, while varying number of degrees of freedom included. Namely, we study the MKVOR*{Hϕ, H$\Delta\phi$,

H$\phi\rho$, H$\Delta\phi\rho$} models with $\Phi_\rho(f) = 1 - f$. For estimating the effect of limiting the decrease of the ρ-meson effective mass on these curves we consider also the MKVOR*H$\Delta\phi\rho$ model with Φ_ρ given by Equation (7) with $\Phi_{\rho,\min} = 0.5$. For the models without ρ^- condensate we see, that the choice $\eta_\rho = \eta_\rho^{(1)} \equiv \eta_\rho^{MKVOR^*}$ maximizes the maximum NS mass, thus proving the efficiency of our "cut" mechanism of the EoS stiffening. In addition, one should notice that the difference between Hϕ and H$\Delta\phi$ curves is as well minimized by the $\eta_\rho^{(1)}$. It means that this choice of η_ρ plays an important role in resolution of the "Δ puzzle" [23]. With an increase of the model number the impact of Δs on the EoS grows, and for $i = 17$ the maximum NS mass becomes 2.03 $M_\odot$, marginally satisfying the maximum NS mass constraint.

The inclusion of the ρ^- condensate changes this tendency. The H$\phi\rho$ curve monotonously increases with an increase of the model index and the choice $\eta_\rho = \eta_\rho^{(1)}$ minimized the maximum mass and maximizes the ρ^- condensate phase transition strength. A peculiar situation occurs if both Δs and ρ^- are included into the model. As one sees from the H$\Delta\phi\rho$ curve in the right panel of Figure 3, the maximum NS mass is close to 2.03, $M_\odot$ and almost independent on the model index. This happens because for low $i \to 1$ the softening comes from both the ρ^- condensate together with Δs, and for large $i \to 17$ the ρ^- condensate disappears and the softening effect of Δs is increased, as was mentioned above. However, this independence of a maximum NS mass on the model number is accidental and holds only for $\Phi_\rho = 1 - f$. If we limit the decrease of the ρ-meson effective mass (see H$\Delta\phi\rho$, $\Phi_{\rho,\min} = 0.5$ curve in the right panel of Figure 3), such a degeneracy is removed.

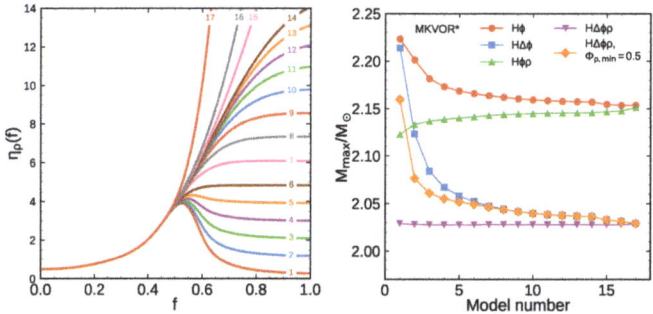

Figure 3. *Left panel*: Scaling functions $\eta_\rho^{(i)}$ for $i = 1\ldots 17$ as functions of the scalar field f. *Right panel*: Maximum NS mass as a function of the model number for MKVOR*{Hϕ, H$\Delta\phi$, H$\phi\rho$, H$\Delta\phi\rho$} models with $\Phi_\rho(f) = 1 - f$ and for MKVOR*H$\Delta\phi\rho$ with Φ_ρ given by Equation (7) with $\Phi_{\rho,\min} = 0.5$.

5. Conclusions

We studied a possibility of charged ρ-meson condensation in a realistic relativistic mean-field model MKVOR* with scaled hadron masses and couplings. The condensation proves to happen by a first-order phase transition and leads to a dramatic reduction of a predicted maximum NS mass, if one uses the universal scaling for masses of all mesons. Nevertheless, the NS maximum mass still passes the observational constraint. We have shown that limiting the decrease of the ρ-meson effective mass allow to reduce the effect of the phase transition on the EoS and increase the maximum neutron star mass. In addition, we demonstrated that our choice of the $\eta_\rho(f)$ scaling function maximizes the neutron star maximum mass in the case without ρ^- condensation. With the condensate included with the universal mass scaling, the maximum mass is almost independent of the choice of $\eta_\rho(f)$. This happens because if the scaling function is chosen to minimize the effect of the ρ^- condensate, the Δ abundance is increased, and vice versa. However, this effect proves to be accidental and does not manifest itself for a different ρ meson mass scaling.

Acknowledgments: The reported study was funded by the Russian Foundation for Basic Research (RFBR) according to the research project No 16-02-00023-A. The work was also supported by the Ministry of Education and Science of the Russian Federation within the state assignment, project No 3.6062.2017/BY, by the Slovak Grant No. VEGA-1/0469/15, and partially (Sections 4.2 and 4.3) by the Russian Science Foundation, Grant No. 17-12-01427. K.A.M. acknowledges the support by the grant of the Foundation for the Advancement of Theoretical Physics "BASIS". E.E.K acknowledges the support by the grant of the Plenipotentiary of the Slovak Government to JINR.

Author Contributions: All authors contributed equally to the formulation of the model, numerical analysis and writing of the paper.

Conflicts of Interest: The authors declare no conflict of interest.

Abbreviations

The following abbreviations are used in this manuscript:

EoS   Equation of state
NS   Neutron star
RMF   Relativistic mean-field
ISM   Isospin-symmetric matter
BEM   Beta-equilibrium matter
PT   Phase transition

References

1. Klahn, T.; Klähn, T.; Blaschke, D.; Typel, S.; Van Dalen, E.N.E.; Faessler, A.; Fuchs, C.; Gaitanos, T.; Grigorian, A.; Ho, A.; et al. Constraints on the high-density nuclear equation of state from the phenomenology of compact stars and heavy-ion collisions. *Phys. Rev. C* **2006**, *74*, 035802.
2. Antoniadis, J.; Freire, P.C.; Wex, N.; Tauris, T.M.; Lynch, R.S.; van Kerkwijk, M.H.; Bassa, C.; Dhillon, V.S.; Hessels, J.W.T.; Kaspi, V.M.; et al. A Massive Pulsar in a Compact Relativistic Binary. *Science* **2013**, *340*, 1233232.
3. Danielewicz, P.; Lacey, R.; Lynch, W.G. Determination of the equation of state of dense matter. *Science* **2002**, *298*, 1592–1596.
4. Schaffner-Bielich, J. Hypernuclear Physics for Neutron Stars. *Nucl. Phys. A* **2008**, *804*, 309–321.
5. Voskresensky, D.N. On the possibility of the condensation of the charged rho meson field in dense isospin asymmetric baryon matter. *Phys. Lett. B* **1997**, *392*, 262–266.
6. Kolomeitsev, E.E.; Maslov, K.A.; Voskresensky, D.N. Delta isobars in relativistic mean-field models with sigma-scaled hadron masses and couplings. *Nucl. Phys. A* **2017**, *961*, 106–141.
7. Kolomeitsev, E.E.; Maslov, K.A.; Voskresensky, D.N. Charged ρ-meson condensation in neutron stars. *Nucl. Phys. A* **2018**, *970*, 291–315.
8. Kolomeitsev, E.E.; Voskresensky, D.N. Relativistic mean-field models with effective hadron masses and coupling constants, and rho- condensation. *Nucl. Phys. A* **2005**, *759*, 373–413.
9. Maslov, K.A.; Kolomeitsev, E.E.; Voskresensky, D.N. Relativistic Mean-Field Models with Scaled Hadron Masses and Couplings: Hyperons and Maximum Neutron Star Mass. *Nucl. Phys. A* **2016**, *950*, 64–109.
10. Riek, F.; Lutz, M.F.M.; Korpa, C.L. Photoabsorption off nuclei with self consistent vertex corrections. *Phys. Rev. C* **2009**, *80*, 024902.
11. Ohnishi, A.; Kawamoto, N.; Miura, K. Brown-Rho Scaling in the Strong Coupling Lattice QCD. *Mod. Phys. Lett. A* **2008**, *23*, 2459–2464.
12. Brown, G.E.; Rho, M. Double decimation and sliding vacua in the nuclear many body system. *Phys. Rept.* **2004**, *396*, 1–39.
13. Paeng, W.G.; Kuo, T.T.S.; Lee, H.K.; Ma, Y.L.; Rho, M. Scale-invariant hidden local symmetry, topology change, and dense baryonic matter. II. *Phys. Rev. D* **2017**, *96*, 014031.
14. Van Straaten, S.; Ford, E.C.; van der Klis, M.; Méndez, M.; Kaaret, P. Relations between timing features and colors in the x-ray binary 4u 0614+09. *Astrophys. J.* **2000**, *540*, 1049–1061.
15. Ozel, F.; Psaltis, D.; Guver, T.; Baym, G.; Heinke, C.; Guillot, S. The Dense Matter Equation of State from Neutron Star Radius and Mass Measurement. *Astrophys. J.* **2016**, *820*, 28.

16. Suleimanov, V.F.; Poutanen, J.; Nättilä, J.; Kajava, J.J.E.; Revnivtsev, M.G.; Werner, K. The direct cooling tail method for X-ray burst analysis to constrain neutron star masses and radii. *Mon. Not. Roy. Astron. Soc.* **2017**, *466*, 906–913.
17. Bogdanov, S. The Nearest Millisecond Pulsar Revisited with XMM-Newton: Improved Mass-Radius Constraints for PSR J0437-4715. *Astrophys. J.* **2013**, *762*, 96.
18. Trümper, J.E.; Burwitz, V.; Haberl, F.; Zavlin, V.E. The puzzles of RX J1856.5-3754: Neutron star or quark star? *Nucl. Phys. B (Proc. Suppl.)* **2004**, *132*, 560–565.
19. Lattimer, J.M. The nuclear equation of state and neutron star masses. *Ann. Rev. Nucl. Part. Sci.* **2012**, *62*, 485–515.
20. Lattimer, J.M.; Steiner, A.W. Neutron Star Masses and Radii from Quiescent Low-Mass X-ray Binaries. *Astrophys. J.* **2014**, *784*, 123.
21. Steiner, A.W.; Lattimer, J.M.; Brown, E.F. Neutron Star Radii, Universal Relations, and the Role of Prior Distributions. *Eur. Phys. J. A* **2016**, *52*, 18.
22. Maslov, K.A.; Kolomeitsev, E.E.; Voskresensky, D.N. Making a soft relativistic mean-field equation of state stiffer at high density. *Phys. Rev. C* **2015**, *92*, 052801.
23. Drago, A.; Lavagno, A.; Pagliara, G.; Pigato, D. Early appearance of Delta isobars in neutron stars. *Phys. Rev. C* **2014**, *90*, 065809.

© 2017 by the authors. Licensee MDPI, Basel, Switzerland. This article is an open access article distributed under the terms and conditions of the Creative Commons Attribution (CC BY) license (http://creativecommons.org/licenses/by/4.0/).

Article

Vector-Interaction-Enhanced Bag Model

Mateusz Cierniak [1,*], **Thomas Klähn** [2], **Tobias Fischer** [1] **and Niels-Uwe F. Bastian** [1]

[1] Institute of Theoretical Physics, University of Wrocław, pl. M. Borna 9, 50-204 Wrocław, Poland; tobias.fischer@ift.uni.wroc.pl (T.F.); niels-uwe@bastian.science (N.-U.F.B.)
[2] Department of Physics and Astronomy, California State University, Long Beach, CA 90840, USA; thomas.klaehn@csulb.edu
* Correspondence: mateusz.cierniak@ift.uni.wroc.pl

Received: 4 December 2017; Accepted: 24 January 2018; Published: 8 February 2018

Abstract: A commonly applied quark matter model in astrophysics is the thermodynamic bag model (tdBAG). The original MIT bag model approximates the effect of quark confinement, but does not explicitly account for the breaking of chiral symmetry, an important property of Quantum Chromodynamics (QCD). It further ignores vector repulsion. The vector-interaction-enhanced bag model (vBag) improves the tdBAG approach by accounting for both dynamical chiral symmetry breaking and repulsive vector interactions. The latter is of particular importance to studies of dense matter in beta-equilibriumto explain the two solar mass maximum mass constraint for neutron stars. The model is motivated by analyses of QCD based Dyson-Schwinger equations (DSE), assuming a simple quark-quark contact interaction. Here, we focus on the study of hybrid neutron star properties resulting from the application of vBag and will discuss possible extensions.

Keywords: Quantum Chromodynamics; dense matter; vector interaction; neutron stars

PACS: 12.39.Ba; 26.60.Kp

---

1. Introduction

The theory of strong interactions, Quantum Chromodynamics (QCD), describes hadrons as bound states of quarks and gluons. These basic degrees of freedom carry the chromodynamic charge, color. Given the QCD feature of a running coupling, i.e., rapidly growing quark–gluon interaction strength with increasing distance (cf. [1] and references therein), and the fact that the net color charge of any observable particle is 0, it is believed that color charged particles in fact cannot be separated. This feature is known as confinement. Besides the running coupling, QCD also exhibits the phenomenon of dynamical chiral symmetry breaking ($D\chi SB$) and its restoration at large densities and high temperatures, believed to be the source of most of the visible mass in the universe.

To date, the only way to address QCD directly is the ab initio lattice QCD approach (cf. [2,3] and references therein). The results of this approach are accurate in the vicinity of vanishing chemical potentials (or equivalently at low densities). They predict a smooth cross-over phase transition at 154 ± 9 MeV (cf. [4–7] and references therein). This is in qualitative agreement with heavy-ion collision experiments [8]. However, at moderate and low collision energies, one encounters finite chemical potentials above this range. In astrophysical systems, e.g., neutron stars and core collapse supernovae, we encounter even larger chemical potentials with densities above normal nuclear density and high isospin asymmetries, far beyond the reach of current generation heavy-ion collision experiments. In both cases, these conditions are inaccessible to lattice QCD.

In fact, currently, no consistent approach exists to simultaneously describe hadron matter and deconfined quark matter at the level of quarks and gluons at high density. Hence, the deconfinement phase transition (i.e., the transition from confined hadron matter to free quarks and gluons) is usually constructed from a given hadronic equation of state (EoS) with baryons and mesons as the basic degrees of freedom and an independently computed quark matter EoS, although there are studies focused on improving this situation (cf.) [9,10]. A general review of recent developments concerning the EoS in astrophysical applications can be found in [11,12].

The two most commonly used effective quark matter models in astrophysics are the thermodynamic bag model (tdBag) of [13] and models of the Nambu–Jona-Lasino type (NJL), cf. [14–17]. The former mimics quark confinement via a phenomenological shift to the EoS, but keeps the quark masses constant. On the other hand, the NJL model exhibits $D\chi SB$, but without modifications does not take confinement into account. Both models do not include repulsive vector interactions, and provide a momentum-independent description of quark properties.

The novel vBag was introduced recently [18] as an effective model for astrophysical studies. It explicitly accounts for $D\chi SB$ and repulsive vector interactions. The latter is of particular importance for studies of neutron star phenomenology, as it allows a hybrid quark-hadron neutron star to reach the limit of 2 solar masses (2 $M_\odot$) in agreement with the recent observations of PSR J1614−2230 and PSR J0348+0432 with masses of 1.928 ± 0.017 $M_\odot$ [19,20] and 2.01 ± 0.04 [21] PSR J0348−0432 with masses of 1.97 ± 0.04 $M_\odot$ [19] and 1.928 ± 0.017 [20,21] respectively. Moreover, vBag mimics deconfinement via a correction to the quark EoS based on the hadron EoS chosen for the construction of the phase transition. This leads to a built-in simultaneous restoration of chiral symmetry and deconfinement. Different Dyson–Schwinger studies suggest that this might be the case in the cross-over domain; the situation is less clear at densities beyond the triple point (cf. [22,23]). vBag has been extended to finite temperatures and arbitrary isospin asymmetry to study the resulting phase diagram [24–26].

The manuscript is organized as follows. In Section 2, we introduce vBag and its derivation from the DSE formalism and present the derived EoS and neutron star mass-radius relations in Section 3. In Section 4, we will discuss the introduction and possible impact of momentum dependence of the single flavor quark properties via the DSE formalism. We will end with a brief summary in Section 5.

2. vBag, an Extended Bag Model

The general in-medium single flavor quark propagator has the form [27,28]

$$S^{-1}(p^2, \tilde{p}_4) = i\vec{\gamma}\vec{p} A(p^2, \tilde{p}_4) + i\gamma_4 \tilde{p}_4 C(p^2, \tilde{p}_4) + B(p^2, \tilde{p}_4), \qquad (1)$$

with $\tilde{p}_4 = p_4 + i\mu$, where μ denotes the chemical potential. Evidently, the gap functions A, B and C account for non–ideal behaviour due to interactions. They follow as solutions of the quark Dyson–Schwinger equation (DSE),

$$S^{-1}(p^2, \tilde{p}_4) = i\vec{\gamma}\vec{p} + i\gamma_4 \tilde{p}_4 + m + \Sigma(p^2, \tilde{p}_4), \qquad (2)$$

where the self-energy takes the shape

$$\Sigma(p^2, \tilde{p}_4) = \int \frac{d^4q}{(2\pi)^4} g^2(\mu) D_{\rho\sigma}(p-q, \mu) \frac{\lambda^a}{2} \gamma^\rho S(q^2, \tilde{q}_4) \Gamma^\sigma_a(q, p, \mu). \qquad (3)$$

In this notation, m is the bare mass, $D_{\rho\sigma}(p-q, \mu)$ is the dressed–gluon propagator and $\Gamma^\sigma_a(q, p, \mu)$ is the dressed quark–gluon vertex. By imposing a specific set of approximations [18] to the self energy term $\Sigma(p^2, \tilde{p}_4)$, one can reproduce the standard NJL model. We start from the rainbow truncation [29], the leading order in a systematic, symmetry-preserving DSE truncation scheme [30,31],

$$\Gamma^\sigma_a(q, p, \mu) = \frac{\lambda_a}{2} \gamma^\sigma. \qquad (4)$$

Next, we impose an effective gluon propagator which is constant in momentum space up to a hard cut-off Λ,

$$g^2 D_{\rho\sigma}(p-q,\mu) = \frac{1}{m_G^2}\Theta(\Lambda^2 - \vec{p}^2)\delta_{\rho\sigma}, \tag{5}$$

equivalent to a quark-quark contact-interaction in configuration space. The Heaviside function Θ provides a three-momentum cutoff for all momenta $\vec{p}^2 > \Lambda^2$. Λ represents a regularization mass scale which, in a realistic treatment, would be removed from the model by taking the limit $\Lambda \to \infty$. For the NJL model this procedure fails and Λ is typically used as a simple UV cutoff. Different regularization procedures are available; in fact the regularization scheme does not have to affect UV divergences only, e.g., infra-red (IR) cutoff schemes can remove unphysical implications [32].

The term m_G in the gluon propagator refers to the gluon mass scale and defines the coupling strength. These approximations allow us to derive the gap equations. The A gap function has a trivial $A = 1$ solution, the rest takes the form

$$B(p^2, \tilde{p}_4) = m + \frac{16 N_c}{9 m_G^2} \int_\Lambda \frac{d^4 q}{(2\pi)^4} \frac{B(q^2, \tilde{q}_4)}{\tilde{q}^2 A^2(q^2, \tilde{q}_4) + \tilde{q}_4^2 C^2(q^2, \tilde{q}_4) + B^2(q^2, \tilde{q}_4)}, \tag{6}$$

$$\tilde{p}_4^2 C(p^2, \tilde{p}_4) = \tilde{p}_4^2 + \frac{8 N_c}{9 m_G^2} \int_\Lambda \frac{d^4 q}{(2\pi)^4} \frac{\tilde{p}_4 \tilde{q}_4 C(q^2, \tilde{q}_4)}{\tilde{q}^2 A^2(q^2, \tilde{q}_4) + \tilde{q}_4^2 C^2(q^2, \tilde{q}_4) + B^2(q^2, \tilde{q}_4)}, \tag{7}$$

where $\int_\Lambda = \int \Theta(\vec{p}^2 - \Lambda^2)$. Both equations can be recast in terms of scalar and vector densities of an ideal spin–degenerate fermi gas,

$$B = m + \frac{4 N_c}{9 m_G^2} n_s(\mu^*, B) \tag{8}$$

$$\mu = \mu^* + \frac{2 N_c}{9 m_G^2} n_v(\mu^*, B) \tag{9}$$

where

$$n_s(\mu^*, B) = 2 \sum_\pm \int_\Lambda \frac{d^3 q}{(2\pi)^3} \frac{B}{E}\left(\frac{1}{2} - \frac{1}{1 + exp(E^\pm/T)}\right) \tag{10}$$

$$n_v(\mu^*, B) = 2 \sum_\pm \int_\Lambda \frac{d^3 q}{(2\pi)^3} \frac{\mp 1}{1 + exp(E^\pm/T)} \tag{11}$$

with $E^2 = \vec{p}^2 + B^2$ and $E^\pm = E \pm \mu^*$. The integrals have no explicit external momentum dependence (p), therefore the gap solutions are constant for a given μ. Typically, for DSE calculations, the pressure is determined in the steepest descent approximation. It consists of an ideal fermi gas and interaction contributions.

$$P_{FG} = Tr Ln S^{-1} = 2 N_c \int_\Lambda \frac{d^4 q}{(2\pi)^4} Ln\left(\vec{p}^2 + \tilde{p}_4^2 + B^2\right), \tag{12}$$

$$P_I = -\frac{1}{2} Tr \Sigma S = \frac{3}{4} m_G^2 (\mu - \mu^*)^2 - \frac{3}{8} m_G^2 (B - m)^2. \tag{13}$$

The merit of the NJL model is the ability to describe chiral symmetry breaking as the formation of a scalar condensate and the restoration of chiral symmetry as melting of the same. The chosen hard cutoff scheme reproduces standard NJL model results and allows to describe quarks as a quasi ideal gas of fermions. Note that after the critical chemical potential μ_χ quark matter can be approximated by an ideal gas of fermions (assuming constant mass equal to the quarks bare mass) shifted by a constant factor (denoted as $B_{\chi,f}$), as seen in Figure 1. This is similar to the standard tdBag model approach (cf.) [13].

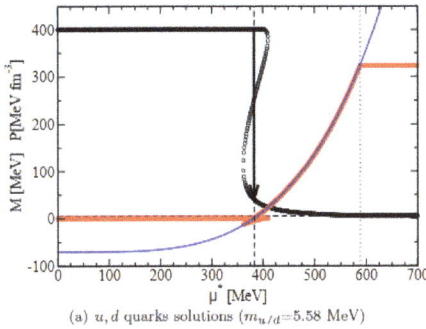
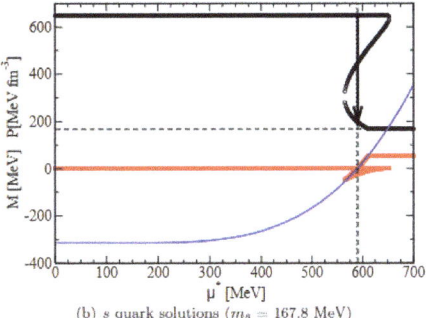

Figure 1. (color online) Single flavor dynamical masses (black) and corresponding pressure (red) computed within the NJL model. The latter is well fitted by the pressure of an ideal Fermi gas (with bare quark mass m_f) shifted by a chiral bag constant B_χ (blue). Figure from [18].

Therefore, we express the single-flavor pressure as

$$P_f(\mu_f) = P_{FG,f}(\mu_f^*) + \frac{K_v}{2} n_{FG,f}^2(\mu_f^*) - B_{\chi,f}. \tag{14}$$

The second term corresponds to the vector condensate, where K_v relates to the vector current–current interaction coupling constant. In our approach, it is defined in terms of the gluon mass scalewith K_v being related to the vector current–current interaction coupling constant, which in combination with the modification of the effective chemical potential μ^* causes stiffening of the EoS with increasing density, as shown in Figure 2.

$$K_v = \frac{2}{9m_G^2}. \tag{15}$$

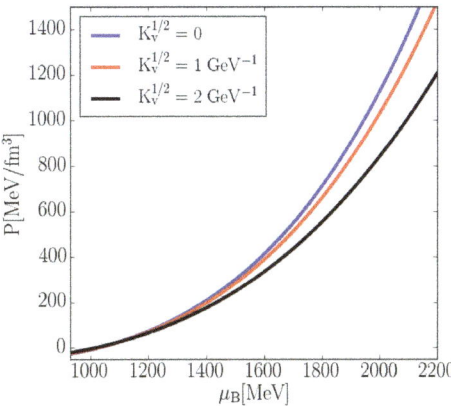

Figure 2. Impact of vector interactions on the stiffness of the EoS. $B_\chi^{1/4} = 150$ MeV.

From Equation (8), it is evident that a corresponding scalar current-current interaction coupling constant is defined as $K_s = 2K_v$. The relation of the coupling constants is consistent with the result obtained after Fierz transformation of the one-gluon exchange interaction [17]. However, we absorbed the effect of scalar interactions in B_χ and vary K_v as an independent model parameter. This procedure is common for NJL-type model studies. Taking the vector interaction into account then results in a modification of the effective chemical potential μ^* and pressure as evident from Equations (9) and (14). This causes stiffening of the EoS with increasing density seen in Figure 2. This term is not included in the standard tdBag model. Chiral symmetry is restored when $P_f(\mu_f) > 0$, and therefore the critical chemical potential can be defined as

$$P_f(\mu_{\chi,f}) = 0. \tag{16}$$

For two-flavor quark matter, this condition is redefined as

$$\sum_f P_f(\mu_{\chi,f}) = 0 \tag{17}$$

to avoid sequential chiral symmetry restoration. This is done so that we can impose simultaneous chiral symmetry restoration and deconfinement at $\mu_{B,\chi}$. This can be achieved by exploiting the fact that the total pressure is fixed only up to a constant factor and therefore we can impose

$$P^Q = \sum_f P_f + B_{dc}. \tag{18}$$

By defining B_{dc} as the hadron pressure at $\mu_{B,\chi}$, we ensure that P^Q and P^H are equal at the point of chiral transition, and therefore it coincides with deconfinement. We can now write the full set of equations that define vBag

$$\mu_f = \mu_f^* + K_v n_{FG,f}(\mu_f^*), \tag{19}$$

$$n_f(\mu_f) = n_{FG,f}(\mu^*), \tag{20}$$

$$P_f(\mu_f) = P_{FG,f}(\mu_f^*) + \frac{K_v}{2} n_{FG,f}^2(\mu_f^*) - B_{\chi,f}, \tag{21}$$

$$\epsilon_f(\mu_f) = \epsilon_{FG,f}(\mu_f^*) + \frac{K_v}{2} n_{FG,f}^2(\mu_f^*) + B_{\chi,f}, \tag{22}$$

$$P^Q = \sum_f P_f + B_{dc}, \tag{23}$$

$$\epsilon^Q = \sum_f \epsilon_f - B_{dc}, \tag{24}$$

where ϵ denotes energy density and n is the particle number density.

3. Neutron Star Mass–Radius Relation

In the left panel of Figure 3, we illustrate the phase transition in β-equilibrated neutron star matter for the chiral bag constants, $B_{\chi,u,d}^{1/4} = 155$ MeV and $B_{\chi,s}^{1/4} = 170$ MeV. B_{dc} has been adjusted so that $\mu_\chi = \mu_{dc}$. Due to the large vacuum mass of the s-quark, the phase transition from hadron to two-flavor quark matter takes place at lower density, followed by the transition to 3f matter at high density. This is the behavior one expects from NJL-type models without flavor coupling channels. It is not accounted for by tdBag which ignores DχSB and consequently predicts a transition from nuclear to three-flavor matter. In contrast, vBag describes a sequential transition from nuclear to two-flavor, and then to three-flavor quark matter. Note that vector interactions are necessary to fulfill the 2 $M_\odot$ constraint. Larger values of $B_\chi^{u,d}$ associated with larger quark masses result in higher critical densities for the phase transition but qualitatively reproduce the above discussed features as long as the transition density does not reach values where already the purely nuclear NS configurations render unstable.

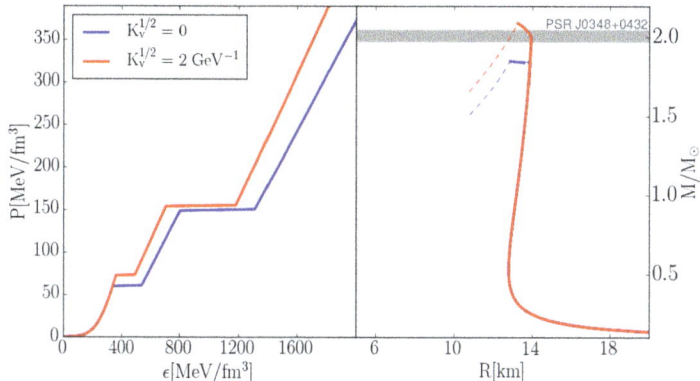

Figure 3. vBag EoS pressure vs. energy density for neutron star matter (**left**); and corresponding mass-radius relations (**right**). The grey band represents the possible masses of the PSR J0348+0432 pulsar [21].

4. Momentum Dependence

As shown in the previous sections, the NJL model can be understood as a particular set of truncations in the quark DSE. The price for the convenient description of chiral symmetry breaking is paid for with the absence of any momentum dependence of the DS gap functions which reflects the well known fact that the NJL model does not exhibit confinement. It does exhibit behavior similar to the tdBag model, which mimics confinement, but none of these two effective models have mass gap solutions with a nontrivial momentum dependence. Consequently, within these models, a confinement criterion that implies the absence of quark mass poles is impossible to account for and the deconfinement transition has to be modeled by imposing additional assumptions. Using a different approximation of the gluon propagator in the quark DSE can however yield a momentum dependent mass-gap, as was shown in the chiral quark model of [33] (the Munczek–Nemirovsky model (MN)) with the gluon propagator

$$g^2 D_{\rho\sigma}(k) = 3\pi^4 \eta^2 \delta_{\rho\sigma} \delta^{(4)}(k). \tag{25}$$

The momentum delta function of the gluon propagator in a crude way mimics the QCD running coupling, a feature absent in the standard NJL model. The model was extended to finite chemical potentials [34] yielding in-medium momentum-dependent solutions

$$A(p^2,\tilde{p}_4) = C(p^2,\tilde{p}_4) = \begin{cases} 2, & \text{if } Re(\tilde{p}^2) < \frac{\eta^2}{4} \\ \frac{1}{2}\left(1 + \sqrt{1 + \frac{2\eta^2}{\tilde{p}^2}}\right), & \text{otherwise} \end{cases} \tag{26}$$

$$B(p^2,\tilde{p}_4) = \begin{cases} \sqrt{\eta^2 - 4\tilde{p}^2} & \text{if } Re(\tilde{p}^2) < \frac{\eta^2}{4} \\ 0 & \text{otherwise} \end{cases} \tag{27}$$

and to non-chiral quarks [35] resulting in a polynomial mass-gap equation

$$B^4 + mB^3 + B^2\left(4\tilde{p}^2 - m^2 - \eta^2\right) - mB\left(4\tilde{p}^2 + m^2 + 2\eta^2\right) - \eta^2 m^2 = 0. \tag{28}$$

Mass-gap solutions can be seen in Figure 4. Note that there is a qualitative change in the behavior of the mass gap of chiral and massive quarks. However, this change is quantitatively small for light quarks. This illustrates the impact of dynamic chiral symmetry breaking on the effective mass of massive quarks and justifies the approximation of light quarks as massless, at the same time showing that such an approximation is increasingly questionable for quarks with masses of the order of 0.1 GeV and above. The key property of this model, however, is the rich momentum-dependent structure of the mass solutions, which shows the impact of IR interactions on quark properties.

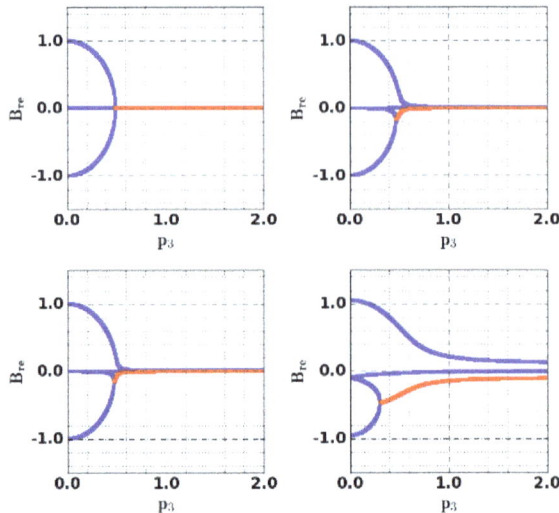

Figure 4. The solution of MN gap equations as a function of momentum. Blue color represents real solutions and red complex for $\eta = 1$ GeV: **(top left)** chiral quark ($m = 0$); **(top right)** up quark ($m = 3$ MeV); **(bottom left)** down quark ($m = 5$ MeV); and **(bottom right)** strange quark ($m = 100$ MeV). All quantities are displayed in units of η. Figure from [35].

5. Conclusions

The vBag model is a novel and easy to implement approach to modeling dense quark matter via a phenomenological Bag approach. It extends the widely used tdBag by taking $D\chi SB$ and repulsive vector interactions into account, and is able to reproduce two solar mass neutron star masses. By connecting B_{dc} to the underlying hadron EoS, it ensures coinciding chiral symmetry restoration and deconfinement. The transition from hadron to quark matter is subsequently introduced via a Maxwell construction, which imposes a 1st order phase transition. This treatment ensures that the effects of $D\chi SB$ are reflected by the EoS, as the hadron EoS is independent of the underlying quark NJL model and therefore a region in which $\mu_\chi < \mu < \mu_{dc}$ might produce unreliable results due to a lack of $D\chi SB$ realization on the hadron side. vBAG is a practical tool for modelers who wish to account for QCD degrees of freedom in complex dense systems—in particular for applications in astrophysics. Furthermore, this model illustrates the power of the DSE approach, explaining standard quark matter models as the NJL and tdBag model in terms of approximations of the quark DSE. At the same time, the DSE approach promises extensions to the widely used effective models, as it accounts naturally for momentum dependent quark properties which might impact the properties of dense stellar objects.

Acknowledgments: The authors acknowledge support from the Bogoliubov–Infeld program and the Polish National Science Center (NCN) under grant numbers UMO–2014/13/B/ST9/02621 (MC and NUB) and UMO-2016/23/B/ST2/00720 (TF).

Author Contributions: T.K. and T.F. developed the model. M.C. and N.-U.F.B. did the numerical calculations. M.C. drafted and finalized the paper after receiving comments from all authors.

Conflicts of Interest: The authors declare no conflict of interest.

Abbreviations

The following abbreviations are used in this manuscript:

QCD	Quantum Chromodynamics
$D\chi SB$	Dynamic Chiral Symmetry Breaking
EoS	Equation of State
MeV	Megaelectronovolt
GeV	Gigaelectronovolt
$M_\odot$	Solar mass
DSE	Dyson–Schwinger equation
NJL	Nambu–Jona-Lasinio
MN	Munczek–Nemirovsky
tdBag	thermodynamic bag
UV	ultra-violet
IR	infra-red

References

1. Roberts, C.D. Three Lectures on Hadron Physics. *J. Phys. Conf. Ser.* **2016**, *706*, 022003.
2. Fodor, Z.; Katz, S.D. Critical point of QCD at finite T and mu, lattice results for physical quark masses. *J. High Energy Phys.* **2004**, *2004*, 529–538.
3. Aoki, Y.; Endrodi, G.; Fodor, Z.; Katz, S.D.; Szabo, K.K. The Order of the quantum chromodynamics transition predicted by the standard model of particle physics. *Nature* **2006**, *443*, 675–678.
4. Borsanyi, S.; Fodor, Z.; Katz, S.D.; Krieg, S.; Ratti, C.; Szabo, K. Fluctuations of conserved charges at finite temperature from lattice QCD. *J. High Energy Phys.* **2012**, *2012*, 138.
5. Bazavov, A.; Bhattacharya, T.; Cheng, M.; Detar, C.; Ding, H.T. The chiral and deconfinement aspects of the QCD transition. *Phys. Rev. D* **2012**, *85*, 054503.
6. Bazavov, A.; Ding, H.T.; Hegde, P.; Kaczmarek, O.; Karsch, F. Freeze-out Conditions in Heavy Ion Collisions from QCD Thermodynamics. *Phys. Rev. Lett.* **2012**, *109*, 192302.
7. Borsanyi, S.; Fodor, Z.; Hoelbling, C.; Katz, S.D.; Krieg, S.; Szabo, K.K. Full result for the QCD equation of state with 2+1 flavors. *Phys. Lett. B* **2014**, *730*, 99–104.
8. Braun-Munzinger, P.; Kalweit, A.; Redlich, K.; Stachel, J. Confronting fluctuations of conserved charges in central nuclear collisions at the LHC with predictions from Lattice QCD. *Phys. Lett. B* **2015**, *747*, 292–298.
9. Dexheimer, V.A.; Schramm, S. A Novel Approach to Model Hybrid Stars. *Phys. Rev. C* **2010**, *81*, 045201.
10. Steinheimer, J.; Schramm, S.; Stocker, H. An Effective chiral Hadron-Quark Equation of State. *J. Phys. G* **2011**, *38*, 035001.
11. Oertel, M.; Hempel, M.; Klähn, T.; Typel, S. Equations of state for supernovae and compact stars. *Rev. Mod. Phys.* **2017**, *89*, 015007.
12. Fischer, T.; Bastian, N.U.; Blaschke, D.; Cierniak, M.; Hempel, M.; Klähn, T.; Martínez-Pinedo, G.; Newton, W.G.; Röpke, G.; Typel, S. The state of matter in simulations of core-collapse supernovae—Reflections and recent developments. *Publ. Astron. Soc. Aust.* **2017**, *34*, e067.
13. Farhi, E.; Jaffe, R.L. Strange Matter. *Phys. Rev. D* **1984**, *30*, 2379–2390.
14. Nambu, Y.; Jona-Lasinio, G. Dynamical Model of Elementary Particles Based on an Analogy with Superconductivity. I. *Phys. Rev.* **1961**, *122*, 345–358.
15. Nambu, Y.; Jona-Lasinio, G. Dynamical model of elementary particles based on an analogy with superconductivity. II. *Phys. Rev.* **1961**, *124*, 246–254.
16. Klevansky, S.P. The Nambu-Jona-Lasinio model of quantum chromodynamics. *Rev. Mod. Phys.* **1992**, *64*, 649–708.
17. Buballa, M. NJL model analysis of quark matter at large density. *Phys. Rep.* **2005**, *407*, 205–376.
18. Klahn, T.; Fischer, T. Vector interaction enhanced bag model for astrophysical applications. *Astrophys. J.* **2015**, *810*, 134.
19. Demorest, P.; Pennucci, T.; Ransom, S.; Roberts, M.; Hessels, J. Shapiro Delay Measurement of a Two Solar Mass Neutron Star. *Nature* **2010**, *467*, 1081–1083.

20. Fonseca, E.; Pennucci, T.T.; Ellis, J.A.; Stairs, I.H.; Nice, D.J.; Ransom, S.M.; Demorest, P.B.; Arzoumanian, Z.; Crowter, K.; Dolch, T.; et al. The NANOGrav Nine-year Data Set: Mass and Geometric Measurements of Binary Millisecond Pulsars. *Astrophys. J.* **2016**, *832*, 167.
21. Antoniadis, J.; Whelan, D.G. A Massive Pulsar in a Compact Relativistic Binary. *Science* **2013**, *340*, 1233232.
22. Qin, S.X.; Chang, L.; Chen, H.; Liu, Y.X.; Roberts, C.D. Phase diagram and critical endpoint for strongly-interacting quarks. *Phys. Rev. Lett.* **2011**, *106*, 172301.
23. Fischer, C.S.; Luecker, J.; Welzbacher, C.A. Phase structure of three and four flavor QCD. *Phys. Rev. D* **2014**, *90*, 034022.
24. Klahn, T.; Fischer, T.; Hempel, M. Simultaneous chiral symmetry restoration and deconfinement—Consequences for the QCD phase diagram. *Astrophys. J.* **2017**, *836*, 89.
25. Fischer, T.; Klähn, T.; Hempel, M. Consequences of simultaneous chiral symmetry breaking and deconfinement for the isospin symmetric phase diagram. *Eur. Phys. J. A* **2016**, *52*, 225.
26. Klähn, T.; Fischer, T.; Cierniak, M.; Hempel, M. Phase Diagram of (Proto)Neutron Star Matter in an Extended Bag Model. *J. Phys. Conf. Ser.* **2017**, *861*, 012026.
27. Rusnak, J.J.; Furnstahl, R.J. Two point fermion correlation functions at finite density. *Z. Phys. A* **1995**, *352*, 345–350.
28. Roberts, C.D.; Schmidt, S.M. Dyson-Schwinger equations: Density, temperature and continuum strong QCD. *Prog. Part. Nucl. Phys.* **2000**, *45*, S1–S103.
29. Gutierrez-Guerrero, L.X.; Bashir, A.; Cloet, I.C.; Roberts, C.D. Pion form factor from a contact interaction. *Phys. Rev. C* **2010**, *81*, 065202.
30. Munczek, H.J. Dynamical chiral symmetry breaking, Goldstone's theorem and the consistency of the Schwinger-Dyson and Bethe-Salpeter Equations. *Phys. Rev. D* **1995**, *52*, 4736–4740.
31. Bender, A.; Roberts, C.D.; Von Smekal, L. Goldstone theorem and diquark confinement beyond rainbow ladder approximation. *Phys. Lett. B* **1996**, *380*, 7–12.
32. Ebert, D.; Feldmann, T.; Reinhardt, H. Extended NJL model for light and heavy mesons without q anti-q thresholds. *Phys. Lett. B* **1996**, *388*, 154–160.
33. Munczek, H.J.; Nemirovsky, A.M. The Ground State q anti-q Mass Spectrum in QCD. *Phys. Rev. D* **1983**, *28*, 181.
34. Klahn, T.; Roberts, C.D.; Chang, L.; Chen, H.; Liu, Y.X. Cold quarks in medium: An equation of state. *Phys. Rev. C* **2010**, *82*, 035801.
35. Cierniak, M.; Klähn, T. Exploring the in-Medium Momentum Dependence of the Dynamical Quark Mass. *arXiv* **2017**, arXiv:nucl-th/1705.06182.

© 2018 by the authors. Licensee MDPI, Basel, Switzerland. This article is an open access article distributed under the terms and conditions of the Creative Commons Attribution (CC BY) license (http://creativecommons.org/licenses/by/4.0/).

Article

Two Novel Approaches to the Hadron-Quark Mixed Phase in Compact Stars

Vahagn Abgaryan [1], David Alvarez-Castillo [2,*], Alexander Ayriyan [1,*], David Blaschke [2,3,4] and Hovik Grigorian [1,5]

1. Laboratory for Information Technologies, Joint Institute for Nuclear Research, Joliot-Curie Street 6, Dubna 141980, Russia; vahagnab@gmail.com (V.A.); hovik.grigorian@gmail.com (H.G.)
2. Bogoliubov Laboratory for Theoretical Physics, Joint Institute for Nuclear Research, Joliot-Curie Street 6, Dubna 141980, Russia; david.blaschke@ift.uni.wroc.pl
3. Institute of Theoretical Physics, University of Wroclaw, Max Born Place 9, 50-204 Wroclaw, Poland
4. National Research Nuclear University (MEPhI), Kashirskoe Shosse 31, Moscow 115409, Russia
5. Department of Physics, Yerevan State University, Alek Manukyan Str. 1, Yerevan 0025, Armenia
* Correspondence: alvarez@theor.jinr.ru (D.A.-C.); ayriyan@jinr.ru (A.A.)

Received: 23 July 2018; Accepted: 31 August 2018; Published: 5 September 2018

Abstract: First-order phase transitions, such as the liquid-gas transition, proceed via formation of structures, such as bubbles and droplets. In strongly interacting compact star matter, at the crust-core transition but also the hadron-quark transition in the core, these structures form different shapes dubbed "pasta phases". We describe two methods to obtain one-parameter families of hybrid equations of state (EoS) substituting the Maxwell construction that mimic the thermodynamic behaviour of pasta phase in between a low-density hadron and a high-density quark matter phase without explicitly computing geometrical structures. Both methods reproduce the Maxwell construction as a limiting case. The first method replaces the behaviour of pressure against chemical potential in a finite region around the critical pressure of the Maxwell construction by a polynomial interpolation. The second method uses extrapolations of the hadronic and quark matter EoS beyond the Maxwell point to define a mixing of both with weight functions bounded by finite limits around the Maxwell point. We apply both methods to the case of a hybrid EoS with a strong first order transition that entails the formation of a third family of compact stars and the corresponding mass twin phenomenon. For both models, we investigate the robustness of this phenomenon against variation of the single parameter: the pressure increment at the critical chemical potential that quantifies the deviation from the Maxwell construction. We also show sets of results for compact star observables other than mass and radius, namely the moment of inertia and the baryon mass.

Keywords: quark-hadron phase transition; pasta phases; speed of sound; hybrid compact stars; mass-radius relation; GW170817

1. Introduction

The understanding of the properties of dense matter in compact star interiors is a subject of current research. Recently, great progress in this direction has been achieved by the detection of the gravitational radiation that emerged from the inspiral phase of two coalescing compact stars, an event named GW170817 [1]. Since it was observed in all other bands of the electromagnetic spectrum, it marked the birth of multi-messenger astronomy. Among the various obtained results, GW170817 has shed light on the properties of the equation of state (EoS) of compact star matter, namely on its stiffness, since through the constraints on the tidal deformability parameter λ [2] from the LIGO-Virgo Collaboration (LVC) results one could estimate the maximum radius of a 1.4 $M_\odot$ compact star to $R_{1.4,\max} = 13.6$ km [3] and maximum mass of nonrotating compacts stars $M_{\text{TOV,max}} = 2.16\ M_\odot$ [4].

Of great scientific interest is the phenomenon of a phase transition from hadronic matter to a deconfined quark phase in hybrid compact stars. Those stars are comprised of a deconfined quark matter core surrounded by a hadronic mantle. The nature of the deconfinement transition is a matter of debate [5,6]. Whether it exhibits a jump in the thermodynamic variables or represents a crossover ([1] is a question that is addressed to both, laboratory experiments as well as compact star observations. The possibility of a mixed phase in neutron stars arises. Standard approaches to describe such a domain of coexistence of competing phases are: (i) the Maxwell construction (for just one chemical potential) which leads to sharp phase boundaries due to constant pressure throughout the mixed phase; (ii) the Gibbs construction (for several chemical potentials corresponding to different conserved quantities) [7], where the pressure changes in the mixed phase which is quasi homogeneous due to the neglect of surface tension effects; and (iii) the constructions with finite size structures of different shapes ("pasta phases" [8]) due to surface tension and Coulomb effects that are mainly modeled with the approximation of sharp surfaces and the surface tension as a free parameter. The adequate description of the letter is a complicated problem where the geometrical properties of the structures, as well as transitions between them, must be taken into account (different methodologies can be found in [9–15]). In the case of the hadron-quark interface, the procedure is well explained in [16] (see also [15] for a recent work); one models several geometrical structures and finds the energetically most favorable ones in different density regions inside compact stars. The occurrence of structures introduces surfaces separating the phases coexisting in the mixed phase. The value of the surface tension determines the size of the structures and thus the amount of surface per volume that can optimally be afforded. While for a vanishing surface tension the mixed phase becomes quasi homogeneous and Δp is largest, a high value of surface tension results in a single surface as for the Maxwell construction that corresponds to $\Delta p = 0$, see Figure 1. The quantitative relation between Δp and the surface tension is under investigation [17].

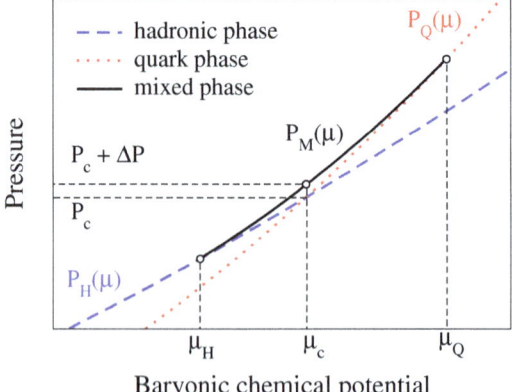

Figure 1. Schematic representation of the interpolation function $P_M(\mu)$ obtained from the mixed phase constructions discussed in this work. For both interpolation methods discussed in the text it has to go though three points: $P_H(\mu_H)$, $P_c + \Delta P$ and $P_Q(\mu_Q)$.

In this work we take a different route and introduce two types of phenomenological interpolations which aim at mimicking the thermodynamic behaviour of those geometrical structures while

---

[1] The word "crossover" is used generically for a transition that does not proceed like in a Maxwell construction at a strictly constant pressure with a jump in (energy) density, but rather by a varying pressure in the transition region. It can thus be a generic crossover transition like in ferromagnetic systems under external magnetic field, but also a first order transition for several globally conserved charges which proceeds via formation of structures of different shapes (pasta phases).

simultaneously exploring the whole corresponding density range in a unified way. A first realization of the idea to describe the transition from the hadronic to the quark matter phase of matter in neutron stars by an interpolation in order to model a crossover-like behaviour was carried out in [18] and followed up in Refs. [19,20], where the jump of the EoS $\varepsilon(P)$ was replaced by a smooth behaviour using as an ansatz a tangens hyperbolicus function.

The mixed phase constructions in this work are developed exclusively for stellar matter where the conditions of charge neutrality and beta equilibrium apply. These constraints allow to express the other chemical potentials in terms of the baryon one. Therefore, only the baryon chemical potential μ remains as the single independent thermodynamic variable. The pressure as a function of μ as shown in Figure 1 can be viewed as a projection from a higher dimensional space spanned by the pressure and several other chemical potentials onto the $P - \mu$ plane where the resulting function $P(\mu)$ is subject to modeling within our simplified approach to mimic the effect of pasta structures in the mixed phase.

A systematic and thermodynamically consistent formulation was recently given in [21,22], where a parabolic interpolation function was introduced to *replace* the behaviour of the hybrid EoS for a Maxwell transition. We shall denote this procedure as the replacement interpolation method (RIM). The resulting hybrid EoS was then used to study the effect of the mixed phase on the properties of compact stars. A second realization of this concept has been worked out recently in [23], where instead of replacing the hadronic and quark matter branches of the hybrid EoS in the limits $\mu_H < \mu < \mu_Q$ (see Figure 1) a *mixing* of these branches is defined using switch functions and a bell-shaped function for the pressure increment with an amplitude $\Delta P = \Delta_P P_c$, where $P_c = P(\mu_c)$ is the critical pressure of the Maxwell construction. This procedure is denoted as the mixing interpolation method (MIM) in [23]. The free parameter Δ_P occurs in both methods with an equivalent influence on the behaviour of the EoS in the mixed phase region, in particular on its extension, see Figure 1. We would like to note that in both methods a negative value of Δ_P would signal that a Maxwell construction using both input EoS $P_H(\mu)$ for hadronic matter and $P_Q(\mu)$ for quark matter would not make sense because it would describe a transition from quark matter at low densities (where $P_Q(\mu)$ is not trustworthy) to hadronic matter at high densities (where $P_H(\mu)$ is not trustworthy). For a discussion of this situation, see Ref. [24].

In this work we present a comparative study of the RIM and MIM approaches to construct mixed phases of the quark-hadron phase transition that mimic the thermodynamic behaviour of pasta phases. We discuss the similarities and differences of these two approaches and apply them to obtain a hybrid EoS under neutron star constraints for which we discuss the resulting hybrid star sequences and their properties. While the first approach (RIM) is rather intuitive and simple to realise as its properties just depend on the order of interpolating polynomial, the second approach (MIM) is based on a procedure of "mixing" the EoS of the two phases in the coexistence region and reminds in its properties on the physics of substitutional compounds as in the crust of compact stars, resulting in an intermediate stiffening effect.

The paper is structured as follows. In Section 2 we start with the reference EOS for the present study, for which a four-polytrope ansatz is employed which features a hadronic phase (first polytrope), a constant pressure polytrope resembling a strong first order phase transition as described by a Maxwell construction (second polytrope) and two polytropes for quark matter phases at high densities. Next, in Section 3, we introduce the RIM and MIM approaches to construct mixed phases when two reference EoS for the low-density (hadronic) and high-density (quark matter) phases are given. We discuss the speed of sound c_s as the key characterizing property of the family of obtained hybrid EoS. Subsequently, in Section 4, we discuss the similarities and differences between the hybrid star EoS of both approaches and show results for the macroscopic properties of compact stars. We motivate these results by the feasibility of detection by multi-messenger astronomy. Consequently, future detections of gravitational wave radiation emitted by of NS–NS or NS–BH mergers shall provide new constraints on both the star mass and radius. Moreover, the determination of the fate of the merger, whether it evolves via a prompt or delayed collapse into a black hole, can be used as an independent estimate on the mass and radius, as proposed in [25]. Up to now, tests for the current compact star models with the at present still single compact star merger event have been performed, e.g., in [22,23,26].

2. Hybrid Star EoS with a Third Family and High-Mass Twins

Compact stars are traditionally divided into white dwarf (first family) and neutron star (second family) branches. Hybrid stars whose equation of state undergoes a sufficiently strong first order phase transition (large jump in energy density $\Delta\varepsilon$) can populate a third family branch in the mass-radius diagram, separated from the second one by a sequence of unstable configurations. As a consequence, there appear so called mass twin configurations: the second and third family solutions overlap within a certain range of masses while the radii of any two stars with the same mass (mass twins) are very different. If the mass-twin phenomenon occurs at high masses $\sim 2~M_\odot$ then one speaks of high-mass twin (HMT) stars [27]. Depending on the critical pressure of the phase transition, the mass-twin phenomenon can occur also at lower masses such as the typical binary radio pulsar mass of $\sim 1.35~M_\odot$, see [22,23,26], so that the corresponding twin star configuration become of relevance for the interpretation of GW170817. In the latter case, a mass ratio $q = m_1/m_2 = 1$ of the merger would not entail that the merging stars have the same radii and internal structure! Would the mass-twin phenomenon (at whatever mass) be observed, this would entail that the QCD phase diagram has to possess at least one critical endpoint since for the study of the cold region of the QCD phase diagram the existence of a first order phase transition between hadron to quark matter had to be concluded. Since the high temperature region of the QCD diagram is known to feature a crossover transition, compact stars can serve as a probe of the existence of a critical end point [28] and provide insight into the properties of matter in heavy ion collision conditions [29].

In order to study the effects of pasta phases at the hadron-quark matter interface in hybrid star interiors, we consider a piecewise polytropic EoS as previously used in various works [3,30–33]. The polytropic representation used in the present work consists of four segments of matter at densities higher than saturation density $n_0 = 0.15~\text{fm}^{-3}$ ($n_0 \ll n_1 < n < n_5$).

$$P(n) = \kappa_i (n/n_0)^{\Gamma_i}, \quad n_i < n < n_{i+1}, \quad i = 1\ldots 4, \tag{1}$$

Each density region is labelled by $i = 1\ldots 4$ with prefactor κ_i and polytropic index Γ_i. HMT stars require a rather stiff nucleonic EoS which here is represented by the first polytrope. The hadron-quark matter first-order phase transition is described by the second polytrope with constant pressure $P_{tr} = \kappa_2$ and vanishing polytropic index ($\Gamma_2 = 0$). At higher densities the polytropes 3 and 4 represent a rather stiff quark matter EoS. The parameters for this HMT realisation are given in Table 1.

Table 1. Parameters for the four-polytrope EoS of Ref. [33], called "ACB4" in Ref. [26]. The corresponding description is presented in Equation (1) of the main text. The last column displays the maximum masses M_{max} on the hadronic (hybrid) branch corresponding to region $i = 1$ ($i = 4$). In addition, the minimal mass M_{min} in region $i = 3$ of the hybrid branch is displayed in that column.

ACB	i	Γ_i	κ_i [MeV/fm^3]	n_i [1/fm^3]	$m_{0,i}$ [MeV]	$M_{max/min}$ [$M_\odot$]
4	1	4.921	2.1680	0.1650	939.56	2.01
	2	0.0	63.178	0.3174	939.56	–
	3	4.000	0.5075	0.5344	1031.2	1.96
	4	2.800	3.2401	0.7500	958.55	2.11

For the present applications to thermodynamically consistent interpolating constructions we need to convert the EoS (1) to the form [33]

$$P(\mu) = \kappa_i \left[(\mu - m_{0,i}) \frac{\Gamma_i - 1}{\kappa_i \Gamma_i} \right]^{\Gamma_i/(\Gamma_i - 1)}, \tag{2}$$

valid for the respective regions (phases) $i = 1\ldots 4$, where for the constant pressure region $i = 2$ this formula collapses to $P(\mu = \mu_c) = P_c = \kappa_2$ because of $\Gamma_2 = 0$. The masses $m_{0,i}$ represent the effective masses of the constituent degrees of freedom in the phase i. For example, in the hadronic region, $m_{0,1} = m_0$, where m_0 is the nucleon mass. At higher densities, this corresponds to effective quark masses. For applying the MIM below, it will be important that the pressure of the hadronic phase ($i = 1$) valid for $\mu < \mu_c$ can be extrapolated to the neighbouring quark matter phase ($i = 3$) where $\mu > \mu_c$ and vice-versa.

HMT star EoS fulfil the Seidov conditions over quantity values at the phase transition [34]

$$\frac{\Delta\varepsilon}{\varepsilon_c} \geq \frac{1}{2} + \frac{3}{2}\frac{P_c}{\varepsilon_c} \tag{3}$$

for the third family of compact stars to exist. These conditions determine the existence of a gap on the mass-radius relation, therefore separating the third family of compact stars from the second one. Once a small region of different matter appears in the centre of the star, the effect can be studied by perturbation theory [34] or by linear response theory [35,36]. The result is that if the Seidov conditions are satisfied, any increase in the central pressure will lead to an instability against oscillations precisely of the same type that happens when the maximum mass is exceeded in the mass-radius relation. The choice of parameters for this EoS corresponds to a sufficiently stiff high-density region in order to prevent gravitational collapse while at the same time not violating the causality condition for the speed of sound $c_s < c$. See [33] for details.

3. Mixed Phase Constructions

In this section we present the details of the interpolation descriptions for the mixed phase between the hadronic and quark matter phases. For this purpose we consider the chemical potential dependent pressures of both the hadronic ($i = 1$) and the neighbouring quark matter ($i = 3$) phases: $P_H(\mu)$, $P_Q(\mu)$, respectively. As mentioned above, our polytropic HMTs EoS features a first order phase transition implemented in the form of a Maxwell construction at a critical chemical potential value μ_c where pressures for both phases are equal:

$$P_Q(\mu_c) = P_H(\mu_c) = P_c, \tag{4}$$

thus both phases are in thermodynamic equilibrium.

3.1. The Replacement Interpolation Method (RIM)

In this mixed phase approach the relevant regions of both, the hadronic and quark matter EoS around the Maxwell critical point (μ_c, P_c) are replaced by a polynomial function defined as

$$P_M(\mu) = \sum_{q=1}^{N} \alpha_q (\mu - \mu_c)^q + (1 + \Delta_P) P_c \tag{5}$$

where Δ_P is a free parameter representing additional pressure of the mixed phase at μ_c. Generally, the ansatz (5) for the mixed phase pressure is an even order ($N = 2k$, $k = 1,2,\ldots$) polynomial and it smoothly matches the EoS at μ_H and μ_Q up to the k-th derivative of the pressure,

$$P_H(\mu_H) = P_M(\mu_H), \quad P_Q(\mu_Q) = P_M(\mu_Q) \tag{6}$$

$$\frac{\partial^q}{\partial \mu^q} P_H(\mu_H) = \frac{\partial^q}{\partial \mu^q} P_M(\mu_H), \quad \frac{\partial^q}{\partial \mu^q} P_Q(\mu_Q) = \frac{\partial^q}{\partial \mu^q} P_M(\mu_Q), \quad q = 1,2,\ldots,k, \tag{7}$$

where $N + 2$ parameter values (μ_H, μ_Q and α_q, for $q = 1,\ldots,N$) can be found by solving the above system of equations, leaving one parameter (ΔP) as a free parameter of this method.

The simplest case of the RIM is the parabolic model for $N = 2$ which has been first introduced in [21,22],

$$P_M(\mu) = \alpha_2 (\mu - \mu_c)^2 + \alpha_1 (\mu - \mu_c) + (1 + \Delta_P) P_c \tag{8}$$

As usual, the parameters α_1, α_2, μ_H and μ_Q are found from the following system of equations involving quantities at the borders of the mixed phase,

$$P_H(\mu_H) = P_M(\mu_H), \quad P_Q(\mu_Q) = P_M(\mu_Q) \tag{9}$$

$$n_H(\mu_H) = n_M(\mu_H), \quad n_Q(\mu_Q) = n_M(\mu_Q). \tag{10}$$

It is evident that the order of the interpolating function (5) will determine whether or not there are discontinuities for the derivatives of the function $P_M(\mu)$.

For instance, the square of the speed of sound,

$$c_s^2 = \frac{\partial P}{\partial \varepsilon} = \frac{\partial \ln \mu}{\partial \ln n}, \tag{11}$$

involves the second derivative of the pressure with respect to μ since $n = \partial P/\partial \mu$, see Figure 2. The result is that for $k = 1$ the function (5) exhibits a clear discontinuity in the speed of sound at ε_c and $\varepsilon_c + \Delta\varepsilon$, whereas in between these borders, the speed of sound slightly increases relative to the case of the Maxwell construction for which $c_s^2 = 0$ in the mixed phase region. For $k = 2$, the mixed phase pressure (5) allows for a continuous speed of sound. However, it is connected at ε_c and $\varepsilon_c + \Delta\varepsilon$ to the speed of sound outside these borders with a jump in its derivative. At the order $k = 3$ and higher the speed of sound behaves smoothly without a jump in its derivative. However, the sharp change in the speed of sound remains as a feature of matter around the transition points μ_H and μ_Q distinctive of a first order phase transition. Moreover, the effect of taking into account the contribution of the higher order polynomials is a softening at the transition that can be associated with crossover–type phase transitions.

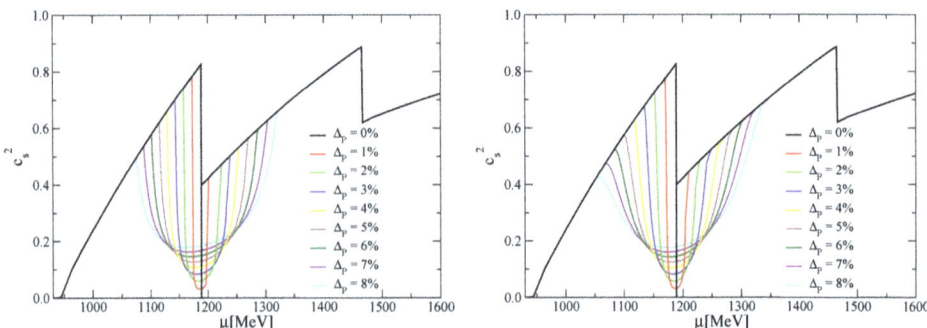

Figure 2. The squared speed of sound as a function of the chemical potential for the RIM construction with $k = 2$ (**left panel**) and $k = 3$ (**right panel**).

3.2. The Mixing Interpolation Method (MIM)

This approach has recently been defined in Ref. [23], where the interpolation ansatz was based on trigonometric functions. Here we will use instead a polynomial ansatz for the interpolation that consists of a pair of functions f_{off} and f_{on} that will switch off and on the hadronic and quark parts of the equation of state, as well as an additional compensating function Δ in order to eliminate thermodynamic instabilities, see Figure 3. This interpolation is applied in the $p - \mu$ plane within the range $\mu_H \leq \mu \leq \mu_Q$.

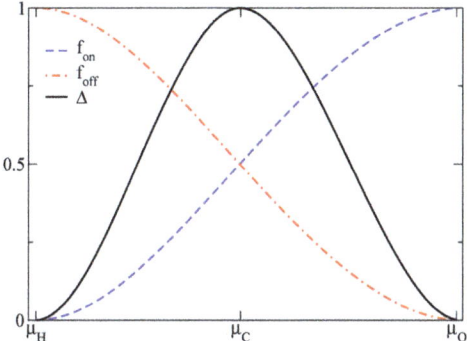

Figure 3. Polynomial switch functions $f_{\text{off/on}}(\mu)$ as well as the function $\Delta(\mu)$.

The pressure that interpolates between the hadron and quark phase at the phase transition reads

$$P(\mu) = P_H(\mu) f_{\text{off}}(\mu) + P_Q(\mu) f_{\text{on}}(\mu) + \Delta(\mu) \Delta P. \qquad (12)$$

Even though f_{off} and f_{on} might be any switching functions, our choice of definition consists of the following pair of left and right side polynomials:

$$f_{>,L} = \alpha_L \left(\frac{\mu - \mu_H}{\mu_Q - \mu_H} \right)^2 + \beta_L \left(\frac{\mu - \mu_H}{\mu_Q - \mu_H} \right)^3 \qquad (13)$$

$$f_{<,R} = \alpha_R \left(\frac{\mu_Q - \mu}{\mu_Q - \mu_H} \right)^2 + \beta_R \left(\frac{\mu_Q - \mu}{\mu_Q - \mu_H} \right)^3 \qquad (14)$$

that together with the complementary functions $f_{>,R}(\mu) = 1 - f_{<,R}(\mu)$ and $f_{<,L}(\mu) = 1 - f_{>,L}(\mu)$ will complete the switch functions. The above coefficients α_L, α_R, β_L and β_R can be determined by the following conditions

$$\left. f_{\lessgtr,L}(\mu) \right|_{\mu=\mu_c} = \left. f_{\lessgtr,R}(\mu) \right|_{\mu=\mu_c} = 1/2$$

$$\left. \frac{\partial f_{\lessgtr,L}(\mu)}{\partial \mu} \right|_{\mu=\mu_c} = \left. \frac{\partial f_{\lessgtr,R}(\mu)}{\partial \mu} \right|_{\mu=\mu_c} \qquad (15)$$

$$\left. \frac{\partial^2 f_{\lessgtr,L}(\mu)}{\partial \mu^2} \right|_{\mu=\mu_c} = \left. \frac{\partial^2 f_{\lessgtr,R}(\mu)}{\partial \mu^2} \right|_{\mu=\mu_c}$$

where the value of $1/2$ is chosen for symmetric convenience. Consequently, the switching functions are defined as

$$f_{\text{on}}(\mu) = \begin{cases} 0, & \mu < \mu_H \\ f_{>,L}, & \mu_H \leq \mu \leq \mu_c \end{cases} \qquad (16)$$

$$f_{\text{off}}(\mu) = \begin{cases} f_{<,R}, & \mu_c \leq \mu \leq \mu_Q \\ 0, & \mu > \mu_Q \end{cases} \qquad (17)$$

and furthermore obey $f_{\text{off/on}}(\mu) = 1 - f_{\text{on/off}}(\mu)$.

In order to construct a proper dimensionless function $\Delta(\mu)$ we introduce

$$\Delta(\mu) = \begin{cases} 0 & \mu < \mu_H \\ g_L(\mu) & \mu_H \leq \mu \leq \mu_C \\ g_R(\mu) & \mu_C \leq \mu \leq \mu_Q \\ 0 & \mu > \mu_Q \end{cases} \qquad (18)$$

consisting of the functions

$$g_L = \delta_L \left(\frac{\mu - \mu_H}{\mu_C - \mu_H} \right)^2 + \gamma_L \left(\frac{\mu - \mu_H}{\mu_C - \mu_H} \right)^3 \qquad (19)$$

$$g_R = \delta_R \left(\frac{\mu_Q - \mu}{\mu_Q - \mu_C} \right)^2 + \gamma_R \left(\frac{\mu_Q - \mu}{\mu_Q - \mu_C} \right)^3 \qquad (20)$$

whose coefficients are determined by the conditions

$$\begin{aligned} g_L(\mu)\big|_{\mu=\mu_C} &= g_R(\mu)\big|_{\mu=\mu_C} = 1 \\ \frac{\partial g_L(\mu)}{\partial \mu}\bigg|_{\mu=\mu_C} &= \frac{\partial g_R(\mu)}{\partial \mu}\bigg|_{\mu=\mu_C} = 0. \end{aligned} \qquad (21)$$

Regarding ΔP as the only free external parameter, up to this moment we have 10 unknowns and eight independent equations which leave us with the possibility to fix the second order derivative of P at μ_H and μ_Q in the following way:

$$\begin{aligned} \frac{\partial^2 P}{\partial \mu^2}\bigg|_{\mu=\mu_H} &= \frac{\partial^2 P_H}{\partial \mu^2}\bigg|_{\mu=\mu_H} \\ \frac{\partial^2 P}{\partial \mu^2}\bigg|_{\mu=\mu_Q} &= \frac{\partial^2 P_Q}{\partial \mu^2}\bigg|_{\mu=\mu_Q}. \end{aligned} \qquad (22)$$

4. Results

4.1. Hybrid Star EoS with Mixed Phases

The two interpolation methods presented above result in a thermodynamically consistent EoS. Knowing that $n = \partial P/\partial \mu$, the thermodynamic identity used to derived all the needed variables at zero temperature reads

$$\varepsilon = -p + \mu n. \qquad (23)$$

Figure 4 shows the resulting mixed phase interpolations for both approaches characterised by the dimensionless pressure increment $\Delta_P = \Delta P/P_c$ that ranges from 1 to 8%, where $\Delta_P = 0\%$ reproduces the Maxwell construction. Figure 5 shows pressure values depending on energy density. The first order phase transition via a Maxwell construction corresponds to the $\Delta_P = 0\%$ case with the pressure being constant in the mixed phase region. Furthermore, Figure 6 shows the squared speed of sound for both approaches where the difference between them becomes obvious: while the MIM shows a peak in the mixed phase region the RIM shows a rather structureless behaviour in this region. This feature is a direct consequence of the functional form of the interpolation implemented by the two methods.

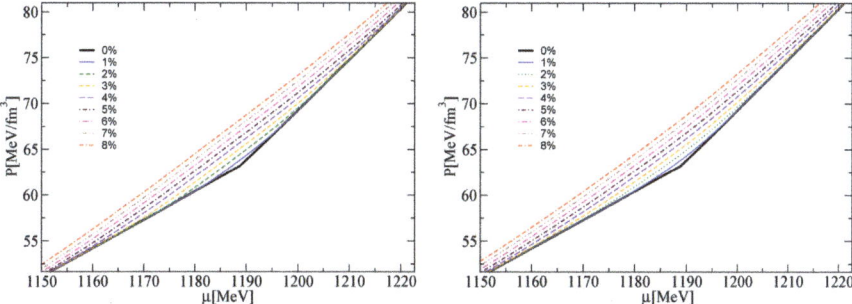

Figure 4. The EoS for pressure P vs. chemical potential μ for both MIM (**left panel**) and RIM for a sixth order polynomial ansatz (**right panel**, $k=3$) approaches to the mixed phase construction. Different curves labelled by percentages correspond to values of $\Delta_P = \Delta P/P_c$, where $\Delta_P = 0$ corresponds to the Maxwell construction.

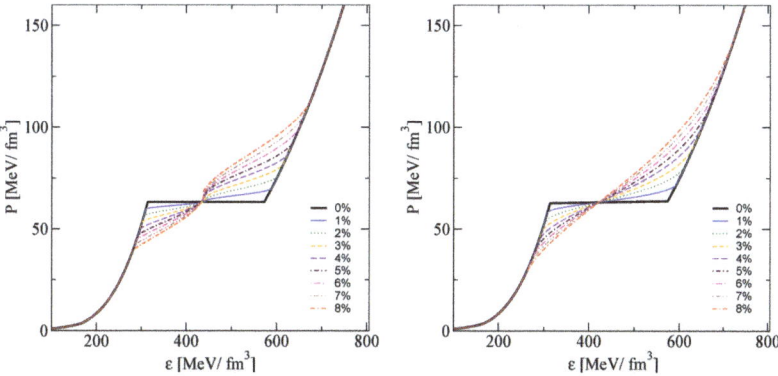

Figure 5. The EoS for pressure P vs. energy density ε for both MIM (**left panel**) and RIM for a sixth order polynomial ansatz (**right panel**, $k=3$) approaches to the mixed phase construction. Different curves labelled by percentages correspond to values of $\Delta_P = \Delta P/P_c$, where $\Delta_P = 0$ corresponds to the Maxwell construction. See [23] for an extended discussion on the MIM approach.

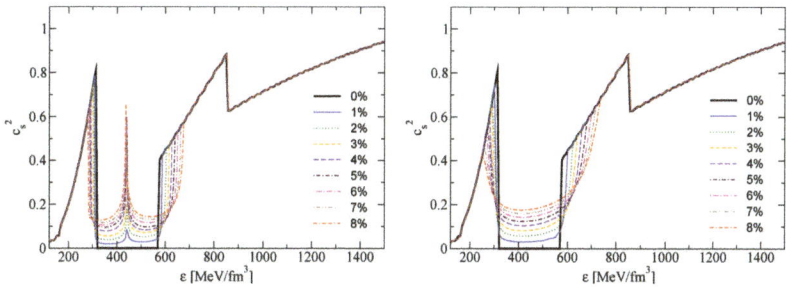

Figure 6. The squared speed of sound c_s^2 against energy density ε for both MIM (**left panel**) and RIM (**right panel**, $k=3$) approaches to the mixed phase construction. Different curves labelled by percentages correspond to values of $\Delta_P = \Delta P/P_c$, where $\Delta_P = 0$ corresponds to the Maxwell construction. A clear feature of the MIM that distinguishes it from the RIM is the intermediate stiffening of the EoS, apparent by the peaked structure inside the mixed phase region. See [23] for an extended discussion on the MIM approach.

4.2. Compact Star Sequences

In order to compute the compact star internal pressure (energy density) profiles leading to mass-radius relations, we solve the Tolman–Oppenheimer–Volkoff (TOV) equations [37,38] derived in the framework of General Relativity for a static, spherically-symmetric compact star

$$\begin{cases} \dfrac{dP(r)}{dr} = -\dfrac{G\left(\varepsilon(r)+P(r)\right)\left(M(r)+4\pi r^3 P(r)\right)}{r\left(r-2GM(r)\right)}, \\ \dfrac{dM(r)}{dr} = 4\pi r^2 \varepsilon(r) \end{cases} \quad (24)$$

with the boundary conditions $P(r = R) = 0$, $M(0) = 0$ and $M(R) = M$ that serve to determine the total stellar mass M and total stellar radius R once a central pressure $P(0) = P(r = 0)$ (and with it the central energy density because $P(\varepsilon)$ is known) is given as input. By increasing the central energy density values, a whole sequence of star configurations up to the one with the maximal mass can be obtained, thus populating the mass-radius diagram. Figure 7 shows compact star sequences for all models characterised by the ΔP value for both, the MIM and RIM approaches together with up-to-date constraints from astrophysical measurements. We can notice that for the lower values of $\Delta P < 6\%$ the HMT phenomenon persists regardless which mixed phase interpolation method has been applied. In Figure 8 we show the mass versus central energy density and the radius versus central pressure for both interpolation methods. For the MIM one observes a trace of the intermediate stiffening effect in the mass versus central energy density which is absent for the RIM.

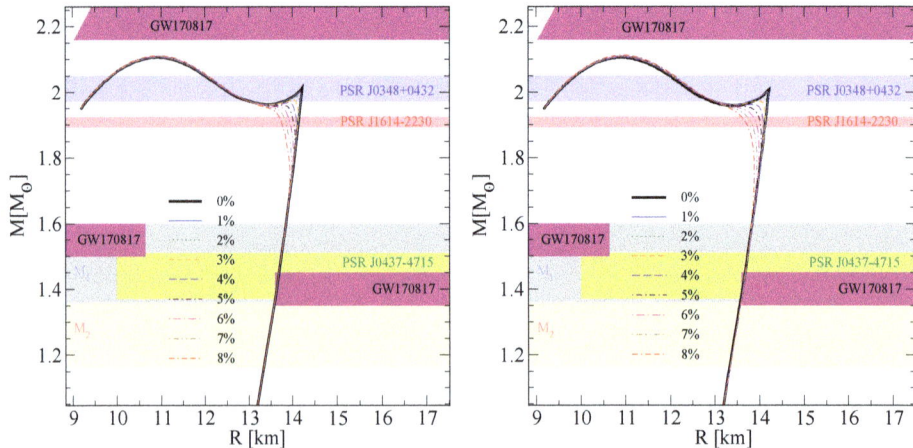

Figure 7. Mass-radius relations for both mixed phase approaches, MIM (**left panel**) and RIM for a sixth order polynomial ansatz (**right panel**). Each curve corresponds to an EoS with a chosen ΔP value given as a percentage of the critical Maxwell pressure P_c represented by alternating line-styles. The shaded areas correspond to compact star measurements: The blue and red horizontal bands correspond to mass measurements of PSR J1614-2230 [39] and PSR J0348+432 [40], respectively. The gray and orange bands denoted by M1 and M2 are the compact star mass windows for the binary merger GW170817. The green band corresponds to the 1.44 ± 0.07 $M_\odot$ mass of PSR J0437-4715 whose radius is expected to be measured by NICER [41]. The hatched regions are excluded by GW170817: the star radius at $1.6\,M_\odot$ cannot be smaller than 10.68 km [25] and for a $1.4\,M_\odot$ the star has to have a radius smaller than 13.6km [3]. The maximum mass of compact stars is estimated to be lower than 2.16 $M_\odot$ [4].

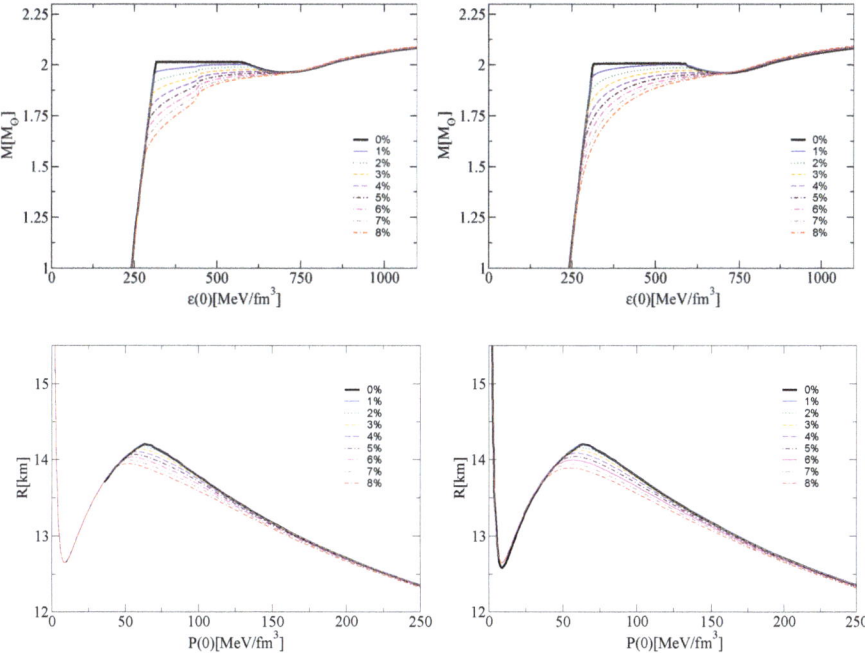

Figure 8. *Upper panel*: Mass as a function of central energy density for both mixed phase approaches, MIM (**left panel**) and RIM (**right panel**). *Lower panel*: Radius as a function of central pressure for all MIM (**left panel**) and RIM (**right panel**, $k = 3$) sequences. Each curve corresponds to an EoS with a chosen ΔP value given as a percentage of the critical Maxwell pressure P_c represented by alternating line-styles. The case $\Delta P = 0$ corresponds to the Maxwell construction which produces a sharp edge in the curves.

In addition, two other quantities of astrophysical interest are the total baryonic mass of the star that results from integrating the following equation

$$\frac{dN_B(r)}{dr} = 4\pi r^2 (1 - \frac{2GM(r)}{r})^{-1/2} n(r), \tag{25}$$

and similarly, its moment of intertia [42]

$$I \simeq \frac{J}{1 + 2GJ/R^3c^2}, \quad J = \frac{8\pi}{3}\int_0^R r^4\left(\rho + \frac{p}{c^2}\right)\Lambda dr, \quad \Lambda = \frac{1}{1 - 2Gm/rc^2}, \tag{26}$$

which are related to observational phenomena as well, like energetic emissions that might conserve baryon mass or moment of inertia dependent pulsar glitches. For a detailed discussion of the moment of inertia in the slow-rotation approximation, and for the hybrid star case see, e.g., [43–45], and references therein. In Figure 9 we show the baryon mass versus radius and and the moment of inertia versus gravitational mass for the compact star sequences obtained in this work with both interpolation methods. When increasing the pressure increment from $\Delta_P = 0$ to 8%, the sharp edges which are obtained for the Maxwell construction case get washed out. One observes no qualitative difference between the MIM and the RIM in the patterns of these families of sequences. For $\Delta_P > 5\%$, the second and third family branches in the M_B versus R diagrams get joined so that neutron star and hybrid star configurations form a connected sequence and the HMT phenomenon get lost. This effect is reflected in the I vs. M diagrams by the loss of multiple values (the lowest branch up to the maximum mass of 2.11 $M_\odot$ shall be ignored because it is unstable). From the M_B versus R diagrams one can read off which configuration on the hybrid star branch would be reached when the maximum mass neutron star configuration would collapse under conservation of baryon number. Comparing the gravitational masses of these two star configurations one can estimate the release of binding energy in this process, see Ref. [45].

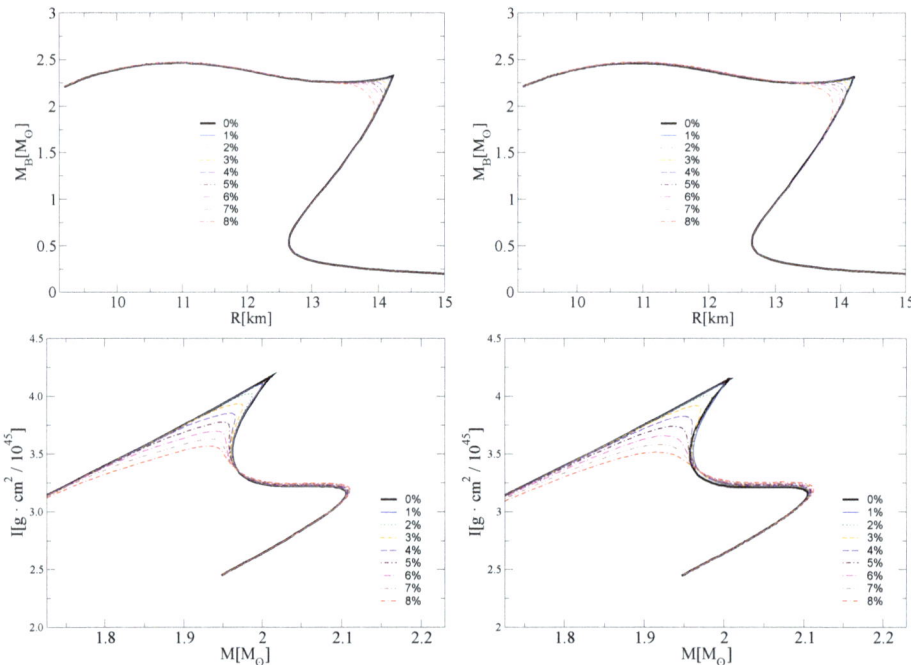

Figure 9. *Upper panel*: Baryonic Mass versus radius for both mixed phase approaches, MIM (**left panel**) and RIM (**right panel**, $k = 3$). *Lower panel*: Moment of inertia as a function of total mass for MIM (**left panel**) and RIM (**right panel**, $k = 3$) approaches. Each curve corresponds to an EoS with a chosen ΔP value given as a percentage of the critical Maxwell pressure P_c represented by alternating line-style values. The case $\Delta P = 0$ corresponds to the Maxwell construction which produces a sharp edge in the curves.

5. Conclusions

In this work we have introduced two interpolation approaches to a mixed phase at the hadron-quark phase transition. An advantage of these two interpolation methods presented here over the construction employing hyperbolic tangent functions [19,20] is the finite extension in chemical potentials of the mixed phase between the hadronic and the quark EoS, whereas the latter strictly converges only at infinity.

While each approach uses a different functional form, both fulfil the same conditions at the border of the mixed phase. We have found that both methods can be distinguished by the behaviour of the speed of sound that they predict. The MIM approach motivated by the analogy with sequential phase transitions occurring for substitutional compounds in the neutron star crust finds an intermediate stiffening of the mixed phase EoS. The RIM approach does not exhibit this feature. In the case of the RIM approach, we have studied both a fourth and sixth order polynomial interpolation. We found that the latter connects the hadron and quark EoS smoothly up to second derivatives, which is visible in the smooth behaviour of the speed of sound. However, the differences in the neutron star properties for both polynomial orders are safely negligible.

The macroscopic properties of compact stars show, for both mixed phase constructions, a very similar systematic behaviour as the pressure increment ΔP is increased: the mass-radius relation smooths out, eliminating the gap between second and third branches, however, we have only considered the highest values. Up to $\Delta_P \sim 5\%$ the HMT phenomenon is robust against the mixed phase construction, regardless whether the MIM or RIM approach is used. For the mass versus central energy density, one observes a trace of the intermediate stiffening effect for the MIM which is absent for the RIM. For the other compact star quantities evaluated here, the baryonic mass and the moment of inertia, both interpolation methods display a similar type of behaviour when the pressure increment is varied.

The methods presented here can potentially be applied to the compact star crust-core transition as well. Just like at the hadron-quark boundary, the transition at the bottom of the crust may proceed via pasta phases dominated by Coulomb forces and surface tension effects [8]. Further astrophysical aspects of mixed phases inside neutron stars include potentially observable effects such as the rotational evolution, pulsar glitches, gravitational wave emission and cooling. They could be sufficiently sensible to the nature of the phase transition, proceeding via pasta phases or not, and thus provide potential signatures of the presence and extension of a mixed phase in compact stars.

Author Contributions: All authors have discussed the results presented in the article and contributed to its final form. The RIM was originally developed by A.A. and H.G. while D.A.-C. and D.B. developed the MIM. A.A. and V.A. generalized RIM for the higher order. V.A. made its realization and called to life H.G's idea of polynomial switching functions. D.A.-C. calculated the compact star configurations.

Funding: This research was funded by the Russian Science Foundation grant number 17-12-01427. D.A.-C. is grateful NCN grant no. UMO-2014/13/B/ST9/02621 and the Bogoliubov-Infeld Program for collaboration between JINR Dubna and Polish Institutes.

Acknowledgments: The authors thank N. Yasutake, K. Maslov and D.N. Voskresensky for fruitful discussions on the features of hadron-quark mixed phase.

Conflicts of Interest: The authors declare no conflict of interest.

References

1. Abbott, B.P.; Abbott, R.; Abbott, T.D.; Acernese, F.; Ackley, K.; Adams, C.; Adams, T.; Addesso, P.; Adhikari, R.X.; Adya, V.B.; et al. GW170817: Observation of gravitational waves from a binary neutron star inspiral. *Phys. Rev. Lett.* **2017**, *119*, 161101. [CrossRef] [PubMed]
2. Hinderer, T.; Lackey, B.D.; Lang, R.N.; Read, J.S. Tidal deformability of neutron stars with realistic equations of state and their gravitational wave signatures in binary inspiral. *Phys. Rev. D* **2010**, *81*, 123016. [CrossRef]
3. Annala, E.; Gorda, T.; Kurkela, A.; Vuorinen, A. Gravitational-wave constraints on the neutron-star-matter Equation of State. *Phys. Rev. Lett.* **2018**, *120*, 172703. [CrossRef] [PubMed]
4. Rezzolla, L.; Most, E.R.; Weih, L.R. Using gravitational-wave observations and quasi-universal relations to constrain the maximum mass of neutron stars. *Astrophys. J.* **2018**, *L25*, 852. [CrossRef]
5. Alford, M.G.; Han, S.; Prakash, M. Generic conditions for stable hybrid stars. *Phys. Rev. D* **2013**, *88*, 083013. [CrossRef]
6. Kojo, T. Phenomenological neutron star equations of state: 3-window modeling of QCD matter. *Eur. Phys. J. A* **2016**, *52*, 51. [CrossRef]
7. Glendenning, N.K. First order phase transitions with more than one conserved charge: Consequences for neutron stars. *Phys. Rev. D* **1992**, *46*, 1274–1287. [CrossRef]
8. Ravenhall, D.G.; Pethick, C.J.; Wilson, J.R. Structure of Matter below Nuclear Saturation Density. *Phys. Rev. Lett.* **1983**, *50*, 2066. [CrossRef]
9. Voskresensky, D.N.; Yasuhira, M.; Tatsumi, T. Charge screening at first order phase transitions and hadron quark mixed phase. *Nucl. Phys. A* **2003**, *723*, 291–339. [CrossRef]
10. Maruyama, T.; Chiba, S.; Schulze, H.J.; Tatsumi, T. Quark deconfinement transition in hyperonic matter. *Phys. Lett. B* **2008**, *659*, 192–196. [CrossRef]

11. Watanabe, G.; Sato, K.; Yasuoka, K.; Ebisuzaki, T. Structure of cold nuclear matter at subnuclear densities by quantum molecular dynamics. *Phys. Rev. C* **2003**, *68*, 035806. [CrossRef]
12. Horowitz, C.J.; Perez-Garcia, M.A.; Berry, D.K.; Piekarewicz, J. Dynamical response of the nuclear 'pasta' in neutron star crusts, *Phys. Rev. C* **2005**, *72*, 035801. [CrossRef]
13. Horowitz, C.J.; Berry, D.K.; Briggs, C.M.; Caplan, M.E.; Cumming, A.; Schneider, A.S. Disordered nuclear pasta, magnetic field decay, and crust cooling in neutron stars. *Phys. Rev. Lett.* **20015**, *114*, 031102. [CrossRef] [PubMed]
14. Newton, W.G.; Stone, J.R. Modeling nuclear 'pasta' and the transition to uniform nuclear matter with the D-3 Skyrme-Hartree-Fock method at finite temperature: Core-collapse supernovae. *Phys. Rev. C* **2009**, *79*, 055801. [CrossRef]
15. Yasutake, N.; Lastowiecki, R.; Benic, S.; Blaschke, D.; Maruyama, T.; Tatsumi, T. Finite-size effects at the hadron-quark transition and heavy hybrid stars. *Phys. Rev. C* **2014**, *89*, 065803. [CrossRef]
16. Maruyama, T.; Chiba, S.; Schulze, H.J.; Tatsumi, T. Hadron-quark mixed phase in hyperon stars. *Phys. Rev. D* **2007**, *76*, 123015. [CrossRef]
17. Maslov, K.; Yasutake, N.; Ayriyan, A.; Blaschke, D.; Grigorian, H.; Maruyama, T.; Tatsumi, T.; Voskresensky, D.N. Hybrid Equation of State with Pasta Phases and Third Family of Compact Stars. Unpublished work, **2018**.
18. Masuda, K.; Hatsuda, T.; Takatsuka, T. Hadron–quark crossover and massive hybrid stars. *Prog. Theor. Exp. Phys.* **2013**, *7*, 073D01. [CrossRef]
19. Alvarez-Castillo, D.E.; Blaschke, D. Mixed phase effects on high-mass twin stars. *Phys. Part. Nucl.* **2015**, *46*, 846–848. [CrossRef]
20. Alvarez-Castillo, D.; Blaschke, D.; Typel, S. Mixed phase within the multi-polytrope approach to high-mass twins. *Astron. Nachr.* **2017**, *338*, 1048–1051. [CrossRef]
21. Ayriyan, A.; Grigorian, H. Model of the Phase Transition Mimicking the Pasta Phase in Cold and Dense Quark-Hadron Matter. *Eur. Phys. J. Web Conf.* **2018**, *173*, 03003. [CrossRef]
22. Ayriyan, A.; Bastian, N.-U.; Blaschke, D.; Grigorian, H.; Maslov, K.; Voskresensky, D.N. Robustness of third family solutions for hybrid stars against mixed phase effects. *Phys. Rev. C* **2018**, *97*, 045802. [CrossRef]
23. Alvarez-Castillo, D.; Blaschke, D. A mixing interpolation method to mimic pasta phases in compact star matter. *arXiv* **2018**, arXiv:1807.03258.
24. Kojo, T.; Powell, P.D.; Song, Y.; Baym, G. Phenomenological QCD equation of state for massive neutron stars. *Phys. Rev. D* **2015**, *91*, 045003. [CrossRef]
25. Bauswein, A.; Just, O.; Janka, H.T.; Stergioulas, N. Neutron-star radius constraints from GW170817 and future detections. *Astrophys. J.* **2017**, *850*, L34. [CrossRef]
26. Paschalidis, V.; Yagi, K.; Alvarez-Castillo, D.; Blaschke, D.B.; Sedrakian, A. Implications from GW170817 and I-Love-Q relations for relativistic hybrid stars. *Phys. Rev. D* **2018**, *97*, 084038. [CrossRef]
27. Benic, S.; Blaschke, D.; Alvarez-Castillo, D.E.; Fischer, T.; Typel, S. A new quark-hadron hybrid equation of state for astrophysics—I. High-mass twin compact stars. *Astron. Astrophys.* **2015**, *577*, A40. [CrossRef]
28. Alvarez-Castillo, D.E.; Blaschke, D. Supporting the existence of the QCD critical point by compact star observations. In Proceedings of the 9th International Workshop on Critical Point and Onset of Deconfinement (CPOD2014), Bielefeld, Germany, 17–21 November 2014. [CrossRef]
29. Alvarez-Castillo, D.; Benic, S.; Blaschke, D.; Han, S.; Typel, S. Neutron star mass limit at 2 $M_\odot$ supports the existence of a CEP. *Eur. Phys. J. A* **2016**, *52*, 232. [CrossRef]
30. Read, J.S.; Lackey, B.D.; Owen, B.J.; Friedman, J.L. Constraints on a phenomenologically parameterized neutron-star equation of state. *Phys. Rev. D* **2009**, *79*, 124032. [CrossRef]
31. Hebeler, K.; Lattimer, K.M.; Pethick, C.J.; Schwenk, A. Equation of state and neutron star properties constrained by nuclear physics and observation. *Astrophys. J.* **2013**, *773*, 11. [CrossRef]
32. Raithel, C.A.; Ozel, F.; Psaltis, D. From Neutron Star Observables to the Equation of State: An Optimal Parametrization. *Astrophys. J.* **2016**, *831*, 44. [CrossRef]
33. Alvarez-Castillo, D.E.; Blaschke, D.B. High-mass twin stars with a multipolytrope equation of state. *Phys. Rev. C* **2017**, *96*, 045809. [CrossRef]
34. Seidov, Z.F. The Stability of a Star with a Phase Change in General Relativity Theory. *Sov. Astron. Lett.* **1971**, *15*, 347–348.

35. Schaeffer, R.; Zdunik, L.; Haensel, P. Phase transitions in stellar cores. I-Equilibrium configurations. *Astron. Astrophys.* **1983**, *126*, 121–145.
36. Zdunik, J.L.; Haensel, P.; Schaeffer, R. Phase transitions in stellar cores. II-Equilibrium configurations in general relativity. *Astron. Astrophys.* **1987**, *172*, 95–110.
37. Tolman, R.C. Static solutions of Einstein's field equations for spheres of fluid. *Phys. Rev.* **1939**, *55*, 364–373. [CrossRef]
38. Oppenheimer, J.R.; Volkoff, G.M. On Massive neutron cores. *Phys. Rev.* **1939**, *55*, 374–381. [CrossRef]
39. Antoniadis, J.; Freire, P.C.; Wex, N.; Tauris, T.M.; Lynch, R.S.; van Kerkwijk, M.H.; Kramer, M.; Bassa, C.; Dhillon, V.S.; Driebe, T.; et al. A Massive Pulsar in a Compact Relativistic Binary, *Science* **2013**, *340*, 1233232. [CrossRef] [PubMed]
40. Arzoumanian, Z.; Brazier, A.; Burke-Spolaor, S.; Chamberlin, S.; Chatterjee, S.; Christy, B.; Cordes, J.M.; Cornish, N.J.; Crowter, K. The NANOGrav 11-year Data Set: High-precision timing of 45 Millisecond Pulsars. *Astrophys. J. Suppl.* **2018**, *235*, 37. [CrossRef]
41. Arzoumanian, Z.; Bogdanov, S.; Cordes, J.; Gendreau, K.; Lai, D.; Lattimer, J.; Link, B.; Lommen, A.; Miller, C.; Ray, P.; et al. X-ray Timing of Neutron Stars, Astrophysical Probes of Extreme Physics. *arXiv* **2009**, arXiv:0902.3264.
42. Ravenhall, D.G.; Pethick, C.J. Neutron star moments of inertia. *Astrophys. J.* **1994**, *424*, 846–851. [CrossRef]
43. Chubarian, E.; Grigorian, H.; Poghosyan, G.S.; Blaschke, D. Deconfinement phase transition in rotating nonspherical compact stars. *Astron. Astrophys.* **2000**, *357*, 968.
44. Zdunik, J.L.; Bejger, M.; Haensel, P.; Gourgoulhon, E. Phase transitions in rotating neutron stars cores: Back bending, stability, corequakes and pulsar timing. *Astron. Astrophys.* **2006**, *450*, 747–758. [CrossRef]
45. Bejger, M.; Blaschke, D.; Haensel, P.; Zdunik, J.L.; Fortin, M. Consequences of a strong phase transition in the dense matter equation of state for the rotational evolution of neutron stars. *Astron. Astrophys.* **2017**, *600*, A39. [CrossRef]

© 2018 by the authors. Licensee MDPI, Basel, Switzerland. This article is an open access article distributed under the terms and conditions of the Creative Commons Attribution (CC BY) license (http://creativecommons.org/licenses/by/4.0/).

Article

A Phenomenological Equation of State of Strongly Interacting Matter with First-Order Phase Transitions and Critical Points

Stefan Typel [1,2,*] and David Blaschke [3,4,5]

[1] Institut für Kernphysik, Technische Universität Darmstadt, Schlossgartenstraße 9, 64289 Darmstadt, Germany
[2] GSI Helmholtzzentrum für Schwerionenforschung GmbH, Planckstraße 1, 64291 Darmstadt, Germany
[3] Institute of Theoretical Physics, University of Wroclaw, Pl. M. Borna 9, 50-204 Wroclaw, Poland; david.blaschke@gmail.com
[4] Bogoliubov Laboratory for Theoretical Physics, Joint Institute of Nuclear Research, Dubna 141980, Russia
[5] National Research Nuclear University (MEPhI), Kashirskoe shosse 31, Moscow 115409, Russia
* Correspondence: s.typel@gsi.de or stypel@ikp.tu-darmstadt.de; Tel.: +49-6151-16-21559

Received: 12 December 2017; Accepted: 29 January 2018; Published: 9 February 2018

Abstract: An extension of the relativistic density functional approach to the equation of state for strongly interacting matter is suggested that generalizes a recently developed modified excluded-volume mechanism to the case of temperature- and density-dependent available-volume fractions. A parametrization of this dependence is presented for which, at low temperatures and suprasaturation densities, a first-order phase transition is obtained. It changes for increasing temperatures to a crossover transition via a critical endpoint. This provides a benchmark case for studies of the role of such a point in hydrodynamic simulations of ultrarelativistic heavy-ion collisions. The approach is thermodynamically consistent and extendable to finite isospin asymmetries that are relevant for simulations of neutron stars, their mergers, and core-collapse supernova explosions.

Keywords: equation of state; QCD matter; phase transition; critical point; modified excluded-volume mechanism

1. Introduction

The simulation of astrophysical phenomena, such as core-collapse supernovae (CCSN) or neutron-star (NS) mergers, requires a careful modeling of strongly interacting matter in a wide range of densities and temperatures. The same applies to the theoretical description of heavy-ion collisions (HIC) that study compressed baryonic matter in the laboratory from low to high beam energies. The properties of such matter are represented by the equation of state (EoS) that provides information on pressure, entropy, energies, and other thermodynamic variables of interest.

A particular feature of QCD matter is the supposed phase transition (PT) from hadronic matter to quark matter when density or temperature increase to sufficiently high values. A strong first-order PT could allow for the existence of a third branch of compact stars and the occurrence of the twin-star phenomenon [1–3]. Signals of the PT might also have direct consequences in dynamical processes when matter in the quark phase expands and cools down, e.g., the release of a second neutrino burst in CCSN [4–6]. For a recent review on the role of the EoS in CCSN simulations, see [7].

The theoretical description of the hadron–quark PT in strongly interacting matter often relies on a construction employing different models for the two phases. With such an approach, the coexistence line of the first-order PT will usually connect a point on the zero-temperature axis at finite baryon chemical potential μ_B with a point at finite temperature T on the zero baryon chemical potential axis.

By this construction, the QCD hadron–quark PT is of first order in the whole temperature–density plane, see for instance [8]. However, from lattice QCD studies, it is known that there is a smooth crossover at $\mu_B = 0$ with increasing T [9,10], so at least one critical point at finite μ_B and T is expected to exist. Other possibilities are that the character of the transition is crossover all over the QCD phase diagram [11] or, as is advocated in studies of the BEC-BCS crossover transition in low-temperature QCD, that a second critical endpoint exists [12–14]. Since lattice QCD studies are presently incapable of exploring the EoS close to the presumed critical point with much confidence, unified models are needed that can account for the existence of these features, see, e.g., [15]. There are dedicated microscopic models available that incorporate the major expected features in the QCD phase diagram, e.g., chiral mean-field models [16] or parity-doublet quark–hadron models [17]. Simulations of CCSN or HIC that are based on a hydrodynamic description of matter during dynamical evolution use the thermodynamic properties of matter encoded in the EoS as an input. Such data can be provided by phenomenological models that do not need to incorporate all of the details of the underlying physics.

In this work, a novel approach is introduced to provide a phenomenological EoS of baryonic matter that exhibits a first-order PT and a critical point at densities and temperatures expected in QCD matter. The parameters of the model can be adjusted to place the coexistence line at arbitrary positions in the phase diagram. The description uses an extension of a relativistic energy density functional for hadronic matter assuming a medium-dependent change in the number of degrees of freedom. This approach employs a recently developed version of a modified excluded-volume (EV) mechanism that gives a thermodynamically consistent EoS with nuclear matter properties that are consistent with present constraints. Here, we concentrate on the hadron–quark transition but not on the liquid–gas PT, which is also contained in our model. The model allows us to study the PT for arbitrary isospin asymmetries; however, only isospin-symmetric matter is considered in this first exploratory study for simplicity. In the present work, no attempt was made to reproduce the EoS of QCD matter at vanishing baryon chemical potential obtained in lattice QCD studies. With appropriately chosen EV parameters, the crossover transition with increasing temperature can be well modeled, even for imaginary chemical potentials, e.g., in a hadron resonance gas model [18]. With improved parametrizations, the structure of the phase diagram in the full space of variables, i.e., temperature, baryon density/chemical potential, and isospin asymmetry, can be investigated in the future.

The theoretical formalism of the model is presented in Section 2, which includes the main equations that define the relevant thermodynamic quantities in Section 2.1. In Section 2.2, details of the parametrization of the interaction and of the effective degeneracy factors are given. They account for the change in the number of degrees of freedom with density and temperature. The phase transitions are explored in Section 3 for isospin-symmetric matter. Conclusions follow in Section 4.

2. Theoretical Model

The theoretical description of strongly interacting matter in the present work is adapted from the model introduced in [19]. It combines a relativistic mean-field (RMF) approach for hadronic matter with density-dependent nucleon–meson couplings and a modified EV mechanism. Here, it is sufficient to provide only the main equations without a detailed derivation. The essential quantities that determine the position of the PT and the critical point in the phase diagram are the effective degeneracy factors that depend on the number densities of the particles and the temperature.

2.1. Relativistic Energy Density Functional with Modified Excluded-Volume Mechanism

The present model assumes neutrons and protons as well as their antiparticles as the basic degrees of freedom. These particles interact by the exchange of mesons, and the model effectively describes the short-range repulsion (ω meson), the intermediate-range attraction (σ meson), and the isospin dependence of the nuclear interaction (ρ and δ mesons), as is common of RMF models. The contribution of leptons or other degrees of freedom like nuclei, hyperons or photons, as required for multi-purpose EoS for astrophysical applications [20], is not considered here.

The nucleons $i = n, p, \bar{n}, \bar{p}$ with rest masses m_i are treated as quasi-particles of energy,

$$E_i(k) = \sqrt{k^2 + (m_i - S_i)^2} + V_i \tag{1}$$

which depends on the particle momentum k and the scalar (S_i) and vector (V_i) potentials. Denoting the particle chemical potentials with μ_i, the contribution of the quasi-particles to the total pressure

$$p = \sum_i p_i + p_{meson} - p^{(r)} \tag{2}$$

of the system can be written as

$$p_i = T g_i^{(eff)} \int \frac{d^3k}{(2\pi)^3} \ln\left[1 + \exp\left(-\frac{E_i(k) - \mu_i}{T}\right)\right] \tag{3}$$

where the medium-dependent effective degeneracy factors

$$g_i^{(eff)} = g_i \Phi_i \tag{4}$$

are a product of the usual degeneracy factor $g_i = 2$ for nucleons and the available-volume fraction Φ_i, which is defined in Section 2.2.

The meson contribution

$$p_{meson} = \frac{1}{2}\left(C_\omega n_\omega^2 + C_\rho n_\rho^2 - C_\sigma n_\sigma^2 - C_\delta n_\delta^2\right) \tag{5}$$

to the total pressure in Equation (2) contains the coupling factors of the mesons

$$C_j = \frac{\Gamma_j^2}{m_j^2} \tag{6}$$

given as a ratio of the density-dependent coupling functions Γ_j and the meson masses m_j. The source densities

$$n_j = \sum_i g_{ij} n_i^{(v)} \tag{7}$$

for vector mesons ($j = \omega, \rho$) and

$$n_j = \sum_i g_{ij} n_i^{(s)} \tag{8}$$

for scalar mesons ($j = \sigma, \delta$) in Equation (5) are obtained from the quasi-particle vector densities

$$n_i^{(v)} = g_i^{(eff)} \int \frac{d^3k}{(2\pi)^3} f_i(k) \tag{9}$$

and scalar densities

$$n_i^{(s)} = g_i^{(eff)} \int \frac{d^3k}{(2\pi)^3} f_i(k) \frac{m_i - S_i}{\sqrt{k^2 + (m_i - S_i)^2}} \tag{10}$$

with the Fermi-Dirac distribution function

$$f_i(k) = \left[\exp\left(\frac{E_i(k) - \mu_i}{T}\right) + 1\right]^{-1}. \tag{11}$$

The scaling factors

$$g_{n\omega} = g_{p\omega} = -g_{\bar{n}\omega} = -g_{\bar{p}\omega} = 1 \quad (12)$$
$$g_{n\rho} = -g_{p\rho} = -g_{\bar{n}\rho} = g_{\bar{p}\rho} = 1 \quad (13)$$
$$g_{n\sigma} = g_{p\sigma} = g_{\bar{n}\sigma} = g_{\bar{p}\sigma} = 1 \quad (14)$$
$$g_{n\delta} = -g_{p\delta} = g_{\bar{n}\delta} = -g_{\bar{p}\delta} = 1 \quad (15)$$

in Equations (7) and (8) determine the coupling between mesons and nucleons. They also appear in the vector potential

$$V_i = C_\omega g_{i\omega} n_\omega + C_\rho g_{i\rho} n_\rho + B_i V^{(r)}_{\text{meson}} + V^{(r)}_i \quad (16)$$

and the scalar potential

$$S_i = C_\sigma g_{i\sigma} n_\sigma + C_\delta g_{i\delta} n_\delta + S^{(r)}_i \quad (17)$$

in the quasi-particle energy (Equation (1)). The rearrangement potential

$$V^{(r)}_{\text{meson}} = \frac{1}{2}\left(C'_\omega n_\omega^2 + C'_\rho n_\rho^2 - C'_\sigma n_\sigma^2 - C'_\delta n_\delta^2\right) \quad (18)$$

contributes to the vector potential (Equation (16)) because the couplings Γ_j in Equation (6) are assumed to depend on the baryon density $n_B = \sum_i B_i n^{(v)}_i$, where $B_i = g_{i\omega}$ is the baryon number of particle i, and the quantities $C'_j = dC_j/dn_B$ are the derivatives of the coupling factors.

The dependence of the available-volume fractions Φ_i in the effective degeneracy factor (Equation (4)) on the vector or scalar quasi-particle densities (9) and (10) also generates rearrangement contributions

$$V^{(r)}_i = -\sum_j p_j \frac{\partial \ln \Phi_j}{\partial n^{(v)}_i} \quad (19)$$

and

$$S^{(r)}_i = \sum_j p_j \frac{\partial \ln \Phi_j}{\partial n^{(s)}_i} \quad (20)$$

in the potentials (Equations (16) and (17)), respectively. Furthermore, there is a rearrangement term

$$p^{(r)} = p^{(r)}_{\text{meson}} + p^{(r)}_\Phi \quad (21)$$

in the total pressure (Equation (2)) with two contributions from the density dependence of the couplings

$$p^{(r)}_{\text{meson}} = -V^{(r)}_{\text{meson}} n_B \quad (22)$$

and

$$p^{(r)}_\Phi = \sum_i \left(n^{(s)}_i S^{(r)}_i - n^{(v)}_i V^{(r)}_i\right) \quad (23)$$

from the EV effects.

The free energy density of the system

$$f = \sum_i \mu_i n^{(v)}_i - p \quad (24)$$

is obtained with the total pressure (Equation (2)) and the chemical potentials of the particles μ_i. They are not independent since, for nucleons with baryon number B_i and charge number Q_i, they are given by

$$\mu_i = B_i \mu_B + Q_i \mu_Q \quad (25)$$

with the baryon chemical potential μ_B and the charge chemical potential μ_Q. Only the latter two are independent quantities. For the internal energy density

$$\varepsilon = f + Ts \tag{26}$$

the entropy density

$$s = -\sum_i g_i^{(\text{eff})} \int \frac{d^3k}{(2\pi)^3} \left[f_i \ln f_i + (1-f_i) \ln (1-f_i) \right] + \sum_i p_i \frac{\partial \ln \Phi_i}{\partial T} \tag{27}$$

is needed. Besides the standard contribution depending on the distribution functions f_i, there is a term from the possible temperature dependence of the available-volume fractions Φ_i. In order to guarantee the third law of thermodynamics, i.e., $\lim_{T \to 0} s = 0$, the temperature derivative of the available-volume fractions has to vanish for $T \to 0$. After solving the equations above for a given T, μ_B, and μ_Q, a fully consistent thermodynamic EoS is obtained. For practical purposes, however, the baryon density n_B and the hadronic charge fraction

$$Y_q = \frac{\sum_i Q_i n_i^{(v)}}{n_B} \tag{28}$$

are used as independent variables instead of μ_B and μ_Q.

A possible shortcoming of models that consider EV effects is the potential appearance of a superluminal speed of sound in certain regions of the space of thermodynamic variables, see, e.g., [21,22]. This causality constraint has to be checked case by case depending on the specific implementation of the EV mechanism.

2.2. Available-Volume Fractions and Model Parameters

For a quantitative evaluation of the EoS in the present approach, the functional forms of the meson–nucleon couplings Γ_j and the available-volume fractions Φ_i have to be specified as well as all parameters. Here, we use the masses of nucleons and mesons and the coupling functions of the DD2 parametrization presented in [23]. It only considers ω, ρ, and σ mesons for the effective description of the nuclear interaction but not the δ mesons. The parameters were obtained by fitting observables (binding energies, radii, etc.) of selected nuclei. With this set, the EoS of nuclear matter at zero temperature exhibits characteristic nuclear matter parameters, e.g., the saturation density ($n_{\text{sat}} = 0.149065$ fm^{-3}), binding energy at saturation ($B = 16.02$ MeV), incompressibility ($K = 242.7$ MeV), symmetry energy ($J = 32.73$ MeV), and slope ($L = 57.94$ MeV), that are consistent with modern constraints from experiment and theory.

EV effects are frequently employed to describe an effective repulsive interaction between particles, in particular in calculations of the EoS in the framework of hadron resonance gas models, see, e.g., [24]. For zero baryon density, a comparison of the resulting EoS with results from lattice QCD calculations can be used to fix the volume parameters. If a finite volume v_i is attributed to each particle i, the available volume for the motion of the particle is reduced from the total system volume V to $V\Phi^{(cl)}$ with the classical available-volume fraction

$$\Phi^{(cl)} = 1 - \sum_i v_i n_i^{(v)} \,. \tag{29}$$

Clearly, there is a limiting density above which a compression of the system becomes impossible. In general, the volumes and available-volume fractions can depend on the particle species, and the EV mechanism can be used to suppress particles, e.g., nuclei, in a mixture, causing them to disappear above a certain density, see, e.g., [25] for applications to the low-temperature and low-density EoS in astrophysical simulations.

In [19], the interpretation of the EV mechanism was changed by moving the factor Φ_i from the system volume V to the degeneracy factor g_i as in Equation (4) and allowing the available-volume fractions to be arbitrary functions of the particle densities and temperature. The medium dependence of the effective degeneracy factors is interpreted as a change in the effective number of degrees of freedom. A decrease in $g_i^{(\text{eff})}$ has the effect of a repulsive interaction between the particles, whereas an increase can be seen as the action of an attractive interaction, c.f., the softening of the nuclear EoS when hyperons are included, see, e.g., [26,27]. This freedom leaves the room to modify the properties of an EoS in a favored way.

In the present application of the modified EV mechanism, the available-volume fraction is defined to be identical for all particles i as

$$\Phi_i = 1 + sg_1(T)\theta(x)\exp\left(-\frac{1}{2x^2}\right) \tag{30}$$

depending on the temperature T and an auxiliary quantity

$$x = v\left(\sum_j B_j n_j^{(v)} - g_8(T)n_{\text{cut}}\right) \tag{31}$$

depending on T and the quasi-particle vector densities $n_i^{(v)}$ with parameters s, v, and a cutoff density of n_{cut}. The θ function in Equation (30) guarantees that $\Phi_i = 1$ for $x \leq 0$ and the EV mechanism has no effect on the EoS. The functions g_1 in Equation (30) and g_8 in Equation (31) are defined as

$$g_t(T) = \theta(T_0 - T)\exp\left[-\frac{t}{2}\left(\frac{T}{T_0 - T}\right)^2\right] \tag{32}$$

with parameters t and T_0. In the limit $T \to 0$, the function g_t approaches one and it decreases with increasing temperature. Furthermore, the derivatives $\partial \Phi_i / \partial T$ approach zero for $T \to 0$, as required for the thermodynamic consistency, because of the choice of the function $g_t(T)$. For $T \to T_0$, the function g_t vanishes very smoothly and there are no effects at higher temperatures because $\Phi_i = 1$. In order to reproduce the correct high-temperature limit, given by a Stefan–Boltzmann-type behavior, a modification of the available-volume fractions for temperatures well above T_0 is required. This is left to future extensions of the model. According to Equation (31), the quantity x is only positive for baryon densities larger than $g_8 n_{\text{cut}}$, a quantity that decreases with increasing temperature. There are no artificial singularities due to the presence of the θ functions in Equations (30) and (32) because all derivatives of the exponential functions are zero when the arguments of the θ functions vanish. The actual values of the parameters for the modified EV mechanism used in the present study are $s = 3$, $v = 2$ fm^3, $T_0 = 270$ MeV, and $n_{\text{cut}} = n_{\text{sat}}$ of the DD2 parametrization.

3. Results

In order to illustrate the characteristic effects of the modified EV mechanism on the EoS, we limit the presentation to the case of symmetric matter, i.e., $Y_q = 0.5$. The pressure p and baryon chemical potential μ_B are calculated as a function of the baryon density n_B. Due to the increase in the available-volume fractions Φ_i or the effective degeneracy factors $g_i^{(\text{eff})}$, a considerable softening of the EoS is observed in a certain range of densities. Below the critical temperature T_{crit}, the pressure is not a monotonous function of the baryon chemical potential for an isotherm. A Maxwell construction is used to determine the two densities of the coexisting phases where p and μ_B are identical. The pressure p as a function of the baryon density n_B is depicted in Figure 1 for selected temperatures. Note that an integral part of the underlying RMF model is a detailed description of the liquid–gas PT in nuclear matter and the formation and dissociation of nuclear clusters in compact-star matter. For details, see, e.g., [28].

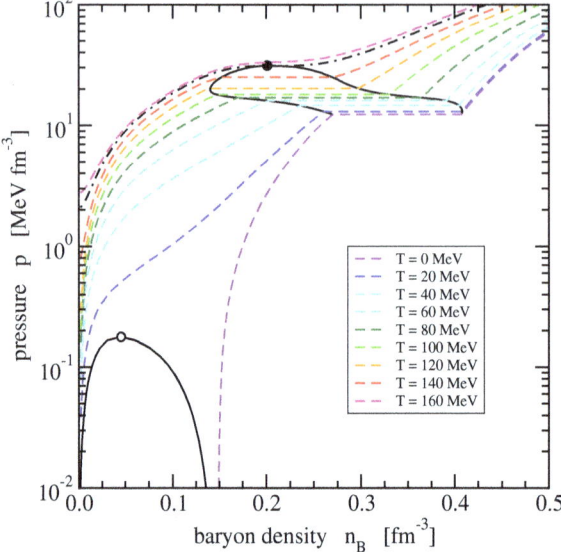

Figure 1. Isotherms in isospin-symmetric strongly interacting matter in the pressure–baryon density diagram at temperatures from 0 to 160 MeV in steps of 20 MeV (dashed colored lines) and at the critical temperature T_{crit} of the pseudo hadron–quark phase transition (black dot-dashed line). The binodals and critical points are denoted by full black lines and a full (open) circle of the pseudo hadron–quark (liquid-gas) phase transition, respectively.

In the coexistence region of the pseudo hadron–quark PT between the low and high density phases, the pressure is constant as typical for a first-order PT. The area of coexistence is enclosed by the binodal, and, above the critical temperature $T_{crit} \approx 155.5$ MeV, there is no PT anymore. The peculiar shape of the binodal is a result of the specific form (30) of the available-volume fractions. It can be adjusted with appropriate changes in the functional form and parameters.

The binodals of the liquid–gas and pseudo hadron–quark PT in the temperature–baryon density plane are shown in panel (a) of Figure 2. At vanishing temperatures, the coexistence region of the pseudo hadron–quark PT covers a density range from 0.270 to 0.408 fm^{-3}, well above the nuclear saturation density n_{sat}. At higher temperatures, it moves to lower densities with an almost constant extension in baryon density except for temperatures close to T_{crit}. Here, the critical density is found as 0.201 fm^{-3}, still above n_{sat}. The dashed line in panel (a) marks the boundary between regions without (lower left) and with (upper right) effects of the modified EV mechanism in the present parametrization. It corresponds to the condition $x = 0$. There is another region in the phase diagram without modified EV effects at temperatures above T_0, outside the figure.

Panel (b) of Figure 2 depicts the lines of the first-order PT in the temperature–baryon chemical potential diagram ending in critical points. With increasing temperature, the baryon chemical potential at the pseudo hadron–quark PT reduces from 979.1 to 591.8 MeV at the critical point. By crossing the transition line, an abrupt change in the density occurs that becomes continuous at the critical point.

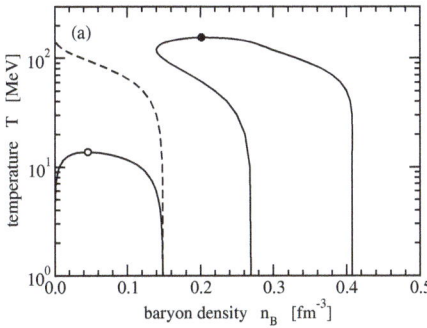 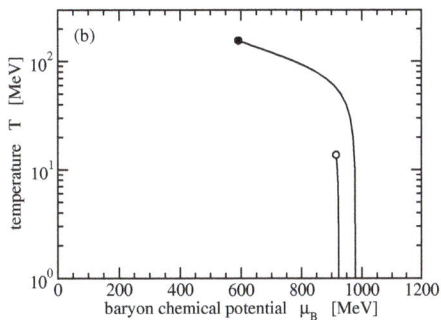

Figure 2. Binodals (full lines) and critical points (full and open circles) of isospin-symmetric strongly interacting matter in (**a**) the temperature–baryon density diagram and (**b**) the temperature–baryon chemical potential diagram. The dashed line in panel (**a**) separates the region without effects of the modified excluded-volume mechanism (lower left) from the region with effects (upper right). Results for the liquid–gas phase transition are shown at subsaturation densities.

4. Conclusions

The extension of the modified EV approach to a density- and temperature-dependent parametrization of the available-volume fractions as introduced in this work was successful in achieving the main goal of this study: As a generic structure of the QCD phase diagram, a first-order pseudo hadron–quark phase transition at low temperatures and a crossover for low baryon densities could be modeled that also includes a critical endpoint at $T_{\mathrm{crit}} = 155.5$ MeV and $\mu_{B,\mathrm{crit}} = 591.8$ MeV. Other patterns of the QCD phase diagram that have been theoretically motivated could also be modeled within the present approach. Further extensions of the model are straightforward. They include the extension to a larger number of components of the hadron resonance gas in the underlying RMF model and an isospin dependence of the EV model. It would be worthwhile to study further thermodynamic quantities such as the speed of sound, heat capacities, or the susceptibilities in such an enlarged model. It would also be interesting to elaborate on a parametrization that would result in a second endpoint at low temperatures, as suggested by Hatsuda et al. [14]. The so-generalized parametrization of the QCD EoS can be used in Bayesian analysis studies for astrophysical applications pertaining to compact stars [29,30], their mergers, and core-collapse supernova explosions, as well as heavy-ion collisions analogous to those studied in [31].

Acknowledgments: This work was supported by the Russian Science Foundation under contract number 17-12-01427. S.T. was supported in part by the DFG through grant No. SFB1245.

Author Contributions: D.B. and S.T. conceived and designed the concept of this work. S.T. worked out the theoretical framework, performed the numerical calculations, and wrote the main body of the paper. D.B. wrote the abstract as well as the discussion, conclusions, and the references sections.

Conflicts of Interest: The authors declare no conflict of interest.

Abbreviations

The following abbreviations are used in this manuscript:

CCSN	core-collapse supernova
EoS	equation of state
EV	excluded-volume
HIC	heavy-ion collision
NS	neutron star
PT	phase transition
QCD	quantum chromodynamics
RMF	relativistic mean-field

References

1. Alford, M.G.; Han, S.; Prakash, M. Generic conditions for stable hybrid stars. *Phys. Rev. D* **2013**, *88*, 083013.
2. Benic, S.; Blaschke, D.; Alvarez-Castillo, D.E.; Fischer, T.; Typel, S. A new quark-hadron hybrid equation of state for astrophysics—I. High-mass twin compact stars. *Astron. Astrophys.* **2015**, *577*, A40.
3. Alvarez-Castillo, D.E.; Blaschke, D.B. High-mass twin stars with a multipolytrope equation of state. *Phys. Rev. C* **2017**, *96*, 045809.
4. Sagert, I.; Fischer, T.; Hempel, M.; Pagliara, G.; Schaffner-Bielich, J.; Mezzacappa, A.; Thielemann, F.K.; Liebendörfer, M. Signals of the QCD phase transition in core-collapse supernovae. *Phys. Rev. Lett.* **2009**, *102*, 081101.
5. Fischer, T.; Whitehouse, S.C.; Mezzacappa, A.; Thielemann, F.K.; Liebendörfer, M. The neutrino signal from protoneutron star accretion and black hole formation. *Astron. Astrophys.* **2009**, *499*, 1–15.
6. Fischer, T.; Sagert, I.; Pagliara, G.; Hempel, M.; Schaffner-Bielich, J.; Rauscher, T.; Thielemann, F.K.; Käppeli, R.; Martinez-Pinedo, G.; Liebendörfer, M. Core-collapse supernova explosions triggered by a quark-hadron phase transition during the early post-bounce phase. *Astrophys. J. Suppl.* **2011**, *194*, 39.
7. Fischer, T.; Bastian, N.U.; Blaschke, D.; Cerniak, M.; Hempel, M.; Klähn, T.; Martínez-Pinedo, G.; Newton, W.G.; Röpke, G.; Typel, S. The state of matter in simulations of core-collapse supernovae—Reflections and recent developments. *Publ. Astron. Soc. Aust.* **2017**, *34*, 67.
8. Blaschke, D.B.; Sandin, F.; Skokov, V.V.; Typel, S. Accessibility of Color Superconducting Quark Matter Phases in Heavy-ion Collisions. *Acta Phys. Pol. Suppl.* **2010**, *3*, 741–746.
9. Bazavov, A.; Bhattacharya, T.; DeTar, C.; Ding, H.; Gottlieb, S.; Gupta, R.; Hegde, P.; Heller, U.M.; Karsch, F.; Laermann, E.; et al. Equation of state in (2 + 1)-flavor QCD. *Phys. Rev. D* **2014**, *90*, 094503.
10. Borsanyi, S.; Fodor, Z.; Hoelbling, C.; Katz, S.D.; Krieg, S.; Szabo, K.K. Full result for the QCD equation of state with 2 + 1 flavors. *Phys. Lett. B* **2014**, *730*, 99–104.
11. Bratovic, N.M.; Hatsuda, T.; Weise, W. Role of Vector Interaction and Axial Anomaly in the PNJL Modeling of the QCD Phase Diagram. *Phys. Lett. B* **2013**, *719*, 131–135.
12. Baym, G.; Hatsuda, T.; Kojo, T.; Powell, P.D.; Song, Y.; Takatsuka, T. From hadrons to quarks in neutron stars. *arXiv* **2017**, arXiv:astro-ph.HE/1707.04966.
13. Abuki, H.; Baym, G.; Hatsuda, T.; Yamamoto, N. The NJL model of dense three-flavor matter with axial anomaly: The low temperature critical point and BEC-BCS diquark crossover. *Phys. Rev. D* **2010**, *81*, 125010.
14. Hatsuda, T.; Tachibana, M.; Yamamoto, N.; Baym, G. New critical point induced by the axial anomaly in dense QCD. *Phys. Rev. Lett.* **2006**, *97*, 122001.
15. Klähn, T.; Fischer, T. Vector interaction enhanced bag model for astrophysical applications. *Astrophys. J.* **2015**, *810*, 134.
16. Dexheimer, V.A.; Schramm, S. A Novel Approach to Model Hybrid Stars. *Phys. Rev. C* **2010**, *81*, 045201.
17. Mukherjee, A.; Steinheimer, J.; Schramm, S. Higher-order baryon number susceptibilities: Interplay between the chiral and the nuclear liquid-gas transitions. *Phys. Rev. C* **2017**, *96*, 025205.
18. Vovchenko, V.; Pasztor, A.; Fodor, Z.; Katz, S.D.; Stoecker, H. Repulsive baryonic interactions and lattice QCD observables at imaginary chemical potential. *Phys. Lett. B* **2017**, *775*, 71–78.
19. Typel, S. Variations on the excluded-volume mechanism. *Eur. Phys. J. A* **2016**, *52*, 16.
20. Oertel, M.; Hempel, M.; Klähn, T.; Typel, S. Equations of state for supernovae and compact stars. *Rev. Mod. Phys.* **2017**, *89*, 015007.
21. Rischke, D.H.; Gorenstein, M.I.; Stoecker, H.; Greiner, W. Excluded volume effect for the nuclear matter equation of state. *Z. Phys. C* **1991**, *51*, 485–490.
22. Satarov, L.M.; Dmitriev, M.N.; Mishustin, I.N. Equation of state of hadron resonance gas and the phase diagram of strongly interacting matter. *Phys. Atom. Nucl.* **2009**, *72*, 1390–1415.
23. Typel, S.; Röpke, G.; Klähn, T.; Blaschke, D.; Wolter, H.H. Composition and thermodynamics of nuclear matter with light clusters. *Phys. Rev. C* **2010**, *81*, 015803.
24. Vovchenko, V.; Gorenstein, M.I.; Stoecker, H. Van der Waals Interactions in Hadron Resonance Gas: From Nuclear Matter to Lattice QCD. *Phys. Rev. Lett.* **2017**, *118*, 182301.
25. Hempel, M.; Schaffner-Bielich, J. Statistical Model for a Complete Supernova Equation of State. *Nucl. Phys. A* **2010**, *837*, 210–254.
26. Chatterjee, D.; Vidaña, I. Do hyperons exist in the interior of neutron stars? *Eur. Phys. J. A* **2016**, *52*, 29.

27. Schaffner-Bielich, J. Hypernuclear Physics for Neutron Stars. *Nucl. Phys. A* **2008**, *804*, 309–321.
28. Typel, S.; Wolter, H.H.; Röpke, G.; Blaschke, D. Effects of the liquid-gas phase transition and cluster formation on the symmetry energy. *Eur. Phys. J. A* **2014**, *50*, 17.
29. Steiner, A.W.; Lattimer, J.M.; Brown, E.F. The Equation of State from Observed Masses and Radii of Neutron Stars. *Astrophys. J.* **2010**, *722*, 33–54.
30. Alvarez-Castillo, D.; Ayriyan, A.; Benic, S.; Blaschke, D.; Grigorian, H.; Typel, S. New class of hybrid EoS and Bayesian M-R data analysis. *Eur. Phys. J. A* **2016**, *52*, 69.
31. Pratt, S.; Sangaline, E.; Sorensen, P.; Wang, H. Constraining the Equation of State of Superhadronic Matter from Heavy-Ion Collisions. *Phys. Rev. Lett.* **2015**, *114*, 202301.

© 2018 by the authors. Licensee MDPI, Basel, Switzerland. This article is an open access article distributed under the terms and conditions of the Creative Commons Attribution (CC BY) license (http://creativecommons.org/licenses/by/4.0/).

Conference Report

Directed Flow in Heavy-Ion Collisions and Its Implications for Astrophysics

Yuri B. Ivanov [1,2,3]

[1] National Research Centre (NRC) "Kurchatov Institute", Moscow 123182, Russia; y.b.ivanov@yandex.ru
[2] National Research Nuclear University "MEPhI", Moscow Engineering Physics Institute, Moscow 115409, Russia
[3] Joint Institute for Nuclear Research (JINR), Dubna 141980, Russia

Received: 17 October 2017; Accepted: 7 November 2017; Published: 14 November 2017

Abstract: Analysis of directed flow (v_1) of protons, antiprotons and pions in heavy-ion collisions is performed in the range of collision energies $\sqrt{s_{NN}}$ = 2.7–39 GeV. Simulations have been done within a three-fluid model employing a purely hadronic equation of state (EoS) and two versions of the EoS with deconfinement transitions: a first-order phase transition and a smooth crossover transition. The crossover EoS is unambiguously preferable for the description of experimental data at lower collision energies $\sqrt{s_{NN}} \lesssim 20$ GeV. However, at higher collision energies $\sqrt{s_{NN}} \gtrsim 20$ GeV, the purely hadronic EoS again becomes advantageous. This indicates that the deconfinement EoS in the quark-gluon sector should be stiffer at high baryon densities than those used in the calculation. The latter finding is in agreement with that discussed in astrophysics in connection with existence of hybrid stars with masses up to about two solar masses.

Keywords: heavy-ion collisions; directed flow; hydrodynamics; deconfinement; hybrid stars

PACS: 25.75.-q; 25.75.Nq; 24.10.Nz

---

1. Introduction

The directed flow [1] is one of the key observables in heavy ion collisions. Nowadays, it is defined as the first coefficient, v_1, in the Fourier expansion of a particle distribution, $d^2N/dy\, d\phi$, in azimuthal angle ϕ with respect to the reaction plane [2,3]

$$\frac{d^2N}{dy\, d\phi} = \frac{dN}{dy}\left(1 + \sum_{n=1}^{\infty} 2\, v_n(y) \cos(n\phi)\right), \tag{1}$$

where y is the longitudinal rapidity of a particle. The directed flow is mainly formed at an early (compression) stage of the collisions and hence is sensitive to early pressure gradients in the evolving nuclear matter [4,5]. As the EoS is harder, stronger pressure is developed. Thus, the directed flow probes the stiffness of the nuclear EoS at the early stage of nuclear collisions [6], which is of prime interest for heavy-ion research and astrophysics.

In Refs. [7–9], a significant reduction of the directed flow in the first-order phase transition to the quark-gluon phase (QGP) (the so-called "softest-point" effect) was predicted, which results from decreasing the pressure gradients in the mixed phase as compared to those in pure hadronic and quark-gluon phases. It was further predicted [10–12] that the directed flow as a function of rapidity exhibits a wiggle near the midrapidity with a negative slope near the midrapidity, when the incident energy is in the range corresponding to onset of the first-order phase transition. Thus, the wiggle near the midrapidity and the wiggle-like behavior of the excitation function of the midrapidity v_1 slope were put forward as a signature of the QGP phase transition. In Ref. [13], it was found that the QGP

EoS is not a necessarily prerequisite for occurrence of the midrapidity v_1 wiggle: A certain combination of space–momentum correlations may result in a negative slope in the rapidity dependence of the directed flow in high-energy nucleus-nucleus collisions. However, this mechanism can be realized only when colliding nuclei become quite transparent so that they pass through each other at the early stage of the collision.

The directed flow of identified hadrons—protons, antiprotons, positive and negative pions—in Au+Au collisions was recently measured in the energy range $\sqrt{s_{NN}}$ = (7.7–39) GeV by the STAR collaboration within the framework of the beam energy scan (BES) program at the BNL Relativistic Heavy Ion Collider (RHIC) [14]. These data have been already discussed in Refs. [15–22]. The Frankfurt group [15] did not succeed to describe the data and to obtain conclusive results. Within a hybrid approach [23], the authors found that there is no sensitivity of the directed flow on the EoS and, in particular, on the occurrence of a first-order phase transition. One of the possible reasons of this result can be that the initial stage of the collision in all scenarios is described within the Ultrarelativistic Quantum Molecular Dynamics (UrQMD) [24] in the hybrid approach. However, this initial stage does not solely determine the final directed flow because the UrQMD results still differ from those obtained within the hybrid approach [23].

In Refs. [16–18], the new STAR data were analyzed within two complementary approaches: kinetic transport approaches of the parton-hadron string dynamics (PHSD) [25] and its purely hadronic version (HSD) [26]), and a hydrodynamic approach of the relativistic three-fluid dynamics (3FD) [27,28]. In contrast to other observables, the directed flow was found to be very sensitive to the accuracy settings of the numerical scheme. Accurate calculations require a very high memory and computation time.

In the present contribution, we refine conclusions on the relevance of used EoSs, in particular, on the stiffness of the EoS at high baryon densities in the QGP sector based on the analysis performed in Refs. [16–18].

2. The 3FD Model

The 3FD approximation is a minimal way to simulate the early-stage nonequilibrium in the colliding nuclei at high incident energies. The 3FD model [27] describes a nuclear collision from the stage of the incident cold nuclei approaching each other, to the final freeze-out stage. Contrary to the conventional one-fluid dynamics, where a local instantaneous stopping of matter of the colliding nuclei is assumed, the 3FD considers inter-penetrating counter-streaming flows of leading baryon-rich matter, which gradually decelerate each other due to mutual friction. The basic idea of a 3FD approximation to heavy-ion collisions is that a generally nonequilibrium distribution of baryon-rich matter at each space–time point can be represented as a sum of two distinct contributions initially associated with constituent nucleons of the projectile and target nuclei. In addition, newly produced particles, populating predominantly the midrapidity region, are attributed to a third, so-called fireball fluid that is governed by the net-baryon-free sector of the EoS.

At the final stage of the collision, the p- and t-fluids are either spatially separated or mutually stopped and unified, while the f-fluid, predominantly located in the midrapidity region, keeps its identity and still overlaps with the baryon-rich fluids to a lesser (at high energies) or greater (at lower energies) extent. The freeze-out is performed accordingly to the procedure described in Ref. [27] and in more detail in Refs. [29,30].

Different EoSs can be implemented in the 3FD model. A key point is that the 3FD model is able to treat a deconfinement transition at the early *nonequilibrium* stage of the collision, when the directed flow is mainly formed. In this work, we apply a purely hadronic EoS [31], an EoS with a crossover transition as constructed in Ref. [32] and an EoS with a first-order phase transition into the QGP [32]. These are illustrated in Figure 1. Note that an onset of deconfinement in the two-phase EoS takes place at rather high baryon densities, above $n \sim 8\, n_0$. In EoSs compatible with constraints on the occurrence of the quark matter phase in massive neutron stars, the phase coexistence starts at about $4\, n_0$ [33]. An example of such an EoS, the DD2 EoS [34], is also displayed in Figure 1. The DD2 EoS will be discussed below. As it will be argued below, this excessive softness of the deconfinement EoSs of Ref. [32] is an obstacle for proper reproduction of the directed flow at high collision energies.

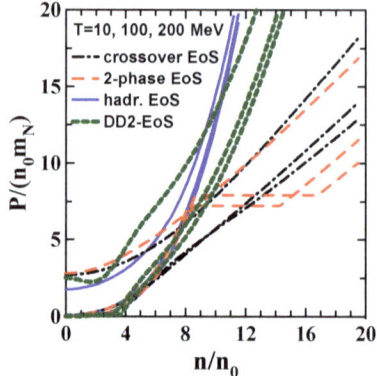

Figure 1. Pressure scaled by the product of normal nuclear density ($n_0 = 0.15$ fm^{-3}) and nucleon mass (m_N) versus baryon density (scaled by n_0) for three EoSs used in the simulations and also for the DD2 EoS [34] that is compatible with astrophysical constraints. Results are presented for three different temperatures $T = 10$, 100 and 200 MeV (bottom-up for corresponding curves).

In recent papers [28,35–40] a large variety of bulk observables has been analyzed with these three EoSs: the baryon stopping [28,37], yields of different hadrons, their rapidity and transverse momentum distributions [35,36,38], the elliptic flow of various species [39,40]. This analysis has been done in the same range of incident energies as that in the present paper. Comparison with available data indicated a definite advantage of the deconfinement scenarios over the purely hadronic one especially at high collision energies. The physical input of the present 3FD calculations is described in detail in Ref. [28].

3. Results

The 3FD simulations were performed for mid-central Au+Au collisions, i.e., at impact parameter $b = 6$ fm. Following the experimental conditions, the acceptance $p_T < 2$ GeV/c for transverse momentum (p_T) of the produced particles is applied to all considered hadrons. This choice is commented in Ref. [18]. In the 3FD model, particles are not isotopically distinguished; i.e., the model deals with nucleons, pions, etc. rather than with protons, neutrons, π^+, π^- and π^0. Therefore, the v_1 values of protons, antiprotons and pions presented below, in fact, are v_1 of nucleons, antinucleons and all (i.e., π^+, π^- and π^0) pions. The directed flow $v_1(y)$ as a function of rapidity y at BES-RHIC bombarding energies is presented in Figure 2 for pions, protons and antiprotons.

As seen, the first-order-transition scenario gives results for the proton v_1, which strongly differ from those in the crossover scenario at $\sqrt{s_{NN}} = 7.7$ and 19.6 GeV. This is in contrast to other bulk observables analyzed so far [28,35–40]. At $\sqrt{s_{NN}} = 39$ GeV, the directed flow of all considered species practically coincides within the first-order-transition and crossover scenarios. It means that the crossover transition to the QGP has been practically completed at $\sqrt{s_{NN}} = 39$ GeV. It also suggests that the region $7.7 \leq \sqrt{s_{NN}} \leq 30$ GeV is the region of the crossover transition.

The crossover EoS is definitely the best in reproduction of the proton $v_1(y)$ at $\sqrt{s_{NN}} \leq 20$ GeV. However, surprisingly, the hadronic scenario becomes preferable for the proton $v_1(y)$ at $\sqrt{s_{NN}} > 20$ GeV. A similar situation takes place in the PHSD/HSD transport approach. Indeed, predictions of the HSD model (i.e., without the deconfinement transition) for the proton $v_1(y)$ become preferable at $\sqrt{s_{NN}} > 30$ GeV [16], i.e., at somewhat higher energies than in the 3FD model. Moreover, the proton v_1 predicted by the UrQMD model, as cited in the experimental paper [14], and, in the recent theoretical work [15], better reproduces the proton $v_1(y)$ data at high collision energies than the PHSD and 3FD-deconfinement models do. Note that the UrQMD model is based on the hadronic dynamics. All these observations could be considered as evidence of a problem in the QGP sector of a EoS. At the same time, the antiproton directed flow at $\sqrt{s_{NN}} > 10$ GeV definitely indicates a preference of the crossover scenario within both the PHSD/HSD and 3FD approaches.

This puzzle has a natural resolution within the 3FD model. The QGP sector of the EoSs with deconfinement [32] was fitted to the lattice QCD data at zero net-baryon density and just extrapolated to nonzero baryon densities. The protons mainly originate from baryon-rich fluids that are governed by the EoS at finite baryon densities. The too strong proton antiflow within the crossover scenario at $\sqrt{s_{NN}} > 20$ GeV is a sign of too soft QGP EoS at high baryon densities. In general, the antiflow or a weak flow indicates softness of an EoS [6–13]. Predictions of the first-order-transition EoS, the QGP sector of which is constructed in the same way as that of the crossover one, fail even at lower collision energies, when the QGP starts to dominate in the collision dynamics, i.e., at $\sqrt{s_{NN}} \gtrsim 15$ GeV. This fact also supports the conjecture on a too soft QGP sector at high baryon densities in the used EoSs.

At the same time, the net-baryon-free (fireball) fluid is governed by the EoS at zero net-baryon density. This fluid is a main source of antiprotons (more than 80% near midrapidity at $\sqrt{s_{NN}} > 20$ GeV and $b = 6$ fm), $v_1(y)$ of which is in good agreement with the data at $\sqrt{s_{NN}} > 20$ GeV within the crossover scenario and even in perfect agreement—within the first-order-transition scenario at $\sqrt{s_{NN}} = 39$ GeV. It is encouraging because at zero net-baryon density the QGP sector of the EoSs is fitted to the lattice QCD data and therefore can be trusted. The crossover scenario, as well as all other scenarios, definitely fails to reproduce the antiproton $v_1(y)$ data at 7.7 GeV. The reason is low multiplicity of produced antiprotons. The antiproton multiplicity in the mid-central ($b = 6$ fm) Au + Au collision at 7.7 Gev is 1 within the deconfinement scenarios and 3 within the hadronic scenario. Therefore, the hydrodynamical approach based on the grand canonical ensemble is certainly inapplicable to the antiprotons in this case. The grand canonical ensemble, with respect to conservation laws, gives a satisfactory description of abundant particle production in heavy ion collisions. However, when applying the statistical treatment to rare probes, one needs to treat the conservation laws exactly, which is the canonical approach. The exact conservation of quantum numbers is known to reduce the phase space available for particle production due to additional constraints appearing through requirements of local quantum number conservation. An example of applying the canonical approach to the strangeness production can be found in [41] and references therein.

The pions are produced from all fluids: near midrapidity ~60% from the baryon-rich fluids and ~40% from the net-baryon-free one at $\sqrt{s_{NN}} > 20$ GeV. Hence, the disagreement of the pion v_1 with data, resulting from redundant softness of the QGP EoS at high baryon densities, is moderate at $\sqrt{s_{NN}} > 20$ GeV. In general, the pion v_1 is less sensitive to the EoS as compared to the proton and antiproton ones. As seen from Figure 2, the deconfinement scenarios are definitely preferable for the pion $v_1(y)$ at $\sqrt{s_{NN}} < 20$ GeV. Though, the hadronic-scenario results are not too far from the experimental data. At $\sqrt{s_{NN}} = 39$ GeV, the hadronic scenario gives even the best description of the pion data because of a higher stiffness of the hadronic EoS at high baryon densities, as compared with that in the considered versions of the QGP EoS.

Thus, all of the analyzed data testify in favor of a harder QGP EoS at high baryon densities than those used in the simulations, i.e., the desired QGP EoS should be closer to the used hadronic EoS at the same baryon densities (see Figure 1). At the same time, a moderate softening of the QGP EoS at moderately high baryon densities is agreement with data at $7.7 \lesssim \sqrt{s_{NN}} \lesssim 20$ GeV.

Here, it is appropriate to mention a discussion on the QGP EoS in astrophysics. In Ref. [42], it was demonstrated that the QGP EoS can be almost indistinguishable from the hadronic EoS at high baryon densities relevant to neutron stars. In particular, this gives a possibility to explain hybrid stars with masses up to about 2 solar masses ($M_\odot$), in such a way that "hybrid stars masquerade as neutron stars" [42]. The discussion of such a possibility has been revived after measurements on two binary pulsars PSR J1614−2230 [43] and PSR J0348l+0432 [44] resulted in the pulsar masses of

$(1.97 \pm 0.04) M_\odot$ and $(2.01 \pm 0.04) M_\odot$, respectively. The obtained results on the directed flow give us another indication of a required hardening of the QGP EoS at high baryon densities.

In this respect, it is instructive to compare the DD2 EoS [34], which is compatible with the existence of hybrid stars with masses up to about 2 solar masses, with those used in the present simulations (see Figure 1). As seen, the DD2 EoS is much closer to the hadronic EoS at high baryon densities as compared to the deconfinement EoSs used in the calculation. This gives hope to the better reproduction of the directed flow at high collision energies $\sqrt{s_{NN}} \gtrsim 20$ GeV with the DD2 EoS.

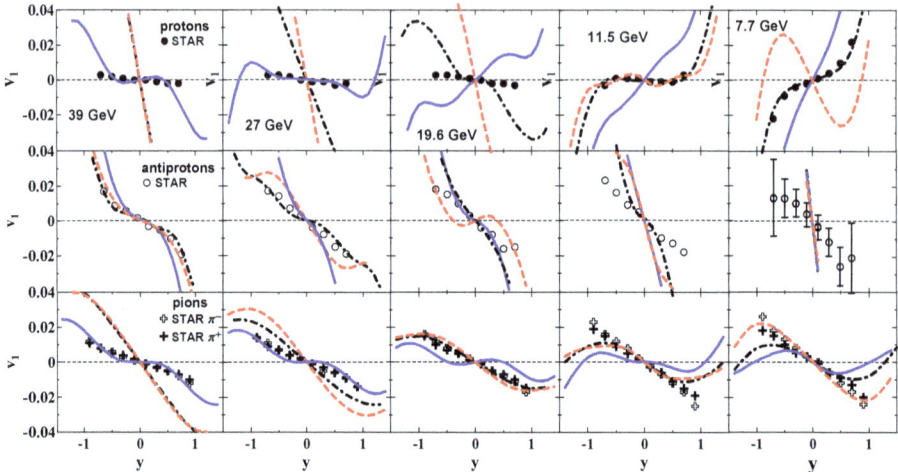

Figure 2. The directed flow $v_1(y)$ for protons, antiprotons and pions from mid-central ($b = 6$ fm) Au+Au collisions at $\sqrt{s_{NN}} = 7.7$–39 GeV calculated with different EoSs. Experimental data are from the STAR collaboration [14].

The slope of the directed flow at the midrapidity is often used to quantify variation of the directed flow with collision energy. The excitation functions for the slopes of the v_1 distributions at midrapidity are summarized in Figure 3, where earlier experimental results from the AGS [45] and SPS [46] are also presented. As noted above, the best reproduction of the data at $\sqrt{s_{NN}} < 20$ GeV is achieved with the crossover EoS. The proton dv_1/dy within the first-order-transition scenario exhibits a wiggle earlier predicted in Refs. [8,9,11,12]. The first-order-transition results demonstrate the worst agreement with the proton and antiproton data on dv_1/dy. The first-order-transition dv_1/dy does not coincide with that for the crossover scenario even at high collision energies (i.e., at 10 GeV $\lesssim \sqrt{s_{NN}} \lesssim 30$ GeV) because the corresponding EoSs are not identical in the region of high baryon densities where the smooth crossover transition is not completed yet, cf. Figure 1, and because of different friction terms, which were separately fitted for each EoS in order to reproduce other bulk observables.

The discrepancies between experiment and the 3FD predictions are smaller for the purely hadronic EoS, however, the agreement for the crossover EoS is definitely better, though it is far from being perfect. However, the poor reproduction of the proton v_1 slope at low energies ($\sqrt{s_{NN}} < 5$ GeV), it is still questionable because the same data, but in terms of the transverse in-plane momentum, $\langle P_x \rangle$, are almost perfectly reproduced by the crossover scenario [17,47]. It is difficult to indicate the beginning of the crossover transition because the crossover results become preferable beginning with relatively low collision energies ($\sqrt{s_{NN}} > 3$ GeV). However, the beginning of the crossover transition can be approximately pointed out as $\sqrt{s_{NN}} \simeq 4$ GeV.

The above discussed problems of the crossover scenario reveal themselves also in the dv_1/dy plot. At high energies ($\sqrt{s_{NN}} > 20$ GeV), the slopes also indicate that the used deconfinement EoSs in the quark-gluon sector at zero baryon chemical potential are quite suitable for reproduction of the antiproton dv_1/dy while those at high baryon densities (proton slope) should be stiffer in order to achieve better description of proton dv_1/dy. A combined effect of this excessive softness of the QGP EoS and the reducing baryon stopping results in more and more negative proton slopes at high collision energies. This is in line with the mechanism discussed in Ref. [13]. The pion flow partially follows the proton pattern, as discussed above. Therefore, the pion v_1 slope also becomes more negative with energy rise.

Of course, the 3FD model does not include all factors determining the directed flow. Initial-state fluctuations, which in particular make the directed flow to be nonzero even at midrapidity, are out of the scope of the 3FD approach. Apparently, these fluctuations can essentially affect the directed flow at high collision energies, when the experimental flow itself is very weak. Another point is so-called afterburner, i.e., the kinetic evolution after the the hydrodynamical freeze-out. This stage is absent in the conventional version of the 3FD. Recently, an event generator THESEUS based on the output of the 3FD model was constructed [48]. Thus, constructed output of the 3FD model can be further evolved within the UrQMD model. Results of Ref. [48] show that such kind of the afterburner mainly affects the pion v_1 at peripheral rapidities and makes it more close to the STAR data [14]. At $\sqrt{s_{NN}} < 5$ GeV, the midrapidity region of the pion v_1 is also affected; however, the pion data are absent at these energies. An additional source of uncertainty is the freeze-out. In Ref. [15], it was demonstrated that the freeze-out procedure and, in particular, its criterion also strongly affect the directed flow. Different freeze-out procedures were not tested within the 3FD model because such a test would amount to the analysis of all other bulk observables that can also be affected by the freeze-out change [29]. Such an extensive test would imply a huge amount of computations. However, this source of uncertainty should be mentioned.

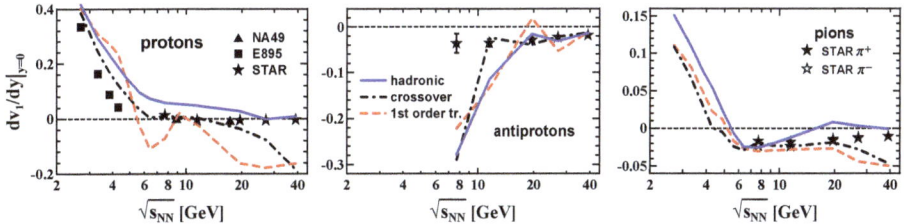

Figure 3. The beam energy dependence of the directed flow slope at midrapidity for protons, antiprotons and pions from mid-central ($b = 6$ fm) Au+Au collisions calculated with different EoSs. The experimental data are from Refs. [14,45,46].

4. Conclusions

In conclusion, the crossover EoS is unambiguously preferable for the most part of experimental data in the considered energy range, though this description is not perfect. Based on the crossover EoS of Ref. [32], the directed flow in semi-central Au+Au collisions indicates that the crossover deconfinement transition takes place in the wide range incident energies $4 \lesssim \sqrt{s_{NN}} \lesssim 30$ GeV. In part, this wide range could be a consequence of that the crossover transition constructed in Ref. [32] is very smooth. In this respect, this version of the crossover EoS certainly contradicts results of the lattice QCD calculations, where a fast crossover, at least at zero chemical potential, was found [49].

At highest computed energies of $\sqrt{s_{NN}} > 20$ GeV, the obtained results indicate that the deconfinement EoSs in the QGP sector should be stiffer at high baryon densities than those used in the calculation, i.e., more similar to the purely hadronic EoS. This observation is in agreement with that discussed in astrophysics, in particular, in connection with a possibility to explain hybrid stars with masses up to about two solar masses. The constraint of existence of such hybrid stars results in the requirement of quite stiff QGP EoS at high baryon densities that is very similar to the hadronic EoS. The obtained results on the directed flow give us another indication of a required hardening of the QGP EoS at high baryon densities. However, this is only an indirect similarity with the astrophysical conjecture because directed-flow simulations are sensitive to the EoS at high temperatures ($T > 100$ MeV) while the hybrid-star calculations are based on zero-temperature EoS.

Acknowledgments: Fruitful discussions with David Blaschke, Hermann Wolter and Dmitri N. Voskresensky are gratefully acknowledged. This work was carried out using computing resources of the federal collective usage center "Complex for simulation and data processing for mega-science facilities" at NRC "Kurchatov Institute", http://ckp.nrcki.ru/. This work was supported by the Russian Science Foundation, Grant No. 17-12-01427.

Conflicts of Interest: The author declares no conflict of interest.

References

1. Danielewicz, P.; Odyniec, G. Transverse Momentum Analysis of Collective Motion in Relativistic Nuclear Collisions. *Phys. Lett. B* **1985**, *157*, 146–150.
2. Voloshin, S.; Zhang, Y. Flow study in relativistic nuclear collisions by Fourier expansion of Azimuthal particle distributions. *Z. Phys. C* **1996**, *70*, 665–671.
3. Voloshin, S.A.; Poskanzer, A.M.; Snellings, R. Collective phenomena in non-central nuclear collisions. In *Relativistic Heavy Ion Physics*; Landolt-Börnstein—Group I Elementary Particles, Nuclei and Atoms; Springer-Verlag: New York, NY, USA, 2010.
4. Sorge, H. Elliptical flow: A signature for early pressure in ultrarelativistic nucleus-nucleus collisions. *Phys. Rev. Lett.* **1997**, *78*, 2309–2312.
5. Herrmann, N.; Wessels, J.P.; Wienold, T. Collective flow in heavy ion collisions. *Ann. Rev. Nucl. Part. Sci.* **1999**, *49*, 581–632.
6. Russkikh, V.N.; Ivanov, Y.B. Collective flow in heavy-ion collisions for Elab = 1 × 160 GeV/nucleon. *Phys. Rev. C* **2006**, *74*, 034904.
7. Hung, C.M.; Shuryak, E.V. Hydrodynamics near the QCD phase transition: Looking for the longest lived fireball. *Phys. Rev. Lett.* **1995**, *75*, 4003–4006.
8. Rischke, D.H.; Pursun, Y.; Maruhn, J.A.; Stöcker, H.; Greiner, W. The phase transition to the quark-gluon plasma and its effects on hydrodynamic flow. *Heavy Ion Phys.* **1995**, *1*, 309–322.
9. Rischke, D.H. Hydrodynamics and collective behaviour in relativistic nuclear collisions. *Nucl. Phys. A* **1996**, *610*, 88–101.
10. Csernai, L.P.; Rohrich, D. Third flow component as QGP signal. *Phys. Lett. B* **1999**, *458*, 454–459.
11. Brachmann, J.; Soff, S.; Dumitru, A.; Stöcker, H.; Maruhn, J.A.; Greiner, W.; Rischke, D.H. Antiflow of nucleons at the softest point of the equation of state. *Phys. Rev. C* **2000**, *61*, 024909.
12. Stöcker, H. Collective flow signals the quark-gluon plasma. *Nucl. Phys. A* **2005**, *750*, 121–147.
13. Snellings, R.J.M.; Sorge, H.; Voloshin, S.A.; Wang, F.Q.; Xu, N. Novel rapidity dependence of directed flow in high energy heavy ion collisions. *Phys. Rev. Lett.* **2000**, *84*, 2803–2805.
14. Adamczyk, L.; Adkins, J.K.; Agakishiev, G.; Aggarwal, M.M.; Ahammed, Z.; Alekseev, I.; Alford, J.; Anson, C.D.; Aparin, A.; Arkhipkin, D.; et al. Beam-Energy Dependence of Directed Flow of Protons, Antiprotons and Pions in Au+Au Collisions. *Phys. Rev. Lett.* **2014**, *112*, 162301.
15. Steinheimer, J.; Auvinen, J.; Petersen, H.; Bleicher, M.; Stöcker, H. Examination of directed flow as a signal for a phase transition in relativistic nuclear collisions. *Phys. Rev. C* **2014**, *89*, 054913.
16. Konchakovski, V.; Cassing, W.; Ivanov, Y.; Toneev, V. Examination of the directed flow puzzle in heavy-ion collisions. *Phys. Rev. C* **2014**, *90*, 014903.
17. Ivanov, Y.B.; Soldatov, A.A. Directed flow indicates a cross-over deconfinement transition in relativistic nuclear collisions. *Phys. Rev. C* **2015**, *91*, 024915.
18. Ivanov, Y.B.; Soldatov, A.A. What can we learn from the directed flow in heavy-ion collisions at BES RHIC energies? *Eur. Phys. J. A* **2016**, *52*, 10.
19. Nara, Y.; Ohnishi, A. Does non-monotonic behavior of directed flow signal the onset of deconfinement? *Nucl. Phys. A* **2016**, *956*, 284–287.
20. Nara, Y.; Niemi, H.; Ohnishi, A.; Stöcker, H. Examination of directed flow as a signature of the softest point of the equation of state in QCD matter. *Phys. Rev. C* **2016**, *94*, 034906.
21. Nara, Y.; Niemi, H.; Steinheimer, J.; Stöcker, H. Equation of state dependence of directed flow in a microscopic transport model. *Phys. Lett. B* **2017**, *769*, 543–548.

22. Singha, S.; Shanmuganathan, P.; Keane, D. The first moment of azimuthal anisotropy in nuclear collisions from AGS to LHC energies. *Adv. High Energy Phys.* **2016**, *2016*, 2836989.
23. Petersen, H.; Steinheimer, J.; Burau, G.; Bleicher, M.; Stöcker, H. A Fully Integrated Transport Approach to Heavy Ion Reactions with an Intermediate Hydrodynamic Stage. *Phys. Rev. C* **2008**, *78*, 044901.
24. Bass, S.; Belkacem, M.; Bleicher, M.; Brandstetter, M.; Bravina, L.; Ernst, C.; Gerland, L.; Hofmann, M.; Hofmann, S.; Konopka, J.; et al. Progress in Particle and Nuclear Physics. *Prog. Part. Nucl. Phys.* **1998**, *41*, 255–369.
25. Cassing, W.; Bratkovskaya, E.L. Parton–hadron–string dynamics: An off-shell transport approach for relativistic energies. *Nucl. Phys. A* **2009**, *831*, 215–242.
26. Cassing, W.; Bratkovskaya, E.L. Hadronic and electromagnetic probes of hot and dense nuclear matter. *Phys. Rep.* **1999**, *308*, 65–233.
27. Ivanov, Y.B.; Russkikh, V.N.; Toneev, V.D. Relativistic heavy-ion collisions within three-fluid hydrodynamics: Hadronic scenario. *Phys. Rev. C* **2006**, *73*, 044904.
28. Ivanov, Y.B. Alternative Scenarios of Relativistic Heavy-Ion Collisions: I. Baryon Stopping. *Phys. Rev. C* **2013**, *87*, 064904.
29. Russkikh, V.N.; Ivanov, Y.B. Vorticity in heavy-ion collisions at the JINR Nuclotron-based Ion Collider fAcility. *Phys. Rev. C* **2007**, *76*, 054907.
30. Ivanov, Y.B.; Russkikh, V.N. On freeze-out problem in relativistic hydrodynamics. *Phys. Atomic Nucl.* **2009**, *72*, 1238–1244.
31. Galitsky, V.M.; Mishustin, I.N. Relativistic effects in collision of heavy ions. *Sov. J. Nucl. Phys.* **1979**, *29*, 363–373. (In Russian)
32. Khvorostukhin, A.S.; Skokov, V.V.; Redlich, K.; Toneev, V.D.; Redlich, K. Lattice QCD constraints on the nuclear equation of state. *Eur. Phys. J. C* **2006**, *48*, 531–543.
33. Klähn, T.; Łastowiecki, R.; Blaschke, D.B. Implications of the measurement of pulsars with two solar masses for quark matter in compact stars and heavy-ion collisions: A Nambu Jona-Lasinio model case study. *Phys. Rev. D* **2013**, *88*, 085001.
34. Typel, S.; Ropke, G.; Klahn, T.; Blaschke, D.; Wolter, H.H. Composition and thermodynamics of nuclear matter with light clusters. *Phys. Rev. C* **2010**, *81*, 015803.
35. Ivanov, Y.B. Alternative Scenarios of Relativistic Heavy-Ion Collisions: II. Particle Production. *Phys. Rev. C* **2013**, *87*, 064905.
36. Ivanov, Y.B. Alternative Scenarios of Relativistic Heavy-Ion Collisions: III. Transverse Momentum Spectra. *Phys. Rev. C* **2014**, *89*, 024903.
37. Ivanov, Y.B. Baryon Stopping as a Probe of Deconfinement Onset in Relativistic Heavy-Ion Collisions. *Phys. Lett. B* **2013**, *721*, 123–130.
38. Ivanov, Y.B. Phase Evolution and Freeze-out within Alternative Scenarios of Relativistic Heavy-Ion Collisions. *Phys. Lett. B* **2013**, *726*, 422–426.
39. Ivanov, Y.B. Elliptic Flow of Protons and Antiprotons in Au+Au Collisions at $\sqrt{s_{NN}}$ = 7.7–62.4 GeV within Alternative Scenarios of Three-Fluid Dynamics. *Phys. Lett. B* **2013**, *723*, 475–480.
40. Ivanov, Y.B.; Soldatov, A.A. Elliptic Flow in Heavy-Ion Collisions at Energies $\sqrt{s_{NN}}$ = 2.7–39 GeV. *Phys. Rev. C* **2015**, *91*, 024914.
41. Hamieh, S.; Redlich, K.; Tounsi, A. Canonical description of strangeness enhancement from p-A to Pb Pb collisions. *Phys. Lett. B* **2000**, *486*, 61–66.
42. Alford, M.; Braby, M.; Paris, M.W.; Reddy, S. Hybrid stars that masquerade as neutron stars. *Astrophys. J.* **2005**, *629*, 969–978.
43. Demorest, P.; Pennucci, T.; Ransom, S.; Roberts, M.; Hessels, J. Shapiro Delay Measurement of A Two Solar Mass Neutron Star. *Nature* **2010**, *467*, 1081–1083.
44. Antoniadis, J.; Freire, P.C.C.; Wex, N.; Tauris, T.M.; Lynch, R.S.; van Kerkwijk, M.H.; Kramer, M.; Bassa, C. A Massive Pulsar in a Compact Relativistic Binary. *Science* **2013**, *340*, 1233232.
45. Ajitanand, N.N.; Alexander, J.; Anderson, M.; Best, D.; Brady, F.P.; Case, T.; Caskey, W.; Cebra, D.; Chance, J.; Cole, B. Sideward Flow in Au+Au Collisions between 2A and 8A GeV. *Phys. Rev. Lett.* **2000**, *84*, 5488.
46. Alt, C.; Anticic, T.; Baatar, B.; Barna, D.; Bartke, J.; Behler, M.; Betev, L.; Bialkowska, H.; Billmeier, A.; Blume, C.; et al. Directed and elliptic flow of charged pions and protons in Pb + Pb collisions at 40 A and 158 A GeV. *Phys. Rev. C* **2003**, *68*, 034903.

47. Ivanov, Y.B.; Soldatov, A.A. Directed flow is a sensitive probe of deconfinement transition. *Eur. Phys. J. A* **2016**, *52*, 246.
48. Batyuk, P.; Blaschke, D.; Bleicher, M.; Ivanov, Y.B.; Karpenko, I.; Merts, S.; Nahrgang, M.; Petersen, H.; Rogachevsky, O. Event simulation based on three-fluid hydrodynamics for collisions at energies available at the Dubna Nuclotron-based Ion Collider Facility and at the Facility for Antiproton and Ion Research in Darmstadt. *Phys. Rev. C* **2016**, *94*, 044917.
49. Aoki, Y.; Endrodi, G.; Fodor, Z.; Katz, S.D.; Szabo, K.K. The Order of the quantum chromodynamics transition predicted by the standard model of particle physics. *Nature* **2006**, *443*, 675–678.

© 2017 by the author. Licensee MDPI, Basel, Switzerland. This article is an open access article distributed under the terms and conditions of the Creative Commons Attribution (CC BY) license (http://creativecommons.org/licenses/by/4.0/).

Article

From Heavy-Ion Collisions to Compact Stars: Equation of State and Relevance of the System Size

Sylvain Mogliacci *, Isobel Kolbé and W. A. Horowitz

Department of Physics, University of Cape Town, Rondebosch 7701, South Africa;
isobel.kolbe@gmail.com (I.K.); wa.horowitz@uct.ac.za (W.A.H.)
* Correspondence: sylvain.mogliacci@uct.ac.za

Received: 30 November 2017; Accepted: 16 January 2018; Published: 23 January 2018

Abstract: In this article, we start by presenting state-of-the-art methods allowing us to compute moments related to the globally conserved baryon number, by means of first principle resummed perturbative frameworks. We focus on such quantities for they convey important properties of the finite temperature and density equation of state, being particularly sensitive to changes in the degrees of freedom across the quark-hadron phase transition. We thus present various number susceptibilities along with the corresponding results as obtained by lattice quantum chromodynamics collaborations, and comment on their comparison. Next, omitting the importance of coupling corrections and considering a zero-density toy model for the sake of argument, we focus on corrections due to the small size of heavy-ion collision systems, by means of spatial compactifications. Briefly motivating the relevance of finite size effects in heavy-ion physics, in opposition to the compact star physics, we present a few preliminary thermodynamic results together with the speed of sound for certain finite size relativistic quantum systems at very high temperature.

Keywords: finite temperature; finite density; quark-gluon plasma; finite size; speed of sound

1. Introduction

The weak coupling expansion of the Quantum Chromodynamics (QCD) grand potential density (which we further call "free energy density", since both are equivalent in the limit of zero density in infinite volume systems), is known to be a central quantity for the thermodynamics of a deconfined system such as those created in Heavy-Ion Collisions (HIC). If computed naively, there are however certain details that need to be improved, and we are going to discuss them in the following.

First of all, such a weak coupling expansion appears not to converge at phenomenologically moderate temperatures [1], relevant to the quark-hadron phase transition which takes place around the pseudo-critical temperature $T_c = 154 \pm 9$ MeV for vanishing quark chemical potentials μ_f [2,3]. Lattice Monte Carlo simulations cannot be used when the chemical potentials, and more specifically the baryon chemical potential, are non-zero due to the so-called sign problem of QCD [4,5]. Consequently, a lot of effort has been put in developing frameworks allowing access to reliable results, from first principles, by continuing the known perturbative outcome toward the non weakly coupled phase transition region, including at non-zero density [6–9]. In the light of the current [10,11] and forthcoming [12,13] experiments, the non perturbative lattice simulations together with these resummation frameworks allow for further insights in the study of the QCD phase diagram [8,14].

Then, a different issue which unfortunately lacks more thorough investigations, is related to the fact that the deconfined systems which are briefly created in collider experiments have finite, and in fact comparatively rather small volumes. For instance, their characteristic sizes are at best of the order of $R \sim 4$–8 Fermis for lead–lead collisions at $\sqrt{s} = 2.76$ TeV [15], and $R \sim 1$–2 Fermis for proton–lead collisions at $\sqrt{s} = 5.02$ TeV [16]. It is then trivial to compare these lengths to a temperature T which is typical, say $T \sim 2T_c$ to be conservative. Moreover, by doing so, one can see that the relevant dimensionless parameter, namely $\Delta = T \times R$, falls into neither of the extreme regimes: It indeed ranges in between a few units to roughly ten (it becomes, however, nearly an order of magnitude smaller if considering an hydrodynamic cell in local equilibrium). For comparison, systems relevant to the description of a compact star are in a complete different regime, with $\Delta = \infty$ being an excellent approximation (accounting for a small, yet non-zero, temperature [17], or evaluating the dimensionless parameter $\mu \times R$ instead). Thus, it appears that in the context of a quark–gluon plasma relevant to the heavy-ion physics, an important question is: Which physical quantities happen to be sensitive to the finite size of the system, and which are not? In this article, we will first briefly recall certain aspects of the QCD thermodynamics for infinite size systems, together with the corresponding fluctuations of globally conserved charges. Next, still relevant to infinite size systems, we will present two state-of-the-art frameworks for resumming the weak coupling expansion of QCD at finite temperature and density. Then, we shall move on to briefly introduce the proposed framework for finite size corrections. After which we will start reviewing results corresponding to the fluctuations and correlations of conserved charges, comparing the resummed perturbative framework results to those of lattice QCD collaborations. Finally, we will present a few preliminary results concerning finite size corrections, using a single non interacting massless scalar field at zero chemical potential as a toy model. A certain number of points will not be emphasized, and we refer the readers to [7,8,18] and references therein for further details on the frameworks as well as all the results for finite density investigations, and to [19] for more details on the finite size preliminary results.

2. Charge Fluctuations in Infinite Size Systems

We start by briefly recalling the link between the Hamiltonian of Quantum Chromodynamics H_{QCD} and its partition function Z_{QCD}, which in infinite size systems reads:

$$Z_{QCD}\left(T, \mu_f; V\right) \equiv \text{Tr} \exp\left[-\frac{1}{T}\left(H_{QCD} - \sum_f \mu_f Q_f\right)\right] = \text{Tr}\left(\rho_{QCD}\right), \quad (1)$$

where Q_f and μ_f respectively denote the conserved charges and the corresponding chemical potentials. Furthermore, in the following, $\langle \vartheta \rangle \equiv \text{Tr}\left(\vartheta \cdot \rho_{QCD}\right)/Z_{QCD}$ will denote thermal averages. While we mainly consider the up, down, and strange quarks with respective chemical potentials μ_u, μ_d, and μ_s, one can also express the partition function in terms of the baryon charge, the electric charge, and the strangeness conserved number, with (μ_B, μ_Q, μ_S) instead. From Equation (1), one can see that the mean and (co)variance of two conserved charges can be expressed in terms of derivatives with respect to the chemical potentials, following:

$$\langle Q_f \rangle = T \frac{\partial}{\partial \mu_f} \log Z_{QCD}, \quad (2)$$

$$\langle (Q_f - \langle Q_f \rangle) \cdot (Q_g - \langle Q_g \rangle) \rangle = T^2 \frac{\partial^2}{\partial \mu_f \partial \mu_g} \log Z_{QCD}, \quad (3)$$

which is straightforwardly related to the first and second order cumulants, respectively. These above quantities, referred to as susceptibilities, are defined for the quark numbers by:

$$\chi_{u_i d_j s_k \ldots}\left(T, \{\mu_f\}\right) \equiv \frac{\partial^n p_{QCD}\left(T, \{\mu_f\}\right)}{\partial \mu_u^i \partial \mu_d^j \partial \mu_s^k \ldots}, \quad (4)$$

with $n = i + j + k + \ldots$, and where we recall that equilibrium thermodynamic quantities such as the pressure follow simple relations in infinite size systems like:

$$p_{\text{QCD}} = \frac{\partial (T \log Z_{\text{QCD}})}{\partial V} \xrightarrow[V \to \infty]{} \frac{T}{V} \log Z_{\text{QCD}}. \quad (5)$$

It should be noted that one may also consider any other conserved charge, instead of the quark numbers. Furthermore, the number susceptibilities are, in general, important as they give information on the correlations and fluctuations of the globally conserved quantum numbers. Therefore, they turn out to be very practical probes for the changes of degrees of freedom across the transition region, specifically when the conserved charge is the baryon number. They are also directly related to the corresponding cumulants, and thus provide some crucial information about the probability distribution of the baryonic degree of freedom, together with insights on the existence and location of a possible critical point on the QCD phase diagram. For more detail on the use of conserved charge cumulants in HIC, we refer the readers to [20,21].

3. Resummed Perturbative Quantum Chromodynamics in Infinite Size Systems

3.1. Resummation Inspired from Dimensional Reduction

The Dimensional Reduction (DR) phenomenon at asymptotically high temperature, which can generally be understood as the appearance of certain effective degrees of freedom in a lower dimension, is well known to account properly for the dynamics of energy scales up to the order gT in QCD [22,23]. Such a procedure is carried out using an effective field theory which is called Electrostatic QCD (EQCD), by integrating out the hard degrees of freedom followed by a careful matching of the effective theory with the original one. EQCD is a three-dimensional SU(N_c) Yang-Mills theory coupled to an adjoint Higgs field [24,25], which accounts for all the infrared divergences encountered in the weak coupling expansions [26]. And as such, EQCD provides a rigorous framework for carrying out higher order loop computations in high temperature perturbative QCD.

Using this knowledge, one can rewrite the QCD pressure as:

$$p_{\text{QCD}} = p_{\text{hard}}(g) + T \, p_{\text{EQCD}}(m_E, g_E, \lambda_E, \zeta), \quad (6)$$

where the parameters $m_E(g), g_E(g), \lambda_E(g)$ and $\zeta(g)$ are also functions of the temperature and the quark chemical potentials, and admit expansions in powers of the four-dimensional gauge coupling g. The contribution p_{hard} is relevant to the hard scale ($\sim T$), and can be computed through a direct loop expansion in QCD. The contribution p_{EQCD} is, on the other hand, relevant to the soft ($\sim gT$) and ultrasoft ($\sim g^2 T$) scales. This contribution is accessible from the partition function of EQCD, and through a partial four-loop order even accessible by means of loop expansion only.

In principle, when computing the EQCD pressure in order to be able to access the full QCD pressure, the entire result (with the EQCD parameters) must be Taylor expanded in powers of g around small values. However, it was suggested in [1] and then first applied at zero chemical potential in [27], that one can simply consider both p_{hard} and p_{EQCD} as functions of the EQCD parameters. Doing so, and not re-expanding them in powers of g resums certain higher order contributions while keeping all correct contributions up to and including the order $g^6 \log(g)$ [28,29]. As a byproduct, the theoretical uncertainties through the renormalization scale dependence of the result is substantially reduced, and the convergence properties are thereby improved.

3.2. Hard-Thermal-Loop Perturbation Theory

The use of a variationally improved perturbation theory framework has been known for decades to allow important higher order resummations as well [30,31]. The introduction of a certain relevant term to be added and subtracted from the Lagrangian density, allows for the treatment of the added/subtracted piece with the non-interacting/interacting part. By doing so, one actually interpolates between the original theory and a theory having appropriately dressed propagators and vertices, while recovering the original theory in the end by setting (see below) $\delta = 1$. For QCD, the procedure is of course more complicated given gauge invariance, and the relevant term is the non-local hard-thermal-loop effective action [32,33]. This procedure is called Hard-Thermal-Loop perturbation theory (HTLpt) [34,35], and the subsequent Lagrangian reorganization reads:

$$\mathcal{L}_{\text{HTLpt}} = \left[\mathcal{L}_{\text{QCD}} + (1 - \delta) \, \mathcal{L}_{\text{HTL}} \right]\Big|_{g \to \sqrt{\delta} g} + \Delta \mathcal{L}_{\text{HTL}}, \qquad (7)$$

where $\mathcal{L}_{\text{HTL}}$ is the gauge invariant HTL improvement term, and δ a formal expansion parameter set to one after the expansion. We notice, in the above, that $\Delta \mathcal{L}_{\text{HTL}}$ is a counter term necessary to cancel all the ultraviolet divergences introduced by the reorganization of the perturbative series, from the ground state of an ideal gas of massless particles to the ground state of an ideal gas of massive quasiparticles.

4. On the Finite Size Corrections

First of all, we would like to refer the readers to the forthcoming article [19], where all the details from the conceptual to computational aspects will be exposed to a greater extent.

We wish now to drastically simplify the approach to the quark–gluon plasma created in high energy collisions, in order to make a first step in accounting for its finite size. To this end, we will disregard the importance of accounting for the interaction, and consider a zero-density toy model for a start: We chose a single non-interacting massless scalar field at zero chemical potential. And indeed, it is worth noticing that such a massless scalar field is, in fact, quite insightful, for it contains information relevant to a gas of non-interacting gluons—albeit a group theory prefactor (to the free energy), which is not present in any of the displayed quantities here, since we conveniently show only the appropriate ratios. We will then solely focus on the corrections due to the small size of our system. Furthermore, we are not discussing here a real finite volume system, leaving it for [19] and referring to works such as [36] for other types of finite volume investigations (this time, relevant to hadronic systems).

Instead, we will simply discuss a quantum relativistic system whose dynamics are governed by the aforementioned field theory, coupled to a heat bath at temperature T and geometrically confined in between two infinite parallel planes separated by a distance L. More precisely, motivated by physical arguments [19], we undertake a spatial compactification by assuming Dirichlet boundary conditions on both infinite planes, and finally arrive, explicitly, to the free-energy density of a neutral, massless, and non-interacting scalar field, which reads:

$$\begin{aligned} f(T,L) &\equiv F(T,L,A)/V \\ &= -\frac{\pi^2 T^4}{90} + \frac{\zeta(3) T^3}{4\pi L} - \frac{\zeta(3) T}{16\pi L^3} \\ &\quad - \frac{T^2}{8L^2} \times \sum_{s=1}^{+\infty} \left[\frac{\operatorname{csch}^2 (2\pi T L \times s)}{s^2} \right] - \frac{T}{16\pi L^3} \times \sum_{s=1}^{+\infty} \left[\frac{\coth (2\pi T L \times s) - 1}{s^3} \right], \end{aligned} \qquad (8)$$

and where $F(T,L,A)$ is the free energy, $A \equiv V/L$ being the area of each of the infinite parallel planes. A few comments are now in order, concerning the Equation (8). The above expression is an exact analytic representation of the free energy density of our system coupled to the heat bath and in between the parallel planes. Moreover, it is resummed to be exponentially fast in terms of convergence for practical numerical evaluations (when the sums are then truncated; see [19] for more detail). Finally, let us notice that the first term in the first line of (8) is the usual (fully) thermal non-compactified result. Next to it, the two simple terms are part of the thermal corrections to the Casimir (geometric) effect due to the presence of boundaries inside the heat bath. The last two terms, each containing an infinite summation, account for both the rest of the thermal corrections to the geometric effect and for the so-called zero-temperature Casimir result. Indeed, it can be checked that applying the (well defined) limit $T \to 0$ to the above will give us:

$$f(T = 0, L) = -\frac{\pi^2}{1440\, L^4}, \qquad (9)$$

which is responsible for the well-known zero-temperature Casimir pressure $p_{Cas} \equiv -\pi^2/(480\,L^4)$. Thus, our expression is not only a very compact one which converges exponentially fast when the sums are truncated for nearly any values of T and L, but it rightfully reproduces the two appropriate limits, namely $L = \infty$ and $T = 0$ respectively.

Let us now present all the results, some concerning the (QCD) infinite volume case at finite density and some being relevant to the (toy model) finite size case at zero density.

5. Results and Discussion

In the present section, we refer the readers to [8] for more detail on the setting of the parameters in the case of the finite density results, the bands corresponding to conservative variations of all the resummed perturbative parameters. We also refer to [19] for a forthcoming thorough investigation concerning the finite size preliminary results which we are presenting. Concerning the finite density results, the blue band corresponds to the DR result while the red and orange bands are the exact one-loop and truncated three-loop HTLpt results. The dashed curves inside the bands correspond to the central values of the renormalization and QCD scales. As for the finite size preliminary results, all quantities which are presented here are relevant to a system in between two infinite parallel planes separated by a distance L, and in contact with a heat bath at temperature T.

5.1. Quantum Chromodynamics Infinite Volume Case at Finite Density

5.1.1. Low Order Susceptibilities

First, we display low order quark and baryon number susceptibilities in Figures 1 and 2 (left), and in Figure 2 (right), respectively. The second- and fourth-order diagonal number susceptibilities are normalized to their non-interacting limits.

From the width of the bands, we clearly see that the DR scale dependence is extremely small for perturbatively relevant temperatures. Moreover, both DR and HTLpt are in accordance with each other, while agreeing quite well with the non perturbative lattice results down to $T \sim 200\text{--}400$ MeV.

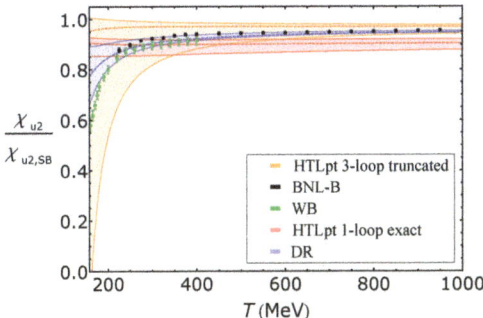

Figure 1. Second-order diagonal quark number susceptibility, normalized to its non-interacting limit. The truncated three-loop HTLpt result is from [37] and the lattice data from the BNL–Bielefeld [38] (BNL–B) as well as from the Wuppertal–Budapest [39] (WB) collaborations.

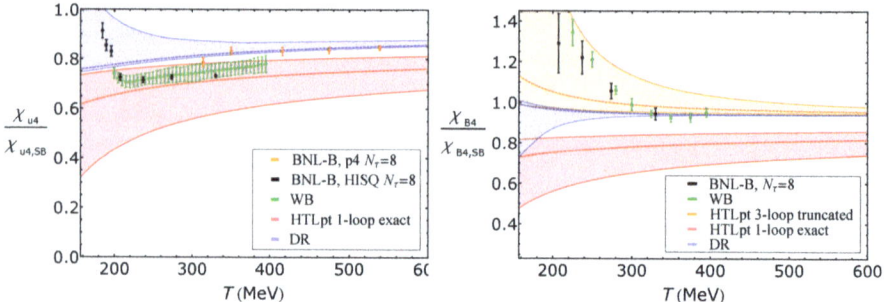

Figure 2. Fourth-order diagonal quark (**left**) and baryon (**right**) number susceptibilities, normalized to the non-interacting limits. The truncated three-loop HTLpt result is from [37] and the lattice data from the BNL–Bielefeld [40–42] (BNL–B) as well as the Wuppertal-Budapest [43] (WB) collaborations.

5.1.2. Kurtoses

Next, we are presenting the kurtosis, a certain ratio of the fourth and second order quark or baryon susceptibilities. It is a measure of how strongly peaked a quantity is, most often used to measure how a critical point is approached during a phase transition [44].

In Figure 3 (left), we plot the DR and HTLpt results together with lattice data which seem to agree with the one-loop HTLpt band at temperatures of $T \sim 300$–400 MeV, however approaching the DR prediction at higher temperatures. The latter reproduces the overall trend of the lattice data better. On the right hand figure, both the three-loop HTLpt and DR predictions seem to agree with the lattice data at around $T \sim 350$ MeV, albeit the DR prediction is much more predictive. Both resummed perturbative results converge to the Stefan Boltzmann limit faster than for the result relevant to quark numbers. This tends to comfort the expectation that the medium should be less sensitive to the hadronic degrees of freedom in this range of temperatures.

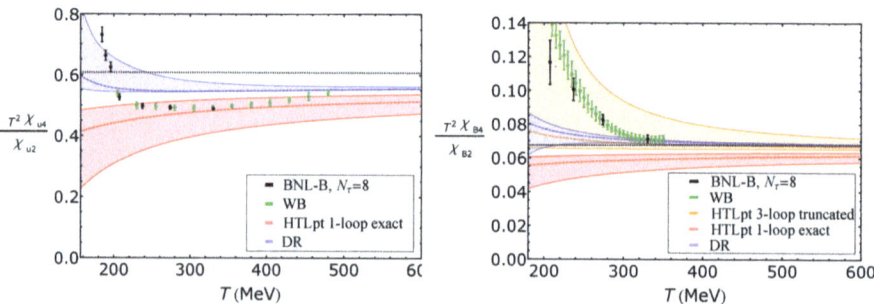

Figure 3. Ratios of low order susceptibilities for the quark (**left**) and baryon (**right**) numbers. The lattice data is from [40,41] (BNL–B) and [43,45] (WB). The three-loop HTLpt result is obtained from the corresponding cumulants of [37]. The black dashed (straight) lines denote the Stefan Boltzmann limits.

5.2. Toy Model Finite Size Case at Zero Density

5.2.1. Finite Size Corrections to the Thermodynamics

We start by plotting the ratio of the free energy with its non-compactified limit (i.e., for which $L \to \infty$). Even though the intrinsic asymmetry of such a finite size system implies that the actual pressure along the planes may not be the same as the pressure across them, we recall [19] that in the non-compactified limit, both pressures reduce to minus the free energy density. Such a quantity is then quite convenient for understanding the effect of finite size corrections.

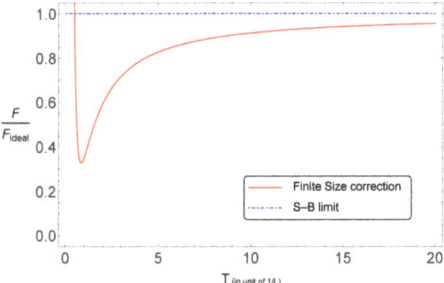

Figure 4. Correction to the free energy of a system at temperature T due to the compactification along one direction with length L. The result is normalized to its non-compactified, non-interacting limit, and plotted as a function of T in unit of $1/L$.

In Figure 4, we notice a sharp increase at low temperature which is simply a consequence of the fact that the function is normalized to the fourth power of temperature. Indeed, the zero temperature limit of the free energy is finite at fixed L: It is the so-called Casimir value (9).

5.2.2. Non Additivity of the Equation of State "Entropy Versus Temperature" in Finite Size Systems

We now wish to bring to the attention of the readers that a finite size system may not only be asymmetric, as mentioned previously, but will also lose some of the property of additivity.

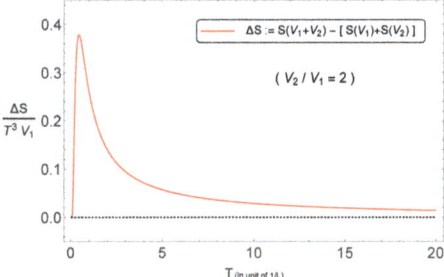

Figure 5. Non additivity of the entropy as a function of the temperature, due to the spatial compactification. The quantity is plotted as a function of T in unit of $1/L$, but goes to zero in the limit $L \to \infty$, and is normalized to appropriate powers of the temperature and volume. One subsystem (V_2) is twice as large as the other.

In Figure 5, we notice that in the large L limit the system becomes fully additive, with for example the equation of state $S(T)$ being additive, as it is expected. However, this happens in the asymptotically small T limit too. But this is merely a consequence of the fact that at zero temperature, the total entropy function vanishes (in the thermodynamic limit, i.e., at present since the volume is infinite, this is called the third law of thermodynamics). We further notice that both subsystems with volumes V_1 and V_2 have the same temperature, in agreement with the zeroth law of thermodynamics.

5.2.3. Finite Size Corrections to the Speed of Sound

Finally, we wish to present the squares of the two possible isochoric speeds of sound in between infinite parallel planes: the one that is transverse to the planes ($c_{s_\perp}^2$), and the one that is longitudinal

($c_{s_{2/3}}^2$). We also refer to [19] for much more detail on the derivation of such a quantity. This sound, propagating in an asymmetric manner, can be understood as due to variations in the pressures (both longitudinal and transverse) and the energy density as a consequence of a certain heat transfer from/toward the system.

Both Figure 6 (left and right) have a number of interesting features, and we also refer to [19] for a more complete interpretation of the result. However, it is already noteworthy that the average of the isochoric speeds of sound in the three directions (c_{s_1}, c_{s_2}, and c_{s_3}; the black line on the left Figure 6) is identically equal to the well known non-compactified limit of $c_s = 1/\sqrt{3}$. For both velocities, the correction seems not to be negligible anymore, starting at about 6–10 % for system sizes relevant to lead–lead collisions at a few hundred MeV. On Figure 6 (right), we display the total energy of such a system, as a function of both isochoric speeds of sound. The negative region for the total energy is simply the consequence of a certain Casimir effect, notifying the fact that the thermal contribution to the energy is not dominant anymore.

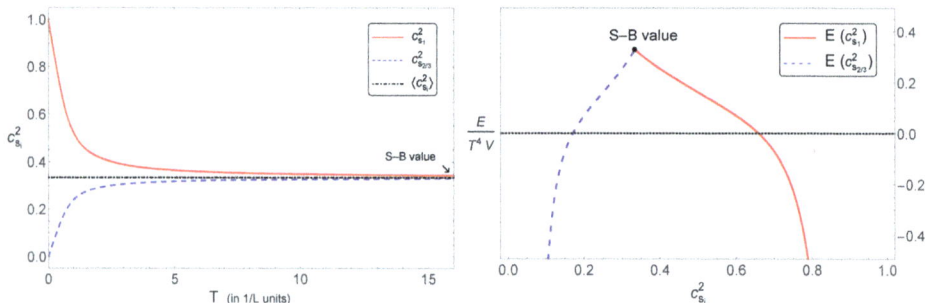

Figure 6. Left figure, we display the two isochoric speeds of sound in between infinite parallel planes, as a function of T in unit of $1/L$. **Right** figure, we display the total energy of the system as a function of each of the isochoric speeds of sound: The transverse one (c_{s_1}), and the longitudinal ones ($c_{s_{2/3}}$).

6. Conclusions

Despite tremendous advances, with various analytic and numerical methods, in the understanding of the thermodynamics of quark–gluon plasmas as created in HIC, we have presented a few preliminary results that suggest the need for a further refinement in the overall picture. This need of improvement seems to be valid, as far as can be understood at present, for quantities related to the thermodynamics of the systems. They seem nevertheless to play an important role in understanding the degrees of freedom at work across the quark–hadron phase transition better. However, the question remains whether or not more dynamical quantities could be less sensitive to the finite size of the system.

Acknowledgments: S.M. would like to acknowledge the financial support from the Claude Leon Foundation, and the South African National Research Foundation (NRF) for supporting his research mobility. S.M. would like to thank the organizers of "Compact stars in the QCD Phase Diagram VI", the Joint Institute for Nuclear Research in Dubna, and in particular David Blaschke and Alexander Sorin. I.K. and W.A.H. wish to thank the SA-CERN Collaboration, the NRF, and the South African National Institute for Theoretical Physics for their support. I. K. wishes to further acknowledge Deutscher Akademischer Austauschdienst for supporting her work.

Conflicts of Interest: The authors declare no conflict of interest.

References

1. Blaizot, J.P.; Iancu, E.; Rebhan, A. On the apparent convergence of perturbative QCD at high temperature. *Phys. Rev. D* **2003**, *68*, 025011.
2. Karsch, F.; Laermann, E.; Peikert, A. Quark mass and flavor dependence of the QCD phase transition. *Nucl. Phys. B* **2001**, *605*, 579–599.

3. Bazavov, A.; Bhattacharya, T.; Cheng, M.; DeTar, C.; Ding, H.-T.; Steven Gottlieb, R.; Gupta, R.; Hegde, P.; Heller, U.M.; Karsch, F.; et al. The chiral and deconfinement aspects of the QCD transition. *Phys. Rev. D* **2012**, *85*, 054503.
4. De Forcrand, P. Simulating QCD at finite density. *PoS LAT 2009*, *2009*, 010
5. Gupta, S. QCD at finite density. *PoS LAT 2010*, *2010*, 007
6. Blaizot, J.P.; Iancu, E.; Rebhan, A. Quark number susceptibilities from HTL resummed thermodynamics. *Phys. Lett. B* **2001**, *523*, 143–150.
7. Andersen, J.O.; Mogliacci, S.; Su, N; Vuorinen, A. Quark number susceptibilities from resummed perturbation theory. *Phys. Rev. D* **2013**, *87*, 074003.
8. Mogliacci, S.; Andersen, J.O.; Strickland, M.; Su, N; Vuorinen, A. Equation of State of hot and dense QCD: Resummed perturbation theory confronts lattice data. *J. High Energy Phys.* **2013**, *1312*, 055.
9. Haque, N.; Bandyopadhyay, A.; Andersen, J.O.; Mustafa, M.G.; Strickland, M.; Su, N. Three-loop HTLpt thermodynamics at finite temperature and chemical potential. *J. High Energy Phys.* **2014**, *1405*, 027.
10. Tannenbaum, M.J. Highlights from BNL-RHIC. *arXiv* **2012**, arXiv:1201.5900.
11. Müller, B.; Schukraft, J.; Wyslouch, B. First Results from Pb+Pb collisions at the LHC. *Ann. Rev. Nucl. Part. Sci.* **2012**, *62*, 361–386.
12. Heuser, J.M. The compressed baryonic matter experiment at FAIR. *Nucl. Phys. A* **2013**, *904–905*, 941c–944c.
13. Kekelidze, V.; Kovalenko, A.; Lednicky, R.; Matveev, V.; Meshkov, I.; Sorin, A.; Trubnikov, G. Project NICA at JINR. *Nucl. Phys. A* **2013**, *904–905*, 945c–948c.
14. Mogliacci, S. Kurtoses and high order cumulants: Insights from resummed perturbation theory. *J. Phys. Conf. Ser.* **2014**, *503*, 012005.
15. Aamodt, K.; Abrahantes Quintana, A.; Adamová, D.; Adare, A.M.; Aggarwal, M.M.; Aglieri Rinella, G.; Agocs, A.G.; Aguilar Salazar, S.; Ahammed, Z.; Ahmad, N.; et al. Two-pion Bose-Einstein correlations in central Pb-Pb collisions at $\sqrt{s_{NN}} = 2.76$ TeV. *Phys. Lett. B* **2011**, *696*, 328–337.
16. Adam, J.; Adamová, D.; Aggarwal, M.M.; Aglieri Rinella, G.; Agnello, M.; Agrawal, N.; Ahammed, Z.; Ahmed, I.; Ahn, S.U.; Aimo, I.; et al. Two-pion femtoscopy in p-Pb collisions at $\sqrt{s_{NN}} = 5.02$ TeV. *Phys. Rev. C* **2015**, *91*, 034906.
17. Vuorinen, A. Quark Matter Equation of State from Perturbative QCD. *EPJ Web Conf.* **2017**, *137*, 09011.
18. Mogliacci, S. Probing the Finite Density Equation of State of QCD via Resummed Perturbation Theory. Ph.D. Thesis, Bielefeld University, Bielefeld, Germany, 2014.
19. Mogliacci, S.; Horowitz, W.A; Kolbé, I. In preparation.
20. Satz, H. Probing the States of Matter in QCD. *Int. J. Mod. Phys. A* **2013**, *28*, 1330043.
21. Koch, V. Hadronic Fluctuations and Correlations. *arXiv* **2008**, arXiv:0810.2520.
22. Appelquist, T.; Pisarski, R.D. High-Temperature Yang-Mills Theories and Three-Dimensional Quantum Chromodynamics. *Phys. Rev. D* **1981**, *23*, 2305.
23. Nadkarni, S. Dimensional reduction in finite-temperature quantum chromodynamics. *Phys. Rev. D* **1983**, *27*, 917–931.
24. Braaten, E.; Nieto, A. Effective field theory approach to high temperature thermodynamics. *Phys. Rev. D* **1995**, *51*, 6990.
25. Kajantie, K.; Laine, M.; Rummukainen, K.; Shaposhnikov, M.E. Generic rules for high temperature dimensional reduction and their application to the standard model. *Nucl. Phys. B* **1996**, *458*, 90–136.
26. Linde, A.D. Infrared Problem in Thermodynamics of the Yang-Mills Gas. *Phys. Lett. B* **1980**, *96*, 289–292.
27. Laine, M.; Schröder, Y. Quark mass thresholds in QCD thermodynamics. *Phys. Rev. D* **2006**, *73*, 085009.
28. Kajantie, K.; Laine, M.; Rummukainen, K.; Schröder, Y. The Pressure of hot QCD up to $g^6 \ln(1/g)$. *Phys. Rev. D* **2003**, *67*, 105008.
29. Vuorinen, A. The Pressure of QCD at finite temperatures and chemical potentials. *Phys. Rev. D* **2003**, *68*, 054017.
30. Kneur, J.-L.; Neveu, A. α_S from F_π and Renormalization Group Optimized Perturbation. *Phys. Rev. D* **2013**, *88*, 074025.
31. Karsch, F.; Patkós, A.; Petreczky, P. Screened perturbation theory. *Phys. Lett. B* **1997**, *401*, 69–73.
32. Frenkel, J.; Taylor, J.C. High Temperature Limit of Thermal QCD. *Nucl. Phys. B* **1990**, *334*, 199–216.
33. Braaten, E.; Pisarski, R.D. Soft Amplitudes in Hot Gauge Theories: A General Analysis. *Nucl. Phys. B* **1990**, *337*, 569–634.

34. Andersen, J.O.; Braaten, E.; Strickland, M. Hard thermal loop resummation of the thermodynamics of a hot gluon plasma. *Phys. Rev. D* **2000**, *61*, 014017.
35. Andersen, J.O.; Braaten, E.; Strickland, M. Hard thermal loop resummation of the free energy of a hot quark-gluon plasma. *Phys. Rev. D* **2000**, *61*, 074016.
36. Karsch, F.; Morita, K.; Redlich, K. Effects of kinematic cuts on net-electric charge fluctuations. *Phys. Rev. C* **2016**, *93*, 034907.
37. Haque, N.; Andersen, J.O.; Mustafa, M.G.; Strickland, M.; Su, N. Three-loop HTLpt Pressure and Susceptibilities at Finite Temperature and Density. *Phys. Rev. D* **2014**, *89*, 061701.
38. Bazavov, A.; Ding, H.-T.; Hegde, P.; Karsch, F.; Miao, C.; Mukherjee, S.; Petreczky, P.; Schmidt, C.; Velytsky, A. Quark number susceptibilities at high temperatures. *Phys. Rev. D* **2013**, *88*, 094021.
39. Borsányi, S.; Fodor, Z.; Katz, S.D.; Krieg, S.; Ratti, C.; Szabó, K.K. Fluctuations of conserved charges at finite temperature from lattice QCD. *J. High Energy Phys.* **2012**, *1201*, 138.
40. Schmidt, C. QCD bulk thermodynamics and conserved charge fluctuations with HISQ fermions. *J. Phys. Conf. Ser.* **2013**, *432*, 012013v.
41. Schmidt, C. Baryon number and charge fluctuations from lattice QCD. *Nucl. Phys. A* **2013**, *904–905*, 865c–868c.
42. Bazavov, A.; Ding, H.-T.; Hegde, P.; Kaczmarek, O.; Karsch, F.; Laermann, E.; Maezawa, Y.; Mukherjee, S.; Ohno, H.; Petreczky, P.; et al. Strangeness at High Temperatures: From Hadrons to Quarks. *Phys. Rev. Lett.* **2013**, *111*, 082301.
43. Borsányi, S. Thermodynamics of the QCD transition from lattice. *Nucl. Phys. A* **2013**, *904–905*, 270c–277c.
44. Stephanov, M.A. On the sign of kurtosis near the QCD critical point. *Phys. Rev. Lett.* **2011**, *107*, 052301.
45. Borsányi, S.; Fodor, Z.; Katz, S.D.; Krieg, S.; Ratti, C.; Szabo, K.K. Freeze-out parameters: lattice meets experiment. *Phys. Rev. Lett.* **2013**, *111*, 062005.

© 2018 by the authors. Licensee MDPI, Basel, Switzerland. This article is an open access article distributed under the terms and conditions of the Creative Commons Attribution (CC BY) license (http://creativecommons.org/licenses/by/4.0/).

Article

Looking for the Phase Transition—Recent NA61/SHINE Results

Ludwik Turko [†]

Institute of Theoretical Physics, University of Wroclaw, pl. M. Borna 9, 50-205 Wroclaw, Poland; ludwik.turko@ift.uni.wroc.pl
† For the NA61/SHINE Collaboration.

Received: 16 January 2018; Accepted: 19 February 2018; Published: 9 March 2018

Abstract: The fixed-target NA61/SHINE experiment at the CERN Super Proton Synchrotron (SPS) seeks to find the critical point (CR) of strongly interacting matter as well as the properties of the onset of deconfinement. The experiment provides a scan of measurements of particle spectra and fluctuations in proton–proton, proton–nucleus, and nucleus–nucleus interactions as functions of collision energy and system size, corresponding to a two-dimensional phase diagram (T-μ_B). New NA61/SHINE results are shown here, including transverse momentum and multiplicity fluctuations in Ar+Sc collisions as compared to NA61 p+p and Be+Be data, as well earlier NA49 A+A results. Recently, a preliminary effect of change in the system size dependence, labelled as the "percolation threshold" or the "onset of fireball", was observed in NA61/SHINE data. This effect is closely related to the vicinity of the hadronic phase space transition region and will be discussed in the text.

Keywords: QCD matter; phase transition; critical point

1. Introduction

The NA61/SHINE, understood as the **S**uper Proton Synchrotron (SPS) **H**eavy **I**on and **N**eutrino **E**xperiment, is a continuation and extension of the NA49 experiment [1,2]. It uses a similar experimental fixed-target setup to NA49 (Figure 1) but with an extended research programme. Beyond an enhanced strong interactions programme, measurements of hadron production for neutrino and cosmic ray experiments are realized. The collaboration involves about 150 physicists from 15 countries and 30 institutions. It is the second largest non-LHC (the Large Hadron Collider) experiment at the CERN.

The strong interaction programme of the NA61/SHINE is dedicated to the study of the onset of deconfinement and the search for the critical point (CR) of hadronic matter, related to the phase transition between hadron gas (HG) and quark–gluon plasma (QGP). The NA49 experiment studied hadron production in Pb+Pb interactions, while the NA61/SHINE collects data varying beam energy within the range of 13A–158A GeV and varying sizes of the colliding systems. This is equivalent to the two-dimensional scan of the hadronic phase diagram in the (T, μ_B) plane, as depicted in Figure 2. The ion collisions research programme was initiated in 2009 with the p+p collisions used later on as reference measurements for heavy ion collisions.

Hadron production measurements for neutrino experiments are just reference measurements of p+C interactions for the T2K experiment, since they are necessary for computing initial neutrino fluxes at J-PARC. It has been extended to measure the production of charged pions and kaons produced in interactions out of thin carbon targets and replicas of the T2K targets what is necessary to test accelerator neutrino beams [3]. Data collection began in 2007.

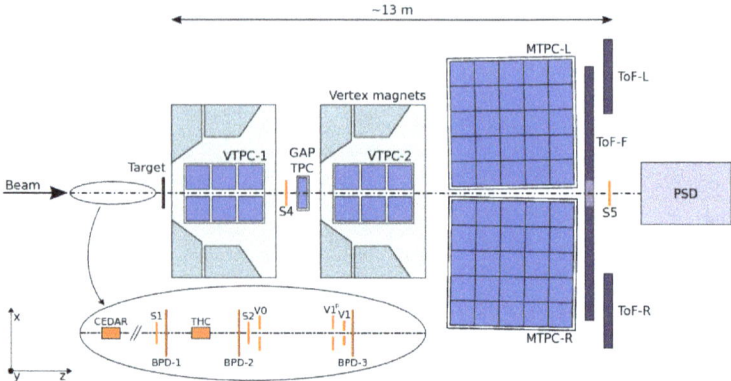

Figure 1. The NA61/SHINE detector consists of a large acceptance hadron spectrometer followed by a set of six Time Projection Chambers (TPCs) as well as Time-of-Flight detectors (ToFs). The high resolution forward calorimeter, the Projectile Spectator Detector (PSD), measures energy flow around the beam direction. For hadron-nucleus interactions, the collision volume is determined by counting low momentum particles emitted from the nuclear target with the Low Momentum Particle Detector (a small TPC) surrounding the target. An array of beam detectors identifies beam particles, secondary hadrons and nuclei as well as primary nuclei, and measures their trajectories precisely.

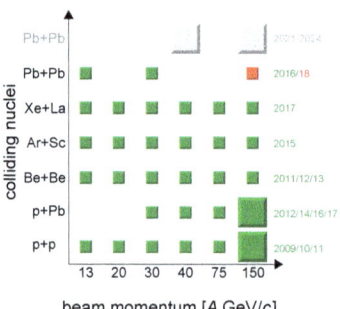

Figure 2. For the programme on strong interactions, NA61/SHINE scans in the system size and beam momentum. In the plot, the recorded data are indicated in green, the approved future data in red, and the proposed extension for the period ⩾2018 is in grey.

Collected p+C data also allow for better understanding of nuclear cascades in the cosmic air showers—necessary in the Pierre Auger and KASCADE experiments [4,5]. These are reference measurements of p+C, p+p, π+C, and K+C interactions for cosmic ray physics. The cosmic ray collisions with the Earth's atmosphere produce secondary air shower radiation. Some of particles produced in such collisions subsequently decay into muons, which are able to reach the surface of the Earth. Cosmic ray-induced muon production can allow the reproduction of primary cosmic ray composition if related hadronic interactions are known [6].

As seen in Figure 3, the phase structure of hadronic matter is involved. Progress in the theoretical understanding of the subject and the collection of more experimental data will allow us to delve further into the subject. While the highest energies achieved at the LHC and RHIC colliders provide data related to the crossover HG/QGP regions, the SPS fixed-target NA61/SHINE experiment is particularly suited to exploring the phase transition line of HG/QGP with the CR included.

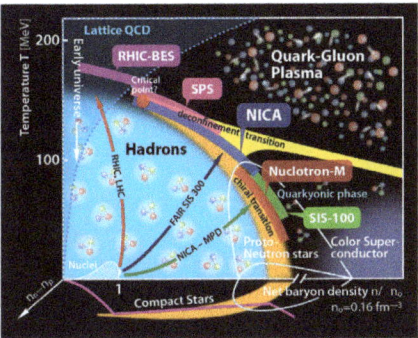

Figure 3. Phase diagram of strongly interacting matter in the temperature and baryonic chemical potential (T, μ_B) plane. Picture taken from this (CSQCD 2017) conference poster.

Heavy ion collision (HIC) experiments reproduce the conditions of the first 10 µs after the Big Bang, when a phase transition from the QGP to a hadron gas would have taken place [7]. It appears, however, that the QGP can be present in the core of massive neutron stars—particularly those with masses exceeding two solar masses [8,9]. That would correspond to the far lower right part of the phase plot, beyond the (T, μ_B) range covered by Figure 3. The CR has been long predicted for thermal quantum chromodynamics (QCDs) at finite μ_B/T [10–12] although this was not unanimously accepted previously [13]. However, lattice QCD calculations are becoming more and more accurate, leading to the present conclusions that the cross-over region occurs at $T_c(\mu_B = 0) = 154 \pm 9$ MeV [14] and the location of a CR is not expected for $\mu_B/T \leqslant 2$ and $T/T_c(\mu_B = 0) > 0.9$ [15]. A more detailed exploration of QCD phase diagram would need both new experimental data with extended detection capabilities and improved theoretical models [16].

Another intriguing and far reaching possibility is the Big Bang phase transition scenario, referred to by Edward Witten as the "cosmic separation of phases" [17]. In the standard approach, the Big Bang QGP is almost matter—antimatter symmetric and evolves to lower temperatures through the crossover region almost vertically to the temperature axis [18]. Edward Witten, using almost "back of envelope" arguments, pointed out the possibility of using the path of universe starting in the QGP phase from the high baryonic chemical potential region reaching almost zeroth temperature, with supercooled QGP. Hadronization then becomes quite an explosive phenomena with a necessary subsequent reheating (see e.g., [19,20]). Corresponding plots, taken from [19], are shown in Figure 4.

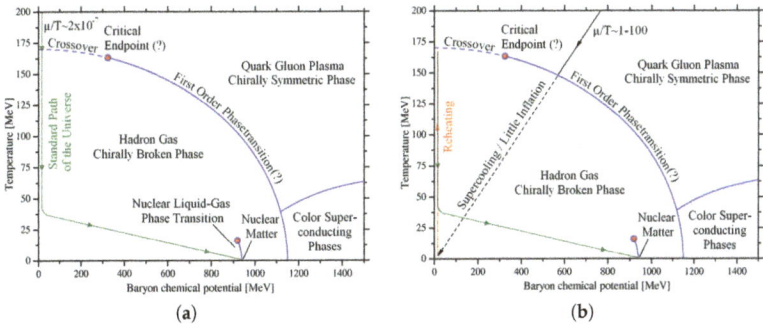

Figure 4. (**a**) Sketch of a possible quantum chromodynamic (QCD)-phase diagram with the commonly accepted standard evolution path of the universe as calculated e.g., in [18]. (**b**) Sketch of a possible QCD phase diagram with the evolution path in the scenario of the cosmic separation of phases.

Beyond cosmological effects (little/tepid inflation) such a possibility would change also our understanding of the hadronization effect in HIC processes.

2. New NA61/SHINE Results

2.1. Irregularities—The Horn

It was expected [21] that ratio K^+/π^+ produced at HIC energies of about $\sqrt{s_{NN}} \approx 10$ GeV should reach a rapid maximum when QGP formation begins. In 1998 there was not enough experimental data to fully check this hypothesis. Present collected results fully confirm the appearance of the horn, although there are still discussions about its relevance to the HG/QGP phase transition.

Recent data from NA61/SHINE [22] show also a strong dependence of the effect on the size of the colliding objects, as seen in Figure 5.

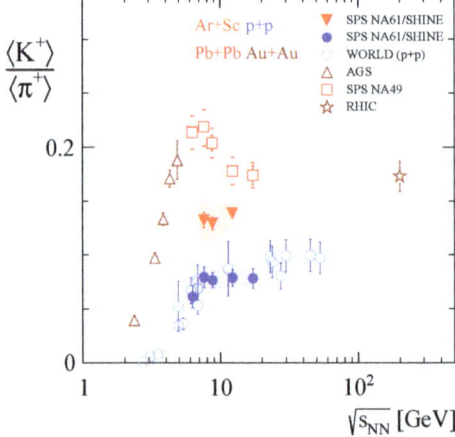

Figure 5. Horn: a strong maximum of the ratio of K^+/π^+ multiplicities. A reduced shadow of the horn structure is visible in p+p reactions.

2.2. Irregularities—The Step

Plateau: A step-in the inverse slope parameter T of the transverse mass spectra m_T at mid-rapidity ($0 < y < 0.2$) plotted against the collision energy per nucleon (Figure 6) is expected for the onset of deconfinement [21]. The effect increases with the size of colliding objects. Qualitatively, a similar structure is visible in p+p collisions, with Be+Be slightly above, consistent with the step structure.

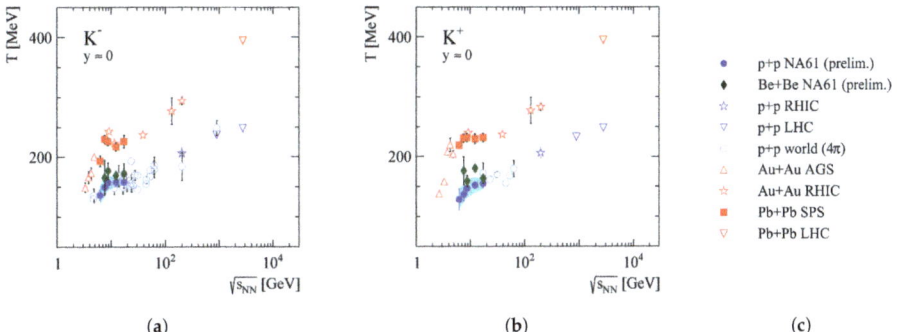

Figure 6. Inverse slope parameters T of negative (**a**) and positive (**b**) kaons exhibit rapid changes in the SPS energy range—also seen in p+p collision. Data collected from all available energy ranges (**c**).

2.3. Fluctuations

The experimental facilities of NA61 allow us to measure fluctuations of various physical quantities which are sensitive to the vicinity of the CR. Indeed, these fluctuations can create a signature of the CR. The analysis of fluctuations of various observables is the main goal of the NA61/SHINE experiment, particularly in a range of energies around 8 GeV per colliding nucleon pair at the center of mass in interactions of light nuclei (corresponding to the beam energy of 30 GeV in the frame of a stationary target). This is just the kinematical region where NA49 data indicate the onset of deconfinement in central Pb+Pb collisions, observing structures in the energy dependence of hadron production in central Pb+Pb collisions which are not observed in hadron interactions [23,24].

Preliminary data presented in Figure 7 do not show any signs of critical behaviour [25].

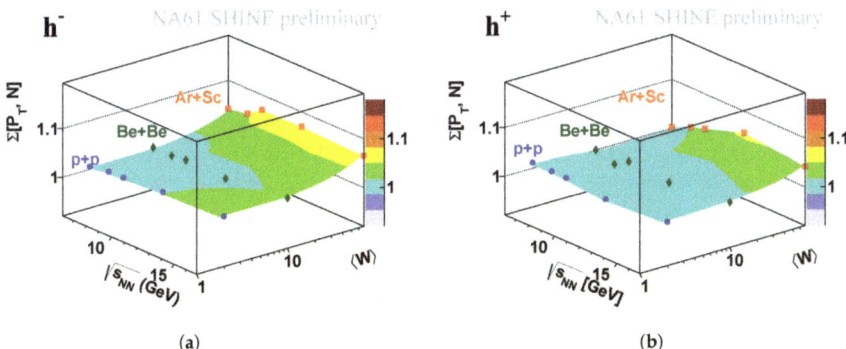

Figure 7. Critical fluctuations in p_T of negative (**a**) and positive (**b**) charged hadrons in ^{40}Ar +^{45}Sc, ^{7}Be + ^{9}Be and p+p collisions.

3. System Size Dependence

In the recent months, some unexpected results were observed by the NA61/SHINE experiment [22,26,27], concerning qualitative differences in system size dependence behaviour. It appears that in particular Be+Be results are very close to p+p at different collision energies. An example of such behaviour is presented in Figure 8.

It looks as if with the increasing size of colliding systems, light clusters are produced more and more copiously, and at some density they start to overlap to reach percolation threshold. This effect would not depend on energy only, but also on the size of the system.

Intensive work now is ongoing to achieve more conclusive results, analyzing recently collected data from Xe+La and Pb+Pb.

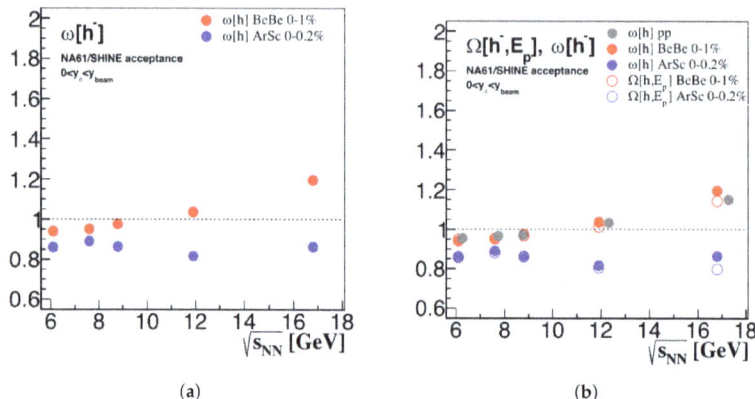

Figure 8. (a) multiplicity fluctuation increases with collision energy in Be+Be but remains constant in Ar+Sc. (b) multiplicity fluctuation in Ar + Sc, ^{7}Be + ^{9}Be and p+p collisions. Be+Be almost identical to p+p fluctuation within statistical errors given by plot's points sizes.

4. Conclusions

The Holy Grail of HIC—Quark Gluon Plasma—still remains an elusive object. Although there are no at present discussions concerning the very existence of this state of matter, there are still open problems connected with experimental signatures of many theoretical ideas and predictions in this field. The NA61/SHINE experiment acts in the energy region particularly suited for the appearance of phase transition effects. Beyond this, the fixed-target technology makes possible to perform 4π geometry measurements which are not accessible in collider-type experiments.

To date, collected and analysed NA61/SHINE data related to theoretical predictions of fluctuations in the presence of CR have not shown any anomalies that could be attributed to this. These data relate to $N - p_T$ fluctuations in p+p, Be+Be, and Ar+Sc central events.

There is clear system size dependence of m_T spectra that differs significantly between p+p and A+A events. This is the effect associated with the transverse collective flow.

The appearance of horn (Figure 5) and step (Figure 6) effects is in accordance with theoretical predictions for the onset of deconfinement in HIC due to mixed phase of HG and QGP [21].

The recent preliminary results of the NA61/SHINE concerning system size dependence may be also a signature for the new physical phenomena. There is a clearly visible jump between light and heavy systems. Be+Be results are very close to the p+p results, independently on the collision energy. In addition, multiplicity fluctuations, close to p+p value in Be+Be collisions, are strongly suppressed in Ar+Sc collisions.

For the CERN long shutdown in 2019–2020, an NA61/SHINE detector upgrade system is foreseen. This would make the precise measurements of open charm and multi-strange hyperon production possible, which are also of great importance both for the neutrino physics programme as well as for the precise measurements of cosmic rays.

Acknowledgments: The author acknowledges support from the Polish National Science Center under contract No. UMO-2014/15/B/ST2/03752 and by the Bogoliubov–Infeld programme for scientific collaboration between JINR Dubna and Polish institutions.

Conflicts of Interest: The author declare no conflict of interest.

Abbreviations

The following abbreviations are used in this manuscript:

AGS	Argonne National Laboratory
CERN	Conseil Europén pour la Recherche Nucléaire
CR	critical point
HG	hadron gas
J-PARC	Japan Proton Accelerator Research Complex
LHC	Large Hadron Collider
HIC	heavy ion collision
QCD	quantum chromodynamics
QGP	quark-gluon plasma
RHIC	Relativistic Heavy Ion Collider
SPS	Super Proton Synchrotron

References

1. Antoniou, N. et al. [NA49-future Collaboration] *Study of Hadron Production in Hadron Nucleus and Nucleus Nucleus Collisions at the CERN SPS*; CERN-SPSC-2006-034; CERN: Genève, Switzerland, 2006.
2. Abgrall, N. et al. [NA61/SHINE Collaboration] *Calibration and Analysis of the 2007 Data*; CERN-SPSC-2008-018; CERN: Genève, Switzerland, 2008.
3. Abgrall, N. et al. [NA61/SHINE Collaboration] Measurements of Cross Sections and Charged Pion Spectra in Proton-Carbon Interactions at 31 GeV/c. *Phys. Rev. C* **2011**, *84*, 034604.
4. Abraham, J. et al. [Pierre Auger Collaboration] Properties and performance of the prototype instrument for the Pierre Auger Observatory. *Nucl. Instrum. Methods Phys. Res. Sect. A Accel. Spectrom. Detect. Assoc. Equip.* **2004**, *523*, 50–95.
5. Antoni, T. et al. [KASCADE Collaboration] The cosmic-ray experiment KASCADE. *Nucl. Instrum. Methods Phys. Res. Sect. A Accel. Spectrom. Detect. Assoc. Equip.* **2003**, *513*, 490.
6. Morison, I. *Introduction to Astronomy and Cosmology*; John Wiley & Sons: Hoboken, NJ, USA, 2008; ISBN 978-0-470-03333-3.
7. Satz, H. *Ultimate Horizons. Probing the Limits of the Universe*; Springer: Berlin/Heidelberg, Germany, 2013; ISBN 1612-3018.
8. Fischer, T.; Bastian, N.U.F.; Wu, M.R.; Typel, S.; Klähn, T.; Blaschke, D.B. High-density phase transition paves the way for supernova explosions of massive blue-supergiant stars. *arXiv* **2017**, arXiv:1712.08788.
9. Benic, S.; Blaschke, D.; Alvarez-Castillo, D.E.; Fischer, T.; Typel, S. A new quark-hadron hybrid equation of state for astrophysics—I. High-mass twin compact stars. *Astron. Astrophys.* **2015**, *577*, A40.
10. Barducci, A.; Casalbuoni, R.; De Curtis, S.; Gatto, R.; Pettini, G. Chiral Symmetry Breaking in QCD at Finite Temperature and Density. *Phys. Lett. B* **1989**, *231*, 463–470.
11. Halasz, A.M.; Jackson, A.D.; Shrock, R.E.; Stephanov, M.A.; Verbaarschot, J.J.M. Phase diagram of QCD. *Phys. Rev. D* **1998**, *58*, 096007.
12. Berges, J.; Rajagopal, K. Color superconductivity and chiral symmetry restoration at nonzero baryon density and temperature. *Nucl. Phys. B* **1999**, *538*, 215–232.
13. De Forcrand, P.; Philipsen, O. The Chiral critical point of $N_f = 3$ QCD at finite density to the order $(\mu/T)^4$. *J. High Energy Phys.* **2008**, *2008(11)*, 012.
14. Bazavov, A. et al. [HotQCD Collaboration] Equation of state in (2+1)-flavor QCD. *Phys. Rev. D* **2014**, *90*, 094503.
15. Bazavov, A.; Ding, H.-T.; Hegde, P.; Kaczmarek, O.; Karsch, F.; Laermann, E.; Maezawa, Y.; Ohno, H.; Petreczky, P. H.; Wagner, M.; et al. The QCD Equation of State to $\mathcal{O}(\mu_B^6)$ from Lattice QCD. *Phys. Rev. D* **2017**, *95*, 054504.

16. Caines, H. The Search for Critical Behavior and Other Features of the QCD Phase Diagram–Current Status and Future Prospects. *Nucl. Phys. A* **2017**, *967*, 121–128.
17. Witten, E. Cosmic Separation of Phases. *Phys. Rev. D* **1984**, *30*, 272.
18. Fromerth, M.J.; Rafelski, J. Hadronization of the quark Universe. *arXiv* **2002**, arXiv:astro-ph/0211346.
19. Boeckel, T.; Schaffner-Bielich, J. A little inflation in the early universe at the QCD phase transition. *Phys. Rev. Lett.* **2010**, *105*, 041301, Erratum in **2011**, *106*, 069901.
20. McInnes, B. The Trajectory of the Cosmic Plasma Through the Quark Matter Phase Diagram. *Phys. Rev. D* **2016**, *93*, 043544.
21. Gazdzicki, M.; Gorenstein, M.I. On the Early Stage of Nucleus–Nucleus Collisions. *Acta Phys. Polon. B* **1999**, *30*, 2705.
22. Lewicki, M.P. [NA61/SHINE Collaboration] Identified kaon production in Ar+Sc collisions at SPS energies. *arXiv* **2017**, arXiv:1712.02417.
23. Alt, C. et al. [NA49 Collaboration] Pion and kaon production in central **Pb+Pb** collisions at **20A** and **30A** GeV: Evidence for the onset of deconfinement. *Phys. Rev. C* **2008**, *77*, 024903.
24. Gazdzicki, M. [NA49 Collaboration] Report from NA49. *J. Phys. G* **2004**, *30*, S701.
25. Aduszkiewicz, A. [NA61/SHINE Collaboration]. Recent results from NA61/SHINE. *Nucl. Phys. A* **2017**, *967*, 35–42.
26. Seryakov, A. [NA61/SHINE Collaboration] Rapid change of multiplicity fluctuations in system size dependence at SPS energies. *arXiv* **2017**, arXiv:1712.03014.
27. Gazdzicki, M. [NA61/SHINE Collaboration] Fluctuations and correlations from NA61/SHINE. *arXiv* **2017**, arXiv:1801.00178.

© 2018 by the author. Licensee MDPI, Basel, Switzerland. This article is an open access article distributed under the terms and conditions of the Creative Commons Attribution (CC BY) license (http://creativecommons.org/licenses/by/4.0/).

Communication

Hadron–Quark Combustion as a Nonlinear, Dynamical System

Amir Ouyed [1,*], Rachid Ouyed [1] and Prashanth Jaikumar [2]

[1] Department of Physics and Astronomy, University of Calgary, Calgary, AB T2N 1N4, Canada; rouyed@ucalgary.ca
[2] Department of Physics and Astronomy, California State University Long Beach, Long Beach, CA 90840, USA; prashanth.jaikumar@csulb.edu
* Correspondence: ahouyedh@ucalgary.ca

Received: 17 January 2018; Accepted: 2 March 2018; Published: 7 March 2018

Abstract: The hadron–quark combustion front is a system that couples various processes, such as chemical reactions, hydrodynamics, diffusion, and neutrino transport. Previous numerical work has shown that this system is very nonlinear, and can be very sensitive to some of these processes. In these proceedings, we contextualize the hadron–quark combustion as a nonlinear system, subject to dramatic feedback triggered by leptonic weak decays and neutrino transport.

Keywords: quarks; combustion; neutron star

1. Introduction

The hypothesis of absolutely stable quark matter [1–3] has very important phenomenological consequences in high energy astrophysics (e.g., [4,5]). For example, it quickly became evident that the conversion of a whole neutron star into a quark star could release thermal and mechanical energy that would be in the same order of magnitude than the energy released by a core collapse supernova [6,7]. An important reason behind the relevance of alternative hypothesis such as the one of absolutely stable quark matter is the fact that computational simulations and modelling cannot recreate the energetic events of many explosive astrophysics phenomena. For example, computational models for core collapse supernovae are still unable to provide robust explosions [8]. Furthermore, the engines of even more energetic phenomena, such as super-luminous supernova [9] and gamma ray bursts [10] remain elusive. The conversion of hadronic to absolutely stable quark matter could give the extra "push" necessary to realize some of these energetic events. Given the potential of this hypothesis of explaining at least in part some of the more mysterious explosive phenomena in astrophysics, the inclusion of this conjecture in models of explosive astrophysics should remain an active research program (e.g., [11]).

Although simple energetics reveal the potential of the hypothesis of absolutely stable quark matter, more sophisticated studies are necessary to prove whether this conversion would be dynamically significant in, for example, powering a supernova-like explosion [6] or a gamma ray burst [12]. Given the high densities and the temperatures of the conversion process, the most apt framework to study the dynamics of this conversion is hydrodynamics, where a fluid of neutrons is "burnt" into a fluid of quarks [13,14]. In the late 1980s, a couple of papers [6,15] appeared that pioneered a semi-analytic method of describing the conversion of neutrons into quarks as a hydrodynamic, combustion process. However, the exact equations that govern this process, the reaction–diffusion–advection equations, are quite complicated. These equations couple various processes, such as radiative transfer, chemical reactions, fluid dynamics, and diffusion, forcing these early papers to simplify considerably the equations in order to find a tractable solution. Nevertheless, due to the nonlinear nature of these equations, simplifications that may appear to be minor could actually have dramatic consequences in the equations' solutions. Nonlinear coupling of various processes could create dramatic outcomes that could lead to orders of magnitude of difference, in, for example, the speed of the conversion.

Later on, a numerical way of solving directly these reaction–diffusion–advection equations was pioneered by Niebergal et al. [16] (hereby Paper I). Their numerical results lead to large differences to much of the previous semi-analytic work. For example, the calculated burning speed of about 0.002c–0.04c, where c is the speed of light, was orders of magnitude faster than what was calculated previously by some semi-analytic models. Their results also hinted at important feedback effects that might arise from coupling neutrino transport into the combustion front. In fact, they found, by solving the semi-analytic, hydrodynamic jump conditions, that, for a given neutrino cooling rate, thermal pressure gradients could slow down the combustion front by orders of magnitude. Ouyed et al. [17] (hereby Paper II) later confirmed this initial intuition numerically, by incorporating neutrino transport and electron pressure into the reaction–diffusion–advection equations. The authors of Paper II discovered that feedback effects triggered by various leptonic processes could affect the burning timescale by orders of magnitude. These two papers showed then that lepton micro-physics are at the very least as important as other parts of the combustion process that have already been deemed important for decades, such as the high density equation of state. The importance of leptons follows from the nonlinear nature of the combustion process—parts of the system that may appear at first glance insignificant may give rise to extreme feedback effects. In the case of neutrinos, we have shown that their omission in simulations would lead to very inadequate results, given that they can dramatically affect the conversion speed.

These proceedings will therefore focus on the importance of lepton micro-physics as a source of nonlinear, feedback effects, by summarizing and contextualizing previously published work, and detailing possible future avenues of research. By micro-physics, we mean the processes that are important at a length-scale of a centimeter, rather than macroscopic processes that appear at the length scale of a compact star (about ten kilometers). Given that, at least to the extent of our knowledge, previous work has never contextualized the issues of hadron–quark combustion using the framework of nonlinear dynamics, we feel that these proceedings could act as a brief introduction to a new way of thinking about the hadron–quark combustion. In particular, we find the concept of feedback loops to be very relevant and illuminating the micro-physics of the flame. In nonlinear dynamics, a feedback loop implies that the output of a system is fed back into input, creating a circuit of cause and effect that can lead to dramatic consequences. The processes coupled in the combustion front can lead to feedback loops, where processes that slow down the burning front could in turn trigger other processes (e.g., the magnifying of pressure gradients) that would slow down the burning front further.

We structure these proceedings in the following way. In Section 2, we describe the intricate structure of the flame and the processes that are coupled in it, and how these processes may lead to feedback. In Section 3, we focus on the effect of neutrinos and electrons, which are the source of the main feedback effects described in Paper I and II. In Section 4, we finish with some concluding remarks.

2. Feedback Effects and the Reaction Zone

The reactions that drive the burning front are:

$$u + e^- \leftrightarrow s + \nu_e, \tag{1}$$

$$u + e^- \leftrightarrow d + \nu_e, \tag{2}$$

$$u + d \leftrightarrow u + s. \tag{3}$$

These reactions are coupled to the reaction–diffusion–advection equations that govern the burning [16,17]:

$$\frac{\partial n_i}{\partial t} = -\nabla \cdot (n_i v - D_i \nabla n_i) + R_i, \tag{4}$$

$$\frac{\partial (hv)}{\partial t} = -\nabla(hv \cdot v) - \nabla P, \tag{5}$$

$$\frac{\partial s}{\partial t} = -\nabla \cdot (sv) - \frac{1}{T}\sum_i \mu_i \frac{dn_i}{dt} + \frac{1}{T}\frac{d\epsilon_{\nu_e}}{dt}, \tag{6}$$

where the index i runs through the different particle species (u, d, s, ν_e). The definition of the variables are: n_i is number density, v is the fluid velocity, ϵ_{ν_e} is electron neutrino energy density, h is enthalpy density, s is entropy density, T is temperature, R_i is the reaction source term, D_i is the diffusion coefficient, and P is pressure.

We enforce charge neutrality by equating the electron number density with $n_e = n_u - n_B$. This is a good assumption given that electrons are degenerate and relativistic, so they move with speeds close to the speed of light in order wash out any charge imbalances.

These equations lead to a reaction zone that acts as an interface between the two flavoured quark "fuel" and the three flavoured quark "ash". The reaction zone is very complex, given that various particles and processes participate in it. Figure 1, which is a snapshot of numerical simulations performed in Paper II, is included to illustrate the complexity of the reaction zone. Figure 1a shows the Fermi momenta of various particles and the temperature gradient along the flame, with the various force gradients caused by the processes. Figure 1b shows the various pressure gradients caused by the different particles. Ultimately, the reactions will be constrained by the transport of s-quarks into the fuel, given that the s-quark acts as an "oxidant" that triggers the conversion. These transport processes shape the width of the reaction zone, which is a function of the nonlinear, hydrodynamic effects related to the distance that fluid velocities carry the u- and d-quarks before they decay into s-quarks. Therefore, much of the processes manifest as force vectors that either slow down the transport of s-quarks into the fuel or accelerate it. These processes may accelerate the burning front or slow it down. We divide the processes along enhancing and quenching, although this division is a simplification because the various processes may be coupled to each other. Enhancing processes accelerate the burning front while quenching processes slow it down. We define the burning speed as the derivative of the interface position versus time.

The enhancing processes are (with more detailed explanations in Paper II):

1. **Flavor equilibration**: The conversion of two flavoured quark matter to three flavoured quark matter through the reactions (1)–(3) releases binding energy in the form of heat, increasing the temperature behind the front. The increase of temperature stiffens the quark EoS, increasing the pressure behind the front and therefore accelerating the burning speed.
2. **Electron capture**: The transformation of electrons into neutrinos through reactions (1) and (2) releases binding energy in the form of heat. Higher temperature enhances the pressure behind the interface, which increases the burning speed.
3. **Neutrino pressure**: Neutrinos deposit momentum into the reaction zone, accelerating the interface into faster speeds.
4. **Loss of lepton number**: Neutrinos, as they diffuse from higher to lower chemical potentials, deposit the chemical potential difference in the form of heat. This heat increases the temperature and therefore enhances the pressure behind the interface. This phenomenon is very similar to what is referred as Joule heating in papers concerning proto–neutron star evolution (e.g., [18]).

The quenching factors are:

1. **Electron pressure**: Electron capture "eats up" the electrons behind the interface, generating a large electron gradient (see the electron Fermi momentum distribution in Figure 1a). These electron gradients generate a degeneracy pressure that pushes the interface backwards, decelerating the burning front. See Figure 1b for a graphical representation of the electron pressure gradient.

2. **Neutrino cooling**: Neutrinos that escape from the burning front carry energy away from the reaction zone, which reduces the temperature and therefore the pressure behind the interface. This quenching effect was first detailed in Paper I.

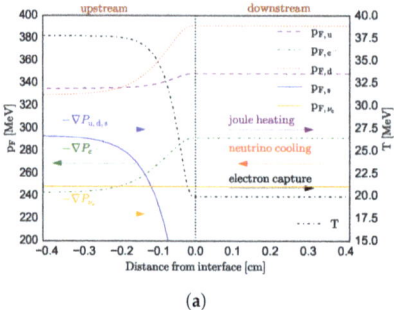 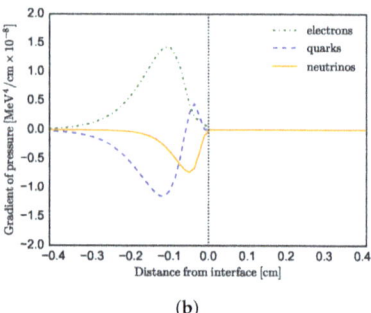

(a) (b)

Figure 1. (**a**): Simulation snapshot of the burning interface. $p_{F,i}$ are the Fermi momenta for particles i, and T is the temperature. In both panels, the interface lies at position zero depicted by the vertical line. The arrows represent the directions of the force vectors and their labels depict the processes that caused them. Upstream is the side behind (left side of the vertical line) the interface, and downstream is the side in front of it (right side of the vertical line). (**b**): The pressure gradients for the leptons and quarks shown in Figure 1a. Figure and caption were taken from Paper II [17].

All these enhancing and quenching processes generate feedback effects. In the case of the hadron–quark combustion front, positive feedback could be how some processes that slow down the burning front in turn lead to other processes (e.g., amplification of certain pressure gradients) that would lead to even more deceleration, generating a nonlinear, exponential effect. Although there are probably many types of feedback loops in the reaction zone given its rich couplings of particles and processes, Papers I and II focused on the positive feedback generated by leptonic weak interactions, which we will explore in the next section.

3. Leptons and Positive Feedback

Electrons and neutrinos are crucial components in the combustion system because they can generate dramatic, positive feedback effects. The authors of Paper I were the first to discover a connection between leptons and positive feedback. They solved the hydrodynamic jump-conditions for the conversion of two flavoured quark matter to three flavoured quark matter, and parameterized neutrino cooling as a small temperature reduction in the three flavoured quark ash. They found that, for a very small amount of cooling—for example, a reduction of 0.1 MeV in the temperature—the thermal pressure would reduce dramatically to the point of almost halting the burning interface.

Paper II discovered more positive feedback effects associated with leptons, and was the first attempt in the literature to incorporate neutrino transport across the reaction zone numerically. Paper II showed, through a combination of semi-analytic studies and numerical simulations, that the leptons themselves can generate huge pressure gradients that can affect dramatically the burning speed (Figure 1b). A key finding is that the quenching process of electron capture could generate positive feedback that could slow down the burning front dramatically if the neutrinos are free streaming, to the point that the burning front halts within the timescales of the simulation (Figure 2).

Much of the source of the dramatic lepton feedback lies in Equation (5) given that the nonlinear momentum is coupled to a lepton degeneracy pressure component in the ∇P term. This lepton pressure term in turn is coupled to reaction source terms, entropy evolution and the transport equation of neutrinos. Given that velocity varies by various orders of magnitude through the simulation timescale, this equation cannot be linearized (e.g., the $\nabla (hv \cdot v)$ term in Equation (5) is strongly nonlinear), as the time-dependent fluid velocity is not merely a perturbation oscillating around an equilibrium point, but instead changes dramatically through time by a forcing due to the ∇P term. In other words, the burning interface is genuinely a nonlinear system out of equilibrium, and linearizing the system would eliminate the dramatic feedback loops, generating an inaccurate solution.

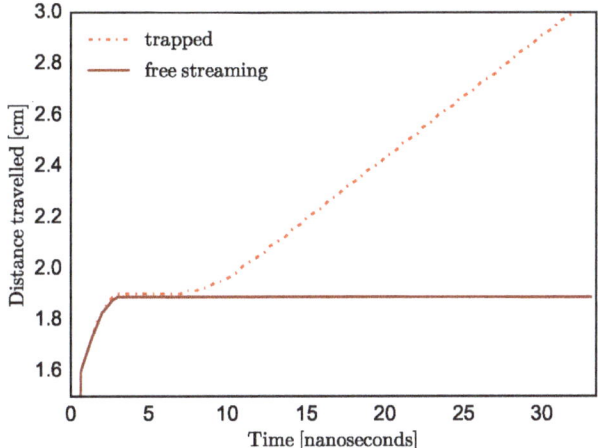

Figure 2. Distance travelled by the combustion front as a function of time from a numerical simulation. The line labeled as "free streaming" represents the burning front with neutrinos free streaming, while the line labeled as "trapped" plots the burning front with trapped neutrinos. Notice how the front halts for the remainder of the simulation for the free streaming case. The thermodynamic parameters for the simulation were an initial temperature of $T = 20$ MeV, an initial lepton fraction of $Y_L = 0.2$, and an initial baryonic density of $n_B = 0.35$ fm^{-3}.

One of the positive pieces of feedback caused by leptons can be described in the following way: processes that slow down the burning front will magnify the electron pressure gradients (see Figure 1b) that oppose the front, given that it gives reactions (1) and (2) more time to "eat up" the electrons behind the front, generating a sharper electron dip. As the electron dip (see Figure 1a for a graphical representation of the electron dip) becomes more dramatic and sharper, this in turn will lead to a slower burning front that would induce an even sharper dip, generating a positive feedback effect. In contrast, a faster burning front might move too quickly for flavour equilibration to catch up with the interface, weakening the electron pressure gradient that opposes the front.

Similarly, another positive piece of feedback caused by quenching effects and associated with leptons is that slower burning leads to slower reaction rates and therefore less neutrinos being created, which reduces neutrino enhancing processes such as Joule heating and neutrino pressure. As the rate of production of neutrinos decreases, so does the burning speed.

A key finding is that quenching processes of electron capture could generate positive feedback that could slow down the burning front dramatically if neutrinos are free streaming, to the point that the burning front halts within the timescales of the simulation (Figure 2). This is because neutrino pressure and Joule heating counteracts the quenching effects of the electron pressure gradient, and therefore once the neutrinos free stream and therefore do not deposit heat or momentum in the reaction zone, the electron pressure gradient stands unopposed, slowing down the burning front by various orders of magnitude. Only when neutrinos are trapped, such as is the case of Figure 3 in Paper II, will the burning front revive, given that neutrinos deposit momentum and heat that pushes the interface; otherwise, if the neutrinos free stream, the front will remain halted. In Paper II, we showed, both through self-consistent simulations and confirmed it with semi-analytic results, that, if neutrinos free stream, the front will effectively halt due to electron pressure gradients.

Therefore, a key message of Paper I and II is the realization that the reaction zone of hadron–quark combustion is a nonlinear system that is quite sensitive to its different parts. As an example of this sensitivity, Paper I and II uncovered the fact that leptonic weak interactions generate nonlinearities that could slow down the burning front's speed by various orders of magnitude.

Instabilities in multidimensions: Paper I and II discovered that the behavior of the burning front is very sensitive to the distribution of neutrinos and electrons. It would be interesting to entertain how these lepton sensitivities manifest in more than one dimension. Given that a realistic, multidimensional compact star model would have spatial anisotropies in its lepton distribution, the burning front would probably halt unevenly across its surface, which would lead to wrinkling. This would be a novel and alternative channel of turbulence, which would exist alongside other more classical instabilities, such as Rayleigh–Taylor or Kelvin–Helmholtz. Therefore, a multidimensional simulation would be ultimately the decisive factor for unearthing the final fate of a neutron star converting into a quark star. It could be that the instabilities slow down the burning dramatically, or lead to cataclysmic outcomes such as quark core-collapse or supersonic burning speeds that would lead to an explosion. Analogous cases that demonstrate the significance of multidimensional studies are core-collapse supernova simulations, where multi-dimensional, hydrodynamic instabilities have turned out to be important for the understanding of the explosion mechanism [8].

The important thing to point out is that the nonlinearities unearthed in one dimension simulations hint that the burning speeds extracted from micro-physical, one-dimensional laminar studies are not the end of it all. Therefore, the astrophysical implications can only be fully understood with a multidimensional code.

4. Conclusions

We reviewed the recent literature on the micro-physical simulations of hadron–quark combustion in order to sketch an outline of the reaction zone as a nonlinear system that experiences feedback. Thus, the main objective of these proceedings was to point out that the combustion front, because of its nonlinear nature, can experience significant feedback and coupling between various parts of the system (e.g., quarks, neutrinos, entropy generations), therefore linearizing the problem, and, ignoring certain parts of the system, can generate an inaccurate picture of its behavior. A key finding is that leptonic weak decays are a key part of this nonlinear system, where the coupling of electrons and neutrinos to entropy generation and hydrodynamics can lead to positive feedback loops that can quench the burning speed almost completely.

Given the nonlinear, dynamical nature of the hadron–quark combustion front, we describe the following common pitfalls in the literature that we attempt to remedy by numerical simulation:

1. Assuming the system is steady-state, in other words, equating all temporal derivatives to zero.
2. Assuming that the front is in pressure equilibrium, that is, fixing $\nabla P = 0$.
3. The above two points lead to the cancellation of the important nonlinearities. Pressure equilibrium and a steady-state momentum make the fluid velocity a constant in space and time.
4. Another related pitfall is collapsing the rich structure of the reaction zone into a discontinuity by solving the jump conditions instead of the continuous hydrodynamic equations. This also leads to a steady-state solution, which eliminates the dynamism of the system.

We must reiterate that the micro-physical simulations reviewed are only in one dimension, and therefore they only offer hints to how these nonlinearities might manifest multidimensionally. Nonetheless, the nonlinear effects hint that fluid-dynamical instabilities could be induced by the coupling of leptons to the fluid, which was termed as a deleptonization instability. This potential deleptonization instabilites, which can only be truly probed with multidimensional simulations, leaves the real timescale of the burning of the whole compact star into a quark star an open question.

The deleptonization instability could slow down the burning of the whole compact star to a matter of hours, or accelerate it towards supersonic detonation that would last less than a millisecond. Therefore, the next pressing step is to hopefully extend these microphysical simulations into the multidimentional regime, which may unearth new and very interesting nonlinearities.

Another important question that arises is what more sophisticated and accurate numerical modeling has to offer for the one-dimensional case. For example, in this study, the exact neutrino Boltzmann transport equations are simplified into isotropic, energy averaged flux-limited diffusion equations. However, since the diffusion approximation assumes that the neutrinos are thermalized and therefore strongly interacting with matter, a more accurate approximation will make matter more transparent to neutrinos, and therefore exacerbate the positive feedback described in this paper because less neutrino momentum and heat will be deposited in the reaction zone. Another approach that could increase the accuracy of the simulation is higher resolution. The most recent simulations we ran used a grid of 600 cm with 48,000 zones (size of zone is dx = 0.0125 cm). However, a higher resolution would probably exacerbate the positive feedback given that the pressure gradients may become sharper due to the reduction of numerical viscosity. Higher order spatial, finite difference schemes (in this work, we used a third order scheme for advection and second order scheme for diffusion) would reduce numerical viscosity as well. Our work therefore acts as a lower bound to the effect of leptons on the interface, with more sophisticated numerical approaches probably magnifying their effect.

Acknowledgments: R.O. is funded by the Natural Sciences and Engineering Research Council of Canada under Grant No. RT731073. P.J. is supported by the U.S. National Science Foundation under Grant No. PHY 1608959.

Author Contributions: A.O. performed the calculations, made the figures and drafted the initial version of the paper. R.O. and P.J. provided critical input for mathematical details and interpretation of results, including final approval of the version to be published.

Conflicts of Interest: The authors declare no conflict of interest.

References

1. Bodmer, A. Collapsed nuclei. *Phys. Rev. D* **1971**, *4*, 1601–1606.
2. Witten, E. Cosmic separation of phases. *Phys. Rev. D* **1984**, *30*, 272–285.
3. Terazawa, H. INS-report, 336 (INS, Univ. of Tokyo); 1989. *J. Phys. Soc. Japan* **1979**, *58*, 1979.
4. Horvath, J.; Benvenuto, O.; Vucetich, H. Nucleation of strange matter in dense stellar cores. *Phys. Rev. D* **1992**, *45*, 3865–3868.
5. Bombaci, I.; Datta, B. Conversion of neutron stars to strange stars as the central engine of gamma-ray bursts. *Astrophys. J. Lett.* **2000**, *530*, L69.
6. Benvenuto, O.; Horvath, J. Evidence for strange matter in supernovae? *Phys. Rev. Lett.* **1989**, *63*, 716–719.
7. Pagliara, G.; Herzog, M.; Röpke, F.K. Combustion of a neutron star into a strange quark star: The neutrino signal. *Phys. Rev. D* **2013**, *87*, 103007.
8. Müller, B. The Core-Collapse Supernova Explosion Mechanism. *Proc. Int. Astron. Union* **2016**, *12*, 17–24.
9. Abbott, T.; Cooke, J.; Curtin, C.; Joudaki, S.; Katsianis, A.; Koekemoer, A.; Mould, J.; Tescari, E.; Uddin, S.; Wang, L. Superluminous Supernovae at High Redshift. *arXiv* **2017**, arXiv:1702.02564.
10. Kumar, P.; Zhang, B. The physics of gamma-ray bursts & relativistic jets. *Phys. Rep.* **2015**, *561*, 1–109.
11. Welbanks, L.; Ouyed, A.; Koning, N.; Ouyed, R. Simulating Hadronic-to-Quark-Matter with Burn-UD: Recent work and astrophysical applications. *J. Phys. Conf. Ser.* **2017**, *861*, 012008.
12. Cheng, K.; Dai, Z. Conversion of neutron stars to strange stars as a possible origin of γ-ray bursts. *Phys. Rev. Lett.* **1996**, *77*, 1210–1213.
13. Lugones, G.; Benvenuto, O. Strange matter equation of state and the combustion of nuclear matter into strange matter in the quark mass-density-dependent model at T > 0. *Phys. Rev. D* **1995**, *52*, 1276–1280.
14. Herzog, M.; Röpke, F.K. Three-dimensional hydrodynamic simulations of the combustion of a neutron star into a quark star. *Phys. Rev. D* **2011**, *84*, 083002.
15. Olinto, A.V. On the conversion of neutron stars into strange stars. *Phys. Lett. B* **1987**, *192*, 71–75.
16. Niebergal, B.; Ouyed, R.; Jaikumar, P. Numerical simulation of the hydrodynamical combustion to strange quark matter. *Phys. Rev. C* **2010**, *82*, 062801.

17. Ouyed, A.; Ouyed, R.; Jaikumar, P. Numerical simulation of the hydrodynamical combustion to strange quark matter in the trapped neutrino regime. *Phys. Lett. B* **2017**, *777*, 184–190.
18. Burrows, A.; Lattimer, J.M. The birth of neutron stars. *Astrophys. J.* **1986**, *307*, 178–196.

© 2018 by the authors. Licensee MDPI, Basel, Switzerland. This article is an open access article distributed under the terms and conditions of the Creative Commons Attribution (CC BY) license (http://creativecommons.org/licenses/by/4.0/).

Review

The Merger of Two Compact Stars: A Tool for Dense Matter Nuclear Physics

Alessandro Drago [1], Giuseppe Pagliara [1], Sergei B. Popov [2], Silvia Traversi [1,*] and Grzegorz Wiktorowicz [3,4]

1. Dipartimento di Fisica e Scienze della Terra dell'Università di Ferrara and INFN Sez. di Ferrara, I-44100 Ferrara, Italy; drago@fe.infn.it (A.D.); pagliara@fe.infn.it (G.P.)
2. Sternberg Astronomical Institute, Lomonosov Moscow State University, 119234 Moscow, Russia; sergepolar@gmail.com
3. National Astronomical Observatories, Chinese Academy of Sciences, Beijing 100012, China; gwiktoro@astrouw.edu.pl
4. School of Astronomy & Space Science, University of the Chinese Academy of Sciences, Beijing 100012, China
* Correspondence: silvia.traversi@unife.it

Received: 1 February 2018; Accepted: 20 February 2018; Published: 7 March 2018

Abstract: We discuss the different signals, in gravitational and electromagnetic waves, emitted during the merger of two compact stars. We will focus in particular on the possible contraints that those signals can provide on the equation of state of dense matter. Indeed, the stiffness of the equation of state and the particle composition of the merging compact stars strongly affect, e.g., the life time of the post-merger remnant and its gravitational wave signal, the emission of the short gamma-ray-burst, the amount of ejected mass and the related kilonova. The first detection of gravitational waves from the merger of two compact stars in August 2017, GW170817, and the subsequent detections of its electromagnetic counterparts, GRB170817A and AT2017gfo, is the first example of the era of "multi-messenger astronomy": we discuss what we have learned from this detection on the equation of state of compact stars and we provide a tentative interpretation of this event, within the two families scenario, as being due to the merger of a hadronic star with a quark star.

Keywords: Gravitational waves; Gamma-ray bursts; nuclear matter; neutron stars

1. Introduction

The observation, on 17 August 2017, of the coalescence of two compact objects characterized by masses in the typical neutron star (NS) range has marked the beginning of the so-called "multi-messenger astronomy" [1] (SN1987a, detected both by optical telescopes and by neutrino observatories, constitutes probably the first example of a multi-messenger astronomical event). Indeed, the merger event has provided a signal in gravitational waves (GW170817) detected by Advanced LIGO and Advanced VIRGO, that has allowed us to localize the binary constraining a sky region of 31 deg^2 and a distance of a 40^{+8}_{-8} Mpc. Moreover, Fermi Gamma-ray Burst Monitor has detected a short Gamma- Ray-Burst event (GRB170817A) delayed by 1.7 s with respect to the merger time [2]. These two detections have been followed by multiple observations revealing the existence of electromagnetic (EM) counterpart of the gravitational wave (GW) event covering the entire EM bands, with signals in the X, UV, optical, IR and radio parts of the spectrum. The separated and joined analysis of these different signals can provide physical insights about open problems in theoretical physics and astrophysics which have, for years, been the subject of speculations and simulations. In particular, the study of the optical counterpart of GW170817, called kilonova (AT2017gfo) because of its peak luminosity, has finally confirmed that NS mergers host r-processes responsible of the synthesis of the most heavy nuclei. Moreover, it has provided information about the amount and features of the ejecta and these could finally give constraints about the importance of the different ejection mechanisms and of the features of the progenitors [3,4].

2. State of the Art before 17 August 2017

2.1. Expectations from the GW Signal

The merger of two compact stars represents one of the most powerful sources of GWs [5]. The process of a merger can be divided into three main stages: the inspiral phase, the coalescence phase and the post merger phase; each of these phases has its own specific waveform which in turn is determined by different physical quantities such as the total mass of the binary, the mass asymmetry, the spin of the two stars, the orbital parameters and, finally, the internal structure of the two stars. During the inspiral phase, the GW's signal is characterized by the chirp mass $M = \frac{(m_1 m_2)^{3/5}}{(m_1+m_2)^{1/5}}$ where m_1 and m_2 are the masses of the two stars (it is customary to label with m_2 the smaller of the two masses). The detection of this part of the signal allows therefore to measure M and to determine, with good accuracy, the total mass of the system $M = m_1 + m_2$. This is due to the fact that from astrophysical observations and from supernovae numerical simulations one can infer that $m_2 \geq 1.1\,M_\odot$. Similarly, the asymmetry parameter $q = m_2/m_1$ is likely to be larger than about 0.6. During most of the inspiral phase one can assume the two stars to be point like sources but when they are at a distance comparable to their radius their finite size can significantly modify the GW signal. Indeed, part of the potential energy of the binary is spent in perturbing the structure of the stars. In turn this leads to an acceleration of the inspiral dynamics with respect to the case of point-like sources (or with respect to the case of the merger of two black holes (BHs)) The physical quantity which parametrizes this effect is the tidal deformability Λ of the two stars [6]. In general, at fixed mass, the larger the radius of the star the larger the value of Λ, the larger is the deviation of the GW signal from the case of point-like sources. Potentially, a precise measurement of the final part of the inspiral phase could lead to very interesting constraints on the radii of the merging compact stars.

Finally, let us discuss the outcome of the merger and the corresponding GW signal. A first possibility is that, when the two compact objects merge, a BH is promptly formed, within a time scale of the order of 1 ms. Correspondingly, the GW signal rapidly switches off. There have been many numerical studies on the conditions for obtaining a prompt collapse [7–10]. A remarkable result is that the value of M above which the remnant collapses rapidly to a BH, $M_{\text{threshold}}$, depends strongly on the equation of state of dense matter. In particular in References [9,10], it has been shown that the ratio between $M_{\text{threshold}}$ and the maximum mass of the cold and non-rotating configuration M_{TOV} to good accuracy scales linearly with the compactness of the maximum mass configuration. This implies that once GW will be detected from mergers, this will allow us to measure $M_{\text{threshold}}$, and to obtain precious information on the structure of cold neutron stars and thus on the equation of state of dense nuclear matter.

If the remnant does not collapse immediately there are three possible outcomes of the merger: a hypermassive star (i.e., a configuration which is stable only as long as differential rotation is not completely dissipated), a supramassive star (i.e., a configuration which is stable only as long as rigid rotation is present) and finally an initially differentially rotating star which is stable even without rotation. In all these cases, the remnant, during the so called phase of the ring-down, will also emit a powerful GW signal although with a spectrum qualitatively very different from the inspiral phase signal. In References [11–14], such a spectrum has been studied as obtained from different numerical simulations and the dominant frequencies have been singled out. Again, these frequencies depend strongly on the equation of state: potentially, if at least one of those modes could be detected one could constrain the radius of the $1.6\,M_\odot$ configuration within a few hundreds meters [15]. One should notice, however, that, in general, the ring-down signal lies in a frequency range above the kHz for which the sensitivity of LIGO and VIRGO is reduced.

2.2. Mechanisms Describing the Prompt Emission of Short GRB and the Extended Emission

The problem of finding the inner engine of short GRBs is linked to the need to overcome two difficulties: first, the generation of a jet with a large Lorentz factor implies a clean environment and therefore a mechanism able to reduce the baryonic pollution is needed; second, some but not all of the short GRBs display an Extended Emission (EE), similar to the quasi-plateau emission observed in the case of long GRBs and lasting up to 10^4 s (or even more in a few cases), suggesting that the inner engine does not switch-off completely after a fraction of a second.

Concerning the way to reduce the baryonic pollution, two mechanisms have been suggested: one is based on the formation of a Black-Hole, so that baryonic material stops being ablated from the surface of the stellar object formed immediately after the merger [16]; the other suggested mechanism is based on the formation of a Quark Star (QS) and, in this case also, baryonic material cannot be ablated once the process of quark deconfinement has reached the surface of the star [17].

Concerning the origin of the EE, again two mechanisms have been proposed. One is based on the formation of a proto-magnetar and describes the EE in a way similar to the emission of a pulsar [18–20]. One needs to assume that after the merger a supramassive star (or even a totally stable star) is formed, since the collapse to a BH needs to be delayed at least by the time associated with the duration of the EE. This mechanism is able to reproduce in a very accurate way the light-curves of the EE, just by using two parameters, the strength of the magnetic field (which needs to be of the order of about 10^{15}–10^{16} G, and the rotation period (which needs to be of the order of a few milliseconds, or shorter). The second mechanism is based on the formation of an accretion disk around the BH [21]: although this possibility cannot be ruled out, no attempt at modeling the EE within this scheme has been made up to now.

Since most of the sGRBs do not display any EE, it is quite natural to assume that most of them are associated with the formation of a BH in less than a second and that in those cases no EE, due to an accretion disk, is produced. Assuming that the EE is explained via a protomagnetar model, two possibilities exist for describing the sGRBs with EE. The first possibility assumes that the prompt emission is due to the formation of a BH. Since the EE is observed after the prompt emission, this scenario needs a "time-reversal" mechanism, so that the EE produced before the collapse to a BH is observed after the prompt emission which is produced when the BH forms [22,23]. The time-reversal is associated to the time needed for the soft EE to leak out of the thick cocoon surrounding the protomagnetar. Instead, the strong prompt emission is emitted soon after the BH is formed and the cocoon exists along the rotation axis. The second possibility is that a QS forms instead of a BH. In this case, the prompt emission takes place when the process of quark deconfinement has reached the surface of the star reducing the baryonic pollution and the EE is due to the proto-magnetar that in this case is a QS [17]. Notice that these two possibilities can be easily distinguished by observations. The "time-reversal" mechanism implies that the prompt emission takes place after the protomagnetar collapses to a BH and therefore the time-separation between the moment of the merger (observed in GWs) and the prompt emission (observed in x- and γ-rays) is of the order of the duration of a supramassive star, i.e., it is easily larger than 10^3–10^4 s. Instead the mechanism based on the formation of a QS requests a time separation between merger and prompt emission of the order of about 10 s, needed for the deconfinement front to reach the surface of the star. This a relevant example of multimessenger analysis at the base of proposals such as the THESEUS mission [24].

2.3. Ejected Mass from NS Mergers, R-Processes and EM Signal

The question about the correct astrophysical mechanism that could be at the base of the r-process nucleosynthesis represents one of the subjects on which physicists have been focusing in the last decade. The first attempt to explain the mystery was to indicate the process of core collapse supernovae (CCSN) as the ideal environment in which r-processes could take place [25]. However, recently, detailed calculations have shown that CCSN does not appear to host the right conditions to create the most neutron-rich nuclei [26–29]. In particular, it seems to be especially difficult for core-collapse supernovae to produce what is known as the "third-peak". These results have pushed the researchers to try to find other possible astrophysical sources which can be responsible for a sufficient emission of matter in the right conditions for r-processes to happen. In the following we discuss the possibility that r-processes take place during the merger.

2.3.1. Ejection Mechanism and Features of the Outgoing Fluid

Binary neutron star (BNS) mergers can result in the ejection of neutron-rich matter, by means of several different possible processes. A classification of the different components of the ejecta has already been made in 2015 by Hotoketzaka and Piran [30]: the main sources are a dynamical ejection and a later ejection of part of the disk formed around the remnant because of neutrino or viscous heating.

The dynamical ejection is due to two different physical mechanisms: the first one is the tidal deformation of the NS, a consequence of the gravitational field that is not axisymmetric; the matter gains sufficient angular momentum and the ejection, mostly in the equatorial plane, starts before the collision and ends about 10 ms after the merger [31]. This material is characterized by a very low electron fraction, $Y_e < 0.1$ [32,33] which can eventually be increased by means of weak reactions in few ms after the merger [34]. The second is the shock that is formed at NSs interface, which spreads the crust material. Also, in the envelope of the remnant, a shock is produced by radial oscillations giving to some fraction of matter the sufficient energy to be ejected. The shock component could be dominant in the case of equal-mass binaries and can also be ejected in the polar direction. The electron fraction is predicted to be higher with respect to the tidal component with values in the range $0.2 < Y_e < 0.4$ [32–34]. This difference is caused principally by the higher neutrino flux which characterizes the polar direction with respect to the equatorial one: indeed electron (anti)neutrino, electron and positron captures can have a deep influence on the evolution of the electron fraction of the ejecta [35]. The dynamically ejected fluid is characterized by a velocity that can reach values of $\beta \sim 0.2$–0.3.

After the merger, some of the ashes of the NSs surround the central part forming a disk of mass in the range $10^{-3} M_\odot < M_{disk} < 0.3\, M_\odot$ [8,36,37]. Part of this disk can generate an outflow caused by viscous or neutrino heating, whose features are characteristic of the type of remnant. If the remnant is a NS the outflow depends also on its lifetime. The strong magnetic fields present at this stage can also play a role. The amount of ejecta is estimated to vary from 5 to 20% of the mass of the disk. This ejecta is usually characterized by lower velocities with respect to the dynamical one reaching a maximum value of about 0.1 c [38]. The electron fraction of this type of ejecta, initially quite low (~ 0.1), can be significantly modified by neutrinos, finally spanning in a range 0.05–0.5 with a distribution which depends on the equation of state employed [32].

Many general relativistic (GR) hydrodynamical simulations of the merger have been performed in order to describe the features of these different kind of ejecta, to evaluate the total amount of material expelled during the phases of the merger and to study the dependence of the results on the features of the binary and on the equation of state describing the NSs.

Concerning the dynamically ejected mass, its tidal component depends on the tidal deformability Λ: the stiffer the equation of state (EoS), the larger the value of Λ and the larger the amount of tidally ejected mass. On the other hand, the compactness of the stars can influence the shock produced at the merger, and also the quantity of material that can be spread out at the moment of the merger. Soft EoSs determine a larger impact velocity and so it is plausible that the correspondent shock and ejected mass will be higher [39]. Concerning the amount of mass ejected from the disk, it is limited by the mass of the disk which in turn depends on the life time of the hypermassive star. Therefore, this component is larger for stiffer equations of state [40]. Future detections of kilonovae will allow to disentagle the various components providing crucial information for nuclear physics [24].

Finally, the amount of ejected matter is deeply influenced by the degree of asymmetry of the binary q. For more asymmetric binaries, the unbound material is larger than in the symmetric case. This result can be explained in terms of the bigger effect of the tidal force that cause the lighter star to be deformed to a drop-like object and, after the merger, to be stretched leading to the formation of a pronounced tidal tail. Also the average electron fraction is influenced by the degree of asymmetry: the effect is particularly strong for soft EOS and it manifests itself as a decrease of the electron fraction of the ejecta with the increase of the mass asymmetry [41].

2.3.2. R-Processes

The ejected mass is reprocessed and through r-processes can in principle generate the distribution of heavy nuclei. It is an open question whether NSs mergers eject an amount of matter sufficient to explain the observed abundances. For these reasons, plenty of simulations have been performed in order to reproduce the path of r process nucleosynthesis: the reaction network included nuclear species between the stability valley and the neutron drip line and considered neutron captures, photodisintegration reactions together with fission and β-decay reactions [39,42–46]. The comparison between the solar abundances as a function of mass number A and the results of these simulated nucleosynthesis (in which the quantity of ejected mass is of the order of 10^{-3}–10^{-2} $M_\odot$ and the merger rate for galaxy is set in a range 10^{-5}–10^{-4} yr^{-1}) shows a good agreement in the regime $A > 120$–140, i.e., a region corresponding to the second and the third peak.

The ability of the simulations to reproduce the abundances of the elements lying between the first and the second peak depends on the obtained distribution of the electron fraction of the ejecta and on the inclusion of the entire network of possible weak interactions. Indeed in References [39,42,43] only the dynamically ejecta are considered while in [46], the outflow of material from the disk is also studied, but all the simulations fail to reproduce the abundances for $A < 120$ because of the low electron fraction attributed to ejecta as a result of neglecting the neutrino absorption processes. In [35], the authors include also the weak interaction of free neutrons obtaining a significant fraction of material with $Y_e = 0.3 - 0.4$ responsible for the production of nuclei in the range $A = 90$–140.

2.3.3. EM Counterpart

A probe of the amount of ejected mass and of the realization of the r-process chains in NS mergers can be the analysis of the EM signal predicted to be associate with this phenomenon [47,48]. The maximum of the luminosity takes place just after the photons can escape the expanding ejecta whose density is reducing. A typical timescale is of the order of 1 day while the luminosity $\sim 10^{42}$ erg s^{-1}, three orders of magnitude larger than the Eddington luminosity for a solar mass star: for this reason this EM events are called kilonovae. The spectral peak can vary in the IR/optical/nearUV wavelengths. The timescale t_{peak}, the luminosity L_{peak} and the effective temperature T_{peak} of the signal depend on the amount M_{ej}, the velocity v and the opacity k of the ejecta [48]:

$$t_{peak} \propto \left(\frac{kM_{ej}}{v}\right)^{\frac{1}{2}}, \quad L_{peak} \propto \left(\frac{vM_{ej}}{k}\right)^{\frac{1}{2}}, \quad T_{peak} \propto \left(vM_{ej}\right)^{-\frac{1}{8}} k^{-\frac{3}{8}}$$

These dependences on the features of the ejecta could translate into an influence of the EoS of NS: an EoS which produces more ejecta will lead to a brighter optical counterpart, peaked on longer timescales and with longer peak wavelengths [39].

3. GW170817-GRB170817A-AT2017gfo

3.1. Analysis of the GW Signal

The signal detected by the LIGO-VIRGO collaborations [49] corresponds to the emission of GWs from an inspiral binary with a chirp mass $M = 1.188^{+0.004}_{-0.002}$ $M_\odot$ which implies a total mass $M = 2.74^{+0.04}_{-0.02}$ $M_\odot$ (under the hypothesis that the spins of the two stars are compatible with the ones observed in binary neutron stars, "low spin case"). In turn, the masses of the components are in the range 1.17–1.6 $M_\odot$, strongly suggesting that the merger was between two NSs. Although the source is quite close, 40 Mpc, it has not been possible to follow the GW signal up to the merger and during the ring-down phase [50]. However, a very useful upper limit on the value of the tidal parameter $\tilde{\Lambda}$ (which depends on the tidal deformabilities and the masses of the two stars) has been set: $\tilde{\Lambda} < 800$ at the 90% level in the low spin case. This constraint is basically model-independent and it allows ua to already rule out a few very stiff equations of state such as MS1 and MS1b which are based on relativistic mean field calculations [51].

What happened during the first few milliseconds after the merger is unclear. Even if not completely excluded, the possibility that the merger has led to a prompt collapse seems to be very unlikely because in that case it would be difficult to explain the observation of the electromagnetic counterparts of GW170817. Actually, one can infer that the post-merger remnant is most probably a hypermassive star: a supramassive star or a stable star would inject part of its huge kinetic rotational energy into the GRB or into the kilonova on a long time scale and there is no evidence, in the observed signals, of such an energy injection [52]. This implies that the total energy of the binary, which can be estimated to be of the order of 95% M (assuming the gravitational binding energy of the binary to be $\sim$5% M) is larger than the maximum mass of the supramassive configuration M_{supra}. Several numerical calculations on rotating compact stars have shown that M_{supra} is to good accuracy $\sim$1.2 M_{TOV} [53]. Combining these results one therefore obtains that $M_{TOV} < 2.2\,M_\odot$. This simple estimate is in agreement with the results of References [52,54,55] and it again disfavors very stiff equations of state which predict maximum masses above 2.2 $M_\odot$ such as e.g., DD2 [56].

If the remnant is a hypermassive star, another constraint can be obtained by imposing that the total energy of the binary is lower than the maximum mass of the hypermassive configuration. This study has been performed in [57] and it allows us to rule out extremely soft equations of state: it has been found that the radius of the 1.6 $M_\odot$ configuration must be larger than about 10.7 km.

To summarize, the first detection of GWs from binary neutron stars has already allowed us to exclude a few examples of dense matter equations of state. In particular, very stiff equations of state based only on nucleonic degrees of freedom seem to be unfavored. We will discuss in the last section how this result actually suggests that strange matter must appear in some form in compact stars.

3.2. The Weak Gamma Emission of GRB170817A: Was it a Standard Short GRB?

As already discussed above, short GRBs are assumed to originate from the merger of two NSs. In the case of the event of August 2017 the GW signal clearly indicates that a merger did take place, but, on the other hand, the gamma-ray emission was delayed by approximately two seconds with respect to the moment of the merger and the observed signal was much weaker than the one of a typical short GRB. It is also relevant to stress that no extended emission was observed, likely indicating that a supramassive star did not form after the merger.

There are two main possible interpretations of the event. The first one assumes that the emission was intrinsically sub-luminous and quasi-isotropic [58,59]. The second one assumes instead a standard short GRB emission, that was observed off-axis [60]. While at the moment, about a hundred days after the event, both possibilities can explain the data, the analysis of the future time-evolution of the synchrotron emission will ultimately be able to distinguish between these two scenarios, telling therefore if GRB170817A was a standard short GRB seen off-axis or if it belongs to a new class of phenomena [61].

Even though at the moment the mechanism which launched GRB170817A is still unclear, some strongly energetic emission in γ and in x-rays was produced and this indicates that the merger did not collapse instantaneously to a BH. There are explicit simulations indicating that if a jet needs to be formed, the object produced in the post-merger needs to survive for at least a few tens of milliseconds [62]. As discussed in the following, the analysis of the kilonova emission also indicates that the result of the merger did not collapse immediately to a BH: a relevant amount of matter was likely emitted from the disk on a time-scale incompatible with an almost instantaneous collapse. This is a very important point to take into account when discussing the possible models for the merger, as we will do in the last section.

3.3. Electromagnetic Signal and Mass Ejection

3.3.1. Analysis of the Optical Transient

On 17 August 2017, the observation took place of the first electromagnetic counterpart to a gravitational wave event attributed to a merger of two NSs. The data in UV, optical and NIR bands extend for a time interval from 0.47 to 18.5 days after the merger and are consistent with a kilonova signal predicted to be associated with a NS merger.

The early spectrum is dominated by a blue component. Over the first few days, the spectra shows a rapid evolution to redder wavelengths: at 1.5 days after merger the optical peak is located around 5000 A and already at 2.5 days it shows a shift to 7000 A, evolving to ∼7800 A at 4.5 days and going finally out of the optical regime in the interval between 4.5 and 7.5 days after the merger. By 10 days the wavelength is >15,000 A. [63]. Moreover, the rate of the decline is observed to change for the different bands belonging to the observed kilonova spectrum: while the decline appears to be quick in the ug band (blue) with a rate of ∼2 mag day, the rizY (red) and the HKs (NIR) bands show a smoother decay causing the spectrum to be dominated by red at late time [64]. The initial luminosities, ∼5 × 10^{41} erg s^{-1} at 0.6 days and ∼2 × 10^{41} erg s^{-1} at 1.5 days, and the short timescale (∼1 day) are consistent with the model called Blue KN: this kind of emission was first proposed by Metzger et al. [48] and subsequently developed as the signal associated to different kinds of matter likely to be ejected during or post merger [65,66]: in [65] the authors analyzed the tidal tails formed during the merger while in [66] is presented a study of the outflow from the remnant in the case of a delayed (>100 ms) BH formation. All this analysis have in common the very low opacity attributed to the ejected mass with values in a range from $k = 0.1$ to $k = 1$ cm^{2}s^{-1}, typical of material containing Fe-group or light r-process nuclei characterized by $A < 140$. Therefore, the Blue KN signal is likely to be associated with r-processes responsible of the formation of nuclei lying between the first and the second peak.

Conversely, the late EM emission, which dominates at longer timescales ∼a week and shows a lower luminosity ∼10^{40}–10^{41} erg s^{-1} fits well with the so called Red KN model [64]: in [67–70] it was first presented the study of the effect of higher opacity of the ejecta on the resulting KN emission. This high opacity (k up to 10 cm^{2}s^{-1}) is attributed to the presence of Lanthanide elements, heavy nuclei with $A > 140$, so the Red KN represents an indication of nucleosyntesis reactions filling the third peak of r-processes.

These observational evidences suggest the presence of material characterized by a not unique value of the opacity and therefore a different content of Lanthanides. Despite the fact that a single component ejecta with a power-law velocity distribution and a time-dependent opacity (studied with an analytical model in [71]) cannot be excluded, the most accredited hypothesis is the existence of at least two components of the ejecta, a Lanthanide poor (for the Blue KN) and a Lanthanide rich (for the Red KN) component [66,72]. This conclusion is also suggested by the fact that the blue component is not obscured by the red one, a clue of the need of distinct regions and angles of emission for the material with different opacity values. This also means that these two components can be attributed to distinct sources [64].

The duration and effective temperature of the KN emission have been studied with models outlined in [68,73] allowing to indicate the mass, velocity and opacity of the ejecta as fitting parameters. The result is that the signal is consistent with a two component model. It consists of: a Blue component with $M_{ej}^B \sim 0.01$–few 0.01 $M_\odot$, velocity $v_{ej}^B = 0.27$–0.3 c and opacity $k^B = 0.5$ cm^{2}s^{-1} requiring a Lanthanide fraction of ∼10^{-4} to 10^{-5} in the outermost ejecta [63,64]; a Red component with $M_{ej}^R \sim 0.04$ $M_\odot$, velocity $v_{ej}^R = 0.12$ c and opacity $k^R = 3.3$ cm^{2}s^{-1} requiring a Lanthanide fraction of ∼10^{-2} [64,74].

Another model able to fit the data is characterized by three components: a Blue one with $M_{ej}^B \sim 0.01$ $M_\odot$, velocity $v_{ej}^B = 0.27$ c and opacity $k^B = 0.5$ cm^{2}s^{-1}; a Purple one with $M_{ej}^P \sim 0.03$ $M_\odot$, velocity

$v_{ej}^P = 0.11$c and opacity $k^P = 3$ cm^{2}s^{-1} and a Red one with $M_{ej}^R \sim 0.01$ $M_\odot$, velocity $v_{ej}^R = 0.16$ c and opacity $k^R = 10$ cm^{2}s^{-1} [64].

3.3.2. Role of Different Ejection Mechanisms

The different opacities of the Red and Blue (and eventually Purple) KN, attributed to a different Lanthanide fraction can be directly connected to the electron fraction of the ejected matter, Y_e: a lower electron fraction corresponds to the ability to synthesize heavier nuclei, so to a bigger concentration of Lanthanides and, as a consequence to a greater opacity. The Y_e of the ejecta depends in turn on the direction and of the mechanism at the base of the ejection [72].

The Blue component KN, characterized by a very low opacity, has to find its origin in a Lanthanide-poor material with an electron fraction >0.25–0.3: this kind of matter can be ejected dynamically by means of the shock generated at the contact surfaces of the two stars at the moment of the merger [32,34,35,39]; the ejecta is expected to be be found within an angle between 30° and 45° [33,41], with respect to the polar axis where the neutrino flux is more intense and so neutrino absorption plays a central role in raising the electron fraction above a value of $Y_e > 0.25$ [40,68]. Moreover, if the remnant of the merger survives as an hypermassive NS , a disk is formed around it which reaches a stable configuration in a few tens of milliseconds while a neutrino-driven wind is formed at a time of $\sim$10 ms [75]. This wind is responsible for the ejection of about the 5% of the mass of the disk, mostly in the polar direction. Simulations reveal that in this direction the large neutrino flux raises the electron fraction up to a distribution which peaks at $Y_e = 0.3$–0.4, so wind also gives rise to a low-opacity ejecta [75,76]. The features which mostly distinguish the two different mechanism is the resulting velocities, 0.2–0.3 c for the dynamical ejecta while lower for the wind ejecta, v < 0.1 c [38,75,76].

The velocity attributed to the Blue component, 0.27–0.3 c, represents an important proof of its dynamical origin, but while some authors suggest that the shock represents the exclusive mechanism for this low opacity signal [63,64], others view this component as a possible result of a union of the dynamical and wind ejecta [40]. In the first case, the required amount of ejected matter, $\sim 10^{-2}$ $M_\odot$ implies the need for a soft EoS in order to reach a high velocity at the impact of the two compact objects: this suggest an upper limit on the NS radius of about 12 km or less [63].

The second hypothesis, instead, does not require such a tight limit on the radius.

Regarding the higher opacity of the Lanthanide richer Red component, the largely accepted interpretation indicates the dynamical mechanism of tidal ejection in the equatorial plane (within an angle of 45°–60°). The squeezed out material is indeed characterized by a very low electron fraction <0.1 [31–33] giving rise to a Red-NIR spectrum with a longer timescale [67,68]. The large amount of mass inferred from the data can be an indication of a high degree of asymmetry of the binary [64]. However, to explain the component characterized by an opacity of $\sim$3 cm^{2}g^{-1} and a very large ejected mlies, we need a soft EoS in order to reach a high velocity at the impact of the two compact objects: this suggests an upper limit on the NS radius of about 12 km or less [63]. The second hypothesis, instead, does not require such a tight limit on the radius.

Regarding the higher opacity of the Lanthanide richer Red component, the largely accepted interpretation indicates the dynamical mechanism of tidal ejection in the equatorial plane (within an angle of 45°–60°). The squeezed out material is indeed characterized by a very low electron fraction <0.1 [31–33] giving rise to a Red-NIR spectrum with a longer timescale [67,68]. The large amount of mass inferred from the data can be an indication of an high degree of asymmetry of the binary [64]. However, to explain the component characterized by an opacity of $\sim$3 cm^{2}g^{-1} and a very large ejected mass, which can be considered as part of the Red KN or a distinct Purple KN, it is also necessary to take into account the disk outflow.

First of all, the wind ejecta for angles >30° are less affected by the neutrino flux maintaining an electron fraction $Y_e \sim 0.25$–0.3 and fitting the required opacity [70,75,77]. At the same time, a contribution can also come from the secular ejecta which affects all the solid angles, but which is equatorial dominated: this viscous-driven ejection can result in the expulsion of up to 30% of the mass of the disk and the Y_e of the material depends on the lifetime of the HMNS with respect to that of the disk (∼ten of ms). In the case of a long-lived HMNS the electron fraction can reach values between 0.2–0.5 with peaks at ∼0.3–0.4 while if the collapse to a black hole happens earlier $Y_e < 0.3$–0.4 [78].

In [40] (three component model), and in [64], the authors suggest that the intermediate opacity component of the KN signal can be, indeed, explained by means of the early viscosity driven secular ejection: this will imply a short-lived remnant (∼30 ms) and a massive disk ∼0.08 $M_\odot$. These two statements point to different directions concerning the features of the EoS: on one side a soft EoS will prevent the remnant to form a long-lived massive neutron star, but on the other side, the greater value of tidal deformability associated to a stiffer EoS will determine the formation of more pronounced tidal tales and thus a more massive disk around the remnant. On the other hand, the upper limit imposed on the tidal deformability by the gravitational waves measurement (see Section 3.1 for details) and the absence of a prompt collapse to BH exclude extremely stiff or extremely soft EoS, respectively. This seems to suggest an EoS is characterized by an intermediate softness.

To summarize, the situation concerning the mechanisms at the base of the kilonova (and of the GRB) is still not settled. In the following we will shortly discuss the global interpretation of the event of August 2017 at the light of the two-families scenario.

4. A Different Hypothesis: A Hadronic Star–Quark Star Merger

The event GW170817 and its electromagnetic counterparts have been generated from the coalescence of two compact stars. In the standard scenario, only one family of compact stars exists, namely the family of stars composed entirely by hadronic degrees of freedom. However, there are some phenomenological indications of the possible existence of a second family of compact stars which are entirely composed by deconfined quarks, namely QSs, see References [79–81]. In this scenario, the first family is populated by hadronic stars (HSs) which could be very compact and "light" due to the softness of the hadronic EoS (with hyperons and delta resonances included) while the second family is populated by QSs which, on the other hand, can support large masses due to the stiffness of the quark matter EoS.

In this scheme, a binary system could be composed of two HSs, two QSs or finally an HS and a QS. Let us discuss these three possibilities in connection with the phenomenology of GW170817.

The threshold mass $M_{threshold}$ for an HS–HS, i.e., the limit mass above which a prompt collapse is obtained, has been estimated to be ∼2.7 $M_\odot$ [82], on the base of the the study performed in [15]. This value is smaller than the total binary mass M inferred from GW170817 [49] and therefore the hypothesis that the binary sytems were an HS–HS system is disfavored within the two families scenario. Also, the possibility that the system was a double QS binary system is excluded because in that case it would be difficult to explain the kilonova which is powered by nuclear radioactive decays: even if some material is ejected from the QSs it is not made of ordinary nuclei and therefore cannot be used inside an r-process chain to produce heavy nuclei. Conversely, the case of a HS–QS merger, in which the prompt collapse is avoided by the formation of a hypermassive hybrid configuration, becomes the most plausible suggestion in the context of the two families scenario [82].

Let us briefly discuss which the possible evolutionary paths are that can lead to the formation of such a mixed system.

The formation of a double compact object like the source of GW170817 is more probable in isolated binary evolution than through dynamical interactions in dense stellar systems [83]. In such a case, a common envelope phase [84] is typically necessary to shorten the orbital period and to allow for the merger to occur within the Hubble time. This phase, in a typical double NS formation route, occurs after the formation of the first compact object [85]. Additionally, the companion may fill its Roche lobe again due to an expansion on the Helium main-sequence and commences a mass transfer phase. Therefore, it is natural to expect that the compact object which formed first may obtain a significantly higher mass than its counterpart. If the total mass of a double compact object is about 2.7 $M_\odot$ or higher, the heavier compact object may reach the mass of ~ 1.5–$1.6 M_\odot$., which in the two-families scenario marks the threshold for deconfinement and the formation of a QS [79]. The secondary cannot accrete mass from the primary, which is already a compact star. Moreover, it has lost a large fraction of mass in the interaction. Therefore, its pre-SN mass will be relatively low and, consequently, its post-SN mass will not be significantly different from the lower limit of a newborn NS $\sim 1.1 M_\odot$; see e.g., [86]. This implies that in the two-families scenario, the binary evolution may favor the situation in which GW170817 is a HS–QS binary rather than a HS–HS. This issue will be further investigated in a forthcoming paper.

Under the hypothesis that the event seen in August 2017 is due to the merger of a HS–QS system, we need now to discuss the possible explanations of the different features seen in the gravitational and electromagnetic signals. First, the gravitational wave signal has clearly indicated that extremely stiff EoSs are ruled out: the limit put on $\tilde{\Lambda}$ is fulfilled only if the radii of the two stars are smaller than about 13.4 km (see the analysis of Ref. [87]). Both HSs and QSs satisfy this limit [79,88], see also Ref. [89] where the tidal deformabilities of HSs and QSs have been computed.

Second, the emission of GRB170817A is probably connected with the formation of a relativistic jet which is launched by a BH-accretion torus system. The scenario discussed in [17] concerns short GRB featuring and extended emission which has not been observed in this case. In our scenario, the compact star which forms immediately after the merger is a hypermassive hybrid star in which the burning of hadronic matter is still active. We expect such a system to collapse to a BH once the differential rotation is dissipated. The sGRB would be produced by the same mechanism studied in References [16,90].

Let us finally discuss the properties of the observed kilonova within our scenario. Perego et al. [40] suggest an effective two components model in which the opacity of the secular ejecta is predicted to be very low (~ 1 cm^{2}s^{-1}), comparable to that of the wind component. This hypothesis has two major consequences: the lifetime of the remnant must be sufficiently long in order to allow weak reactions to raise the electron fraction to >0.3 and the tidal ejecta must give a very relevant contribution.

Both these requirements can be fulfilled in the context of the HS–QS merger; indeed the hybrid star configuration predicted by this model can survive as a hypermassive configuration for a time of the order of hundreds of ms. Moreover, for an asymmetric binary, characterized by $q = 0.75$–0.8, the predicted tidal deformability of the lightest star (the hadronic one) can reach value of ~ 500. This quite high value of λ together with the supposed high asymmetry of the binary can result in a relevant contribution of the tidal effect on the total ejected mass. This allows to explain the third peak of r-processes and the Red KN without the need of a high opacity secular ejecta (notice also that the value $\lambda \sim 500$ is largely above the lower limit derived from the analysis of the EM counterpart performed in [91]).

It is worth noticing that another possibility has been proposed which is able to explain the data from GW170817 and which also makes use of the formation of quark matter in compact stars: it is based on the so-called "twin stars scenario" and is discussed in Reference [92]. In that case, the binary system is composed of a purely hadronic star and of a hybrid star, because two hadronic stars would be excluded by the tidal deformability constraint. Notice that these two scenarios provide different predictions on a variety of phenomena. First, in Reference [17] it has been shown the two-families scenario can describe the prompt emission of short GRBs displaying also an extended emission (see Section 2.2). The mechanism described in that paper cannot be adapted to the twin stars scenario. Second, a prediction of the two-families scenario is the possibility of a prompt collapse also for small-mass binaries, if they are both hadronic stars [82]. Again, this possibility does not exist within the twin-stars scenario.

In conclusion, despite the need of hydrodynamical simulations in order to make more quantitative predictions, the HS–QS merger can represent a viable way to explain the features of the GW170817, GRB170817A and AT2017gfo.

Acknowledgments: S.B.P. acknowledges support from the Russian Science Fondation project 14-12-00146. G.W. is partly supported by the President's International Fellowship Initiative (PIFI) of the Chinese Academy of Sciences under grant no.2018PM0017 and by the Strategic Priority Research Program of the Chinese Academy of Science "Multi-waveband Gravitational Wave Universe" (Grant No. XDB23040000).

Author Contributions: All authors contributed equally to this work.

Conflicts of Interest: The authors declare no conflict of interest.

References

1. Abbott, B.P.; Abbott, R.; Adhikari, R.X.; Ananyeva, A.; Anderson, S.B.; Appert, S.; Arai, K.; Araya, M.C.; Barayoga, J.C.; Barish, B.C.; et al. Multi-messenger Observations of a Binary Neutron Star Merger. *Astrophys. J.* **2017**, *848*, L12.
2. Abbott, B.P.; Abbott, R.; Abbott, T.D.; Acernese, F.; Ackley, K.; Adams, C.; Adams, T.; Addesso, P.; Adhikari, R.X.; Adya, V.B.; et al. Gravitational Waves and Gamma-rays from a Binary Neutron Star Merger: GW170817 and GRB 170817A. *Astrophys. J.* **2017**, *848*, L13.
3. Abbott, B.P.; Abbott, R.; Abbott, T.D.; Acernese, F.; Ackley, K.; Adams, C.; Adams, T.; Addesso, P.; Adhikari, R.X.; Adya, V.B.; et al. Estimating the Contribution of Dynamical Ejecta in the Kilonova Associated with GW170817. *Astrophys. J.* **2017**, *850*, L39.
4. Abbott, B.P.; Abbott, R.; Abbott, T.D.; Acernese, F.; Ackley, K.; Adams, C.; Adams, T.; Addesso, P.; Adhikari, R.X.; Adya, V.B.; et al. On the Progenitor of Binary Neutron Star Merger GW170817. *Astrophys. J.* **2017**, *850*, L40.
5. Shibata, M.; Uryu, K. Gravitational waves from the merger of binary neutron stars in a fully general relativistic simulation. *Prog. Theor. Phys.* **2002**, *107*, 265–303.
6. Flanagan, E.E.; Hinderer, T. Constraining neutron star tidal Love numbers with gravitational wave detectors. *Phys. Rev. D* **2008**, *77*, 021502.
7. Baiotti, L.; Giacomazzo, B.; Rezzolla, L. Accurate evolutions of inspiralling neutron-star binaries: Prompt and delayed collapse to black hole. *Phys. Rev. D* **2008**, *78*, 084033.
8. Hotokezaka, K.; Kiuchi, K.; Kyutoku, K.; Muranushi, T.; Sekiguchi, Y.I.; Shibata, M.; Taniguchi, K. Remnant massive neutron stars of binary neutron star mergers: Evolution process and gravitational waveform. *Phys. Rev. D* **2013**, *88*, 044026.
9. Bauswein, A.; Baumgarte, T.W.; Janka, H.T. Prompt merger collapse and the maximum mass of neutron stars. *Phys. Rev. Lett.* **2013**, *111*, 131101.
10. Bauswein, A.; Stergioulas, N. Semi-analytic derivation of the threshold mass for prompt collapse in binary neutron star mergers. *Mon. Not. R. Astron. Soc.* **2017**, *471*, 4956–4965.
11. Bauswein, A.; Janka, H.T. Measuring neutron-star properties via gravitational waves from binary mergers. *Phys. Rev. Lett.* **2012**, *108*, 011101.
12. Bauswein, A.; Janka, H.T.; Hebeler, K.; Schwenk, A. Equation-of-state dependence of the gravitational-wave signal from the ring-down phase of neutron-star mergers. *Phys. Rev. D* **2012**, *86*, 063001.
13. Takami, K.; Rezzolla, L.; Baiotti, L. Constraining the Equation of State of Neutron Stars from Binary Mergers. *Phys. Rev. Lett.* **2014**, *113*, 091104.
14. Maione, F.; De Pietri, R.; Feo, A.; Löffler, F. Spectral analysis of gravitational waves from binary neutron star merger remnants. *Phys. Rev. D* **2017**, *96*, 063011.
15. Bauswein, A.; Stergioulas, N.; Janka, H.T. Exploring properties of high-density matter through remnants of neutron-star mergers. *Eur. Phys. J. A* **2016**, *52*, 56.
16. Rezzolla, L.; Giacomazzo, B.; Baiotti, L.; Granot, J.; Kouveliotou, C.; Aloy, M.A. The missing link: Merging neutron stars naturally produce jet-like structures and can power short Gamma-Ray Bursts. *Astrophys. J.* **2011**, *732*, L6.
17. Drago, A.; Lavagno, A.; Metzger, B.; Pagliara, G. Quark deconfinement and the duration of short Gamma Ray Bursts. *Phys. Rev. D* **2016**, *93*, 103001.
18. Lyons, N.; O'Brien, P.T.; Zhang, B.; Willingale, R.; Troja, E.; Starling, R.L.C. Can X-Ray Emission Powered by a Spinning-Down Magnetar Explain Some GRB Light Curve Features? *Mon. Not. R. Astron. Soc.* **2010**, *402*, 705–712.

19. Dall'Osso, S.; Stratta, G.; Guetta, D.; Covino, S.; De Cesare, G.; Stella, L. GRB Afterglows with Energy Injection from a spinning down NS. *Astron. Astrophys.* **2011**, *526*, A121.
20. Rowlinson, A.; O'Brien, P.T.; Metzger, B.D.; Tanvir, N.R.; Levan, A.J. Signatures of magnetar central engines in short GRB lightcurves. *Mon. Not. R. Astron. Soc.* **2013**, *430*, 1061–1087.
21. Van Putten, M.H.P.M.; Lee, G.M.; Della Valle, M.; Amati, L.; Levinson, A. On the origin of short GRBs with Extended Emission and long GRBs without associated SN. *Mon. Not. R. Astron. Soc.* **2014**, *444*, L58–L62.
22. Rezzolla, L.; Kumar, P. A novel paradigm for short gamma-ray bursts with extended X-ray emission. *Astrophys. J.* **2015**, *802*, 95.
23. Ciolfi, R.; Siegel, D.M. Short gamma-ray bursts in the "time-reversal" scenario. *Astrophys. J.* **2015**, *798*, L36.
24. Stratta, G.; Ciolfi, R.; Amati, L.; Ghirlanda, G.; Tanvir, N.; Bozzo, E.; Gotz, D.; O'Brien, P.; Frontera, F.; Osborne, J.P.; et al. THESEUS: A key space mission for Multi-Messenger Astrophysics. *arXiv* **2017**, arXiv:1712.08153.
25. Burbidge, M.E.; Burbidge, G.R.; Fowler, W.A.; Hoyle, F. Synthesis of the elements in stars. *Rev. Mod. Phys.* **1957**, *29*, 547–650.
26. Hoffman, R.D.; Muller, B.; Janka, H.T. Nucleosynthesis in O-Ne-Mg Supernovae. *Astrophys. J.* **2008**, *676*, L127.
27. Fischer, T.; Whitehouse, S.C.; Mezzacappa, A.; Thielemann, F.K.; Liebendorfer, M. Protoneutron star evolution and the neutrino driven wind in general relativistic neutrino radiation hydrodynamics simulations. *Astron. Astrophys.* **2010**, *517*, A80.
28. Arcones, A.; Janka, H.T.; Scheck, L. Nucleosynthesis-relevant conditions in neutrino-driven supernova outflows. 1. Spherically symmetric hydrodynamic simulations. *Astron. Astrophys.* **2007**, *467*, 1227–1248.
29. Roberts, L.F.; Woosley, S.E.; Hoffman, R.D. Integrated Nucleosynthesis in Neutrino Driven Winds. *Astrophys. J.* **2010**, *722*, 954–967.
30. Hotokezaka, K.; Piran, T. Mass ejection from neutron star mergers: different components and expected radio signals. *Mon. Not. R. Astron. Soc.* **2015**, *450*, 1430–1440.
31. Hotokezaka, K.; Kiuchi, K.; Kyutoku, K.; Okawa, H.; Sekiguchi, Y.I.; Shibata, M.; Taniguchi, K. Mass ejection from the merger of binary neutron stars. *Phys. Rev. D* **2013**, *87*, 024001.
32. Palenzuela, C.; Liebling, S.L.; Neilsen, D.; Lehner, L.; Caballero, O.L.; O'Connor, E.; Anderson, M. Effects of the microphysical Equation of State in the mergers of magnetized Neutron Stars With Neutrino Cooling. *Phys. Rev. D* **2015**, *92*, 044045.
33. Radice, D.; Galeazzi, F.; Lippuner, J.; Roberts, L.F.; Ott, C.D.; Rezzolla, L. Dynamical Mass Ejection from Binary Neutron Star Mergers. *Mon. Not. R. Astron. Soc.* **2016**, *460*, 3255–3271.
34. Sekiguchi, Y.; Kiuchi, K.; Kyutoku, K.; Shibata, M. Dynamical mass ejection from binary neutron star mergers: Radiation-hydrodynamics study in general relativity. *Phys. Rev. D* **2015**, *91*, 064059.
35. Goriely, S.; Bauswein, A.; Just, O.; Pllumbi, E.; Janka, H.T. Impact of weak interactions of free nucleons on the r-process in dynamical ejecta from neutron-star mergers. *Mon. Not. R. Astron. Soc.* **2015**, *452*, 3894–3904.
36. Shibata, M.; Taniguchi, K. Merger of binary neutron stars to a black hole: disk mass, short gamma-ray bursts, and quasinormal mode ringing. *Phys. Rev. D* **2006**, *73*, 064027.
37. Rezzolla, L.; Baiotti, L.; Giacomazzo, B.; Link, D.; Font, J.A. Accurate evolutions of unequal-mass neutron-star binaries: Properties of the torus and short GRB engines. *Class. Quant. Gravity* **2010**, *27*, 114105.
38. Siegel, D.M.; Metzger, B.D. Three-Dimensional General-Relativistic Magnetohydrodynamic Simulations of Remnant Accretion Disks from Neutron Star Mergers: Outflows and r-Process Nucleosynthesis. *Phys. Rev. Lett.* **2017**, *119*, 231102.
39. Bauswein, A.; Goriely, S.; Janka, H.T. Systematics of dynamical mass ejection, nucleosynthesis, and radioactively powered electromagnetic signals from neutron-star mergers. *Astrophys. J.* **2013**, *773*, 78.
40. Perego, A.; Radice, D.; Bernuzzi, S. AT 2017gfo: An Anisotropic and Three-component Kilonova Counterpart of GW170817. *Astrophys. J.* **2017**, *850*, L37.
41. Sekiguchi, Y.; Kiuchi, K.; Kyutoku, K.; Shibata, M.; Taniguchi, K. Dynamical mass ejection from the merger of asymmetric binary neutron stars: Radiation-hydrodynamics study in general relativity. *Phys. Rev. D* **2016**, *93*, 124046.
42. Goriely, S.; Bauswein, A.; Janka, H.T. R-Process Nucleosynthesis in Dynamically Ejected Matter of Neutron Star Mergers. *Astrophys. J.* **2011**, *738*, L32.

43. Korobkin, O.; Rosswog, S.; Arcones, A.; Winteler, C. On the astrophysical robustness of neutron star merger r-process. *Mon. Not. R. Astron. Soc.* **2012**, *426*, 1940–1949.
44. Goriely, S.; Sida, J.L.; Lemaître, J.F.; Panebianco, S.; Dubray, N.; Hilaire, S.; Bauswein, A.; Janka, H.T. New fission fragment distributions and r-process origin of the rare-earth elements. *Phys. Rev. Lett.* **2013**, *111*, 242502.
45. Just, O.; Bauswein, A.; Pulpillo, R.A.; Goriely, S.; Janka, H.T. Comprehensive nucleosynthesis analysis for ejecta of compact binary mergers. *Mon. Not. R. Astron. Soc.* **2015**, *448*, 541–567.
46. Siegel, D.M.; Metzger, B.D. Three-dimensional GRMHD simulations of neutrino-cooled accretion disks from neutron star mergers. *arXiv* **2017**, arXiv:1711.00868.
47. Li, L.X.; Paczynski, B. Transient events from neutron star mergers. *Astrophys. J.* **1998**, *507*, L59.
48. Metzger, B.D.; Martinez-Pinedo, G.; Darbha, S.; Quataert, E.; Arcones, A.; Kasen, D.; Thomas, R.; Nugent, P.; Panov, I.V.; Zinner, N.T. Electromagnetic Counterparts of Compact Object Mergers Powered by the Radioactive Decay of R-process Nuclei. *Mon. Not. R. Astron. Soc.* **2010**, *406*, 2650–2662.
49. Abbott, B.P.; Abbott, R.; Abbott, T.D.; Acernese, F.; Ackley, K.; Adams, C.; Adams, T.; Addesso, P.; Adhikari, R.X.; Adya, V.B.; et al. GW170817: Observation of Gravitational Waves from a Binary Neutron Star Inspiral. *Phys. Rev. Lett.* **2017**, *119*, 161101.
50. Abbott, B.P.; Abbott, R.; Abbott, T.D.; Acernese, F.; Ackley, K.; Adams, C.; Adams, T.; Addesso, P.; Adhikari, R.X.; Adya, V.B.; et al. Search for Post-merger Gravitational Waves from the Remnant of the Binary Neutron Star Merger GW170817. *Astrophys. J.* **2017**, *851*, L16.
51. Mueller, H.; Serot, B.D. Relativistic mean field theory and the high density nuclear equation of state. *Nucl. Phys. A* **1996**, *606*, 508–537.
52. Margalit, B.; Metzger, B.D. Constraining the Maximum Mass of Neutron Stars From Multi-Messenger Observations of GW170817. *Astrophys. J.* **2017**, *850*, L19.
53. Lasota, J.P.; Haensel, P.; Abramowicz, M.A. Fast rotation of neutron stars. *Astrophys. J.* **1996**, *456*, 300.
54. Ruiz, M.; Shapiro, S.L.; Tsokaros, A. GW170817, General Relativistic Magnetohydrodynamic Simulations, and the Neutron Star Maximum Mass. *Phys. Rev. D* **2018**, *97*, 021501.
55. Rezzolla, L.; Most, E.R.; Weih, L.R. Using gravitational-wave observations and quasi-universal relations to constrain the maximum mass of neutron stars. *Astrophys. J. Lett.* **2018**, *852*, L25.
56. Banik, S.; Hempel, M.; Bandyopadhyay, D. New Hyperon Equations of State for Supernovae and Neutron Stars in Density-dependent Hadron Field Theory. *Astrophys. J. Suppl. Ser.* **2014**, *214*, 22.
57. Bauswein, A.; Just, O.; Janka, H.T.; Stergioulas, N. Neutron-star radius constraints from GW170817 and future detections. *Astrophys. J.* **2017**, *850*, L34.
58. Gottlieb, O.; Nakar, E.; Piran, T.; Hotokezaka, K. A cocoon shock breakout as the origin of the γ-ray emission in GW170817. *arXiv* **2017**, arXiv:1710.05896.
59. Kasliwal, M.M.; Nakar, E.; Singer, L.P.; Kaplan, D.L.; Cook, D.O.; Van Sistine, A.; Lau, R.M.; Fremling, C.; Gottlieb, O.; Jencson, J.E.; et al. Illuminating Gravitational Waves: A Concordant Picture of Photons from a Neutron Star Merger. *Science* **2017**, *358*, 1559–1565.
60. Lazzati, D.; Perna, R.; Morsony, B.J.; López-Cámara, D.; Cantiello, M.; Ciolfi, R.; Giacomazzo, B.; Workman, J.C. Late time afterglow observations reveal a collimated relativistic jet in the ejecta of the binary neutron star merger GW170817. *arXiv* **2017**, arXiv:1712.03237.
61. Margutti, R.; Alexander, K.D.; Xie, X.; Sironi, L.; Metzger, B.D.; Kathirgamaraju, A.; Fong, W.; Blanchard, P.K.; Berger, E.; MacFadyen, A.; et al. The Binary Neutron Star event LIGO/VIRGO GW170817 a hundred days after merger: synchrotron emission across the electromagnetic spectrum. *arXiv* **2018**, arXiv:1801.03531.
62. Ruiz, M.; Shapiro, S.L. General relativistic magnetohydrodynamics simulations of prompt-collapse neutron star mergers: The absence of jets. *Phys. Rev. D* **2017**, *96*, 084063.
63. Nicholl, M.; Berger, E.; Kasen, D.; Metzger, B.D.; Elias, J.; Briceño, C.; Alexander, K.D.; Blanchard, P.K.; Chornock, R.; Cowperthwaite, P.S.; et al. The Electromagnetic Counterpart of the Binary Neutron Star Merger LIGO/VIRGO GW170817. III. Optical and UV Spectra of a Blue Kilonova From Fast Polar Ejecta. *Astrophys. J.* **2017**, *848*, L18.
64. Cowperthwaite, P.S.; Berger, E.; Villar, V.A.; Metzger, B.D.; Nicholl, M.; Chornock, R.; Blanchard, P.K.; Fong, W.; Margutti, R.; Soares-Santos, M. The Electromagnetic Counterpart of the Binary Neutron Star Merger LIGO/Virgo GW170817. II. UV, Optical, and Near-infrared Light Curves and Comparison to Kilonova Models. *Astrophys. J.* **2017**, *848*, L17.

65. Roberts, L.F.; Kasen, D.; Lee, W.H.; Ramirez-Ruiz, E. Electromagnetic Transients Powered by Nuclear Decay in the Tidal Tails of Coalescing Compact Binaries. *Astrophys. J.* **2011**, *736*, L21.
66. Metzger, B.D.; Fernández, R. Red or blue? A potential kilonova imprint of the delay until black hole formation following a neutron star merger. *Mon. Not. R. Astron. Soc.* **2014**, *441*, 3444–3453.
67. Barnes, J.; Kasen, D. Effect of a High Opacity on the Light Curves of Radioactively Powered Transients from Compact Object Mergers. *Astrophys. J.* **2013**, *775*, 18.
68. Kasen, D.; Badnell, N.R.; Barnes, J. Opacities and Spectra of the *r*-process Ejecta from Neutron Star Mergers. *Astrophys. J.* **2013**, *774*, 25.
69. Tanaka, M.; Hotokezaka, K. Radiative Transfer Simulations of Neutron Star Merger Ejecta. *Astrophys. J.* **2013**, *775*, 113.
70. Tanaka, M.; Kato, D.; Gaigalas, G.; Rynkun, P.; Radžiūtė, L.; Wanajo, S.; Sekiguchi, Y.; Nakamura, N.; Tanuma, H.; Murakami, I.; et al. Properties of Kilonovae from Dynamical and Post-Merger Ejecta of Neutron Star Mergers. *Astrophys. J.* **2018**, *852*, 109.
71. Waxman, E.; Ofek, E.; Kushnir, D.; Gal-Yam, A. Constraints on the ejecta of the GW170817 neutron-star merger from its electromagnetic emission. *arXiv* **2017**, arXiv:1711.09638.
72. Wollaeger, R.T.; Korobkin, O.; Fontes, C.J.; Rosswog, S.K.; Even, W.P.; Fryer, C.L.; Sollerman, J.; Hungerford, A.L.; van Rossum, D.R.; Wollaber, A.B. Impact of ejecta morphology and composition on the electromagnetic signatures of neutron star mergers. *arXiv* **2017**, arXiv:1705.07084.
73. Villar, V.A.; Berger, E.; Metzger, B.D.; Guillochon, J. Theoretical Models of Optical Transients. I. A Broad Exploration of the Duration-Luminosity Phase Space. *Astrophys. J.* **2017**, *849*, 70.
74. Chornock, R.; Berger, E.; Kasen, D.; Cowperthwaite, P.S.; Nicholl, M.; Villar, V.A.; Alexander, K.D.; Blanchard, P.K.; Eftekhari, T.; Fong, W.; et al. The Electromagnetic Counterpart of the Binary Neutron Star Merger LIGO/VIRGO GW170817. IV. Detection of Near-infrared Signatures of r-process Nucleosynthesis with Gemini-South. *Astrophys. J.* **2017**, *848*, L19.
75. Perego, A.; Rosswog, S.; Cabezón, R.M.; Korobkin, O.; Käppeli, R.; Arcones, A.; Liebendörfer, M. Neutrino-driven winds from neutron star merger remnants. *Mon. Not. R. Astron. Soc.* **2014**, *443*, 3134–3156.
76. Fernández, R.; Metzger, B.D. Delayed outflows from black hole accretion tori following neutron star binary coalescence. *Mon. Not. R. Astron. Soc.* **2013**, *435*, 502–517.
77. Kasen, D.; Fernandez, R.; Metzger, B. Kilonova light curves from the disc wind outflows of compact object mergers. *Mon. Not. R. Astron. Soc.* **2015**, *450*, 1777–1786.
78. Fujibayashi, S.; Kiuchi, K.; Nishimura, N.; Sekiguchi, Y.; Shibata, M. Mass Ejection from the Remnant of Binary Neutron Star Merger: Viscous-Radiation Hydrodynamics Study. *arXiv* **2017**, arXiv:1711.02093.
79. Drago, A.; Lavagno, A.; Pagliara, G.; Pigato, D. The scenario of two families of compact stars. *Eur. Phys. J. A* **2016**, *52*, 40.
80. Drago, A.; Pagliara, G. The scenario of two families of compact stars. *Eur. Phys. J. A* **2016**, *52*, 41.
81. Wiktorowicz, G.; Drago, A.; Pagliara, G.; Popov, S.B. Strange quark stars in binaries: formation rates, mergers and explosive phenomena. *Astrophys. J.* **2017**, *846*, 163.
82. Drago, A.; Pagliara, G. Merger of two neutron stars: predictions from the two-families scenario. *Astrophys. J.* **2018**, *852*, L32.
83. Belczynski, K.; Askar, A.; Arca-Sedda, M.; Chruslinska, M.; Donnari, M.; Giersz, M.; Benacquista, M.; Spurzem, R.; Jin, D.; Wiktorowicz, G.; et al. The origin of the first neutron star—Neutron star merger. *arXiv* **2017**, arXiv:1712.00632.
84. Ivanova, N.; Justham, S.; Chen, X.; De Marco, O.; Fryer, C.L.; Gaburov, E.; Ge, H.; Glebbeek, E.; Han, Z.; Li, X.D.; et al. Common envelope evolution: Where we stand and how we can move forward. *Astron. Astrophys. Rev.* **2013**, *21*, 59.
85. Chruslinska, M.; Belczynski, K.; Klencki, J.; Benacquista, M. Double neutron stars: merger rates revisited. *Mon. Not. R. Astron. Soc.* **2018**, *474*, 2937–2958.
86. Fryer, C.L.; Belczynski, K.; Wiktorowicz, G.; Dominik, M.; Kalogera, V.; Holz, D.E. Compact Remnant Mass Function: Dependence on the Explosion Mechanism and Metallicity. *Astrophys. J.* **2012**, *749*, 91.
87. Annala, E.; Gorda, T.; Kurkela, A.; Vuorinen, A. Gravitational-wave constraints on the neutron-star-matter Equation of State. *arXiv* **2017**, arXiv:1711.02644.
88. Drago, A.; Lavagno, A.; Pagliara, G. Can very compact and very massive neutron stars both exist? *Phys. Rev. D* **2014**, *89*, 043014.

89. Hinderer, T.; Lackey, B.D.; Lang, R.N.; Read, J.S. Tidal deformability of neutron stars with realistic equations of state and their gravitational wave signatures in binary inspiral. *Phys. Rev. D.* **2010**, *81*, 123016.
90. Ruiz, M.; Lang, R.N.; Paschalidis, V.; Shapiro, S.L. Binary Neutron Star Mergers: A jet Engine for Short Gamma-ray Bursts. *Astrophys. J.* **2016**, *824*, L6.
91. Radice, D.; Perego, A.; Zappa, F. GW170817: Joint Constraint on the Neutron Star Equation of State from Multimessenger Observations. *Astrophys. J. Lett.* **2018**, *852*, L29.
92. Paschalidis, V.; Yagi, K.; Alvarez-Castillo, D.; Blaschke, D.B.; Sedrakian, A. Implications from GW170817 and I-Love-Q relations for relativistic hybrid stars. *arXiv* **2017**, arXiv:1712.00451.

© 2018 by the authors. Licensee MDPI, Basel, Switzerland. This article is an open access article distributed under the terms and conditions of the Creative Commons Attribution (CC BY) license (http://creativecommons.org/licenses/by/4.0/).

Article

Prospects of Constraining the Dense Matter Equation of State from Timing Analysis of Pulsars in Double Neutron Star Binaries: The Cases of PSR J0737−3039A and PSR J1757−1854

Manjari Bagchi

The Institute of Mathematical Sciences (IMSc-HBNI), 4th Cross Road, CIT Campus, Chennai 600 113, India; manjari.bagchi@gmail.com

Received: 4 December 2017; Accepted: 8 February 2018; Published: 12 February 2018

Abstract: The Lense-Thirring effect from spinning neutron stars in double neutron star binaries contributes to the periastron advance of the orbit. This extra term involves the moment of inertia of the neutron stars. The moment of inertia, on the other hand, depends on the mass and spin of the neutron star, as well as the equation of state of the matter. If at least one member of the double neutron star binary (better the faster one) is a radio pulsar, then accurate timing analysis might lead to the estimation of the contribution of the Lense-Thirring effect to the periastron advance, which will lead to the measurement of the moment of inertia of the pulsar. The combination of the knowledge on the values of the moment of inertia, the mass and the spin of the pulsar will give a new constraint on the equation of state. Pulsars in double neutron star binaries are the best for this purpose as short orbits and moderately high eccentricities make the Lense-Thirring effect substantial, whereas tidal effects are negligible (unlike pulsars with main sequence or white-dwarf binaries). The most promising pulsars are PSR J0737−3039A and PSR J1757−1854. The spin-precession of pulsars due to the misalignment between the spin and the orbital angular momentum vectors affect the contribution of the Lense-Thirring effect to the periastron advance. This effect has been explored for both PSR J0737−3039A and PSR J1757−1854, and as the misalignment angles for both of these pulsars are small, the variation in the Lense-Thirring term is not much. However, to extract the Lense-Thirring effect from the observed rate of the periastron advance, more accurate timing solutions including precise proper motion and distance measurements are essential.

Keywords: dense matter; equation of state; stars: neutron; pulsars: general, pulsars: PSR J0737−3039A; pulsars: PSR J1757−1854

---

1. Introduction

Timing analysis of binary pulsars leads to the measurement of pulsar's spin, Keplerian orbital parameters (the orbital period P_b, the orbital eccentricity e, the longitude of periastron ω, the projected semi-major axis of the orbit $x_p = a_p \sin i$, where a_p is the semi-major axis of the pulsar orbit and i is the angle between the orbit and the sky-plane, and the epoch of the periastron passage[1]), as well as post-Keplerian (PK) parameters like the Einstein parameter (γ), Shapiro range (r) and shape (s) parameters, the rate of the periastron advance $\dot\omega$, the rate of change of the orbital period $\dot P_b$, the relativistic deformation of the orbit δ_θ, etc. [4]. Sometimes, the Shapiro delay is parametrized

---

[1] The remaining Keplerian parameter, the longitude of the ascending node φ, does not come in the standard pulsar timing algorithm. It can be measured via proper motion only in very special cases; see [1–3] for details.

differently, with parameters h_3 and ς [5], or with r and z_s [6], and these parameters can easily be expressed in terms of conventional parameters s and r.

Measurement of PK parameters leads to estimation of masses of the pulsars and their companions, as well as tests of various theories of gravity [7,8]. Note that, in principle, measurements of only two PK parameters are enough to extract two unknowns, i.e., the masses of the pulsar and the companion, while measurements of more than two PK parameters lead to tests of gravity theory through consistency. However, the uncertainty in measurement is not equal for every PK parameter, and usually the two most accurate parameters are used to obtain the best mass estimates (e.g., Figure 1 of [9]). Additionally, relativistic binary pulsars have the potential to constrain the Equation of State (EoS) of matter at extreme densities. Measurements of masses of two pulsars being around 2 $M_\odot$ have already ruled out many soft EsoS [10,11]. On the other hand, the recent analysis of the gravitational wave event GW170817 from the merger of two neutron stars ruled out some extremely stiff EsoS [12,13]. Still, a large number of EsoS is allowed, including several hybrid [14] and strange quark matter [15] EsoS in addition to standard hadronic ones. Therefore, further progress on this issue is essential, and that is expected in the near future as, on the one hand more, neutron star-neutron star mergers are expected to be detected in future runs of the advanced LIGO, and on the other hand, many binary pulsars are being timed regularly and accurately. Besides, upcoming radio telescopes like MeerKAT and the Square Kilometre Array (SKA) will lead to significant improvement in pulsar timing.

Among all of the PK parameters, in the present article, I concentrate mainly on the periastron advance, which has the potential to constrain the dense matter equation of state due to the Lense-Thirring effect. Neutron star-neutron star binaries or 'Double Neutron Star' (DNS) systems are the best for this purpose as these systems are compact enough to display effects of strong field gravity, yet wide enough to have negligible tidal effects. Moreover, due to the high compactness of neutron stars, spin-induced quadrupole moments are also negligible. On the other hand, as these neutron stars are rapidly spinning (spin periods of most of the pulsars in DNSs are less than 100 milliseconds), the Lense-Thirring effect is significant. That is why I concentrate on DNSs in this article, although the mathematical formulations are valid for any kind of general relativistic binaries with the possibility of additional effects depending on the nature of the objects in the binaries. There are about seventeen DNSs known at the present time, including the first discovered binary pulsar: the Hulse–Taylor binary, PSR B1913+16. For one DNS, PSR J0737−3039A/B, both members are radio pulsars, and the system is known as a double pulsar [2].

For a long period of time, the double pulsar was the most relativistic binary. However, the recently-discovered DNS, PSR J1757−1854 shows even stronger effects of general relativity in some aspects, i.e., $\dot{P}$ and γ [17], due to its high eccentricity combined with a small orbital period. In fact, its eccentricity is 6.9-times and orbital period 1.8-times larger than those of PSR J0737−3039A/B. It is noteworthy that $\dot{\omega}$ is larger for PSR J0737−3039A/B than PSR J1757−1854 where the contribution of eccentricity is less dominant; see Lorimer and Kramer [4] for expressions for PK parameters.

PSR J1906+0746 is another DNS having the second smallest value of P_b, just after PSR J0737−3039A/B. However, its P_b is slightly (1.1-times) and e is significantly (7.1-times) smaller than those of PSR J1757−1854. That is why although PSR J1906+0746 has the third largest value of $\dot{\omega}$, other PK parameters are small, even smaller than most of the other DNSs. Finally, the latest discovered DNS, PSR J1946+2052, has broken all the records by having the smallest value of P_b and the largest value of $\dot{\omega}$ [18].

---

[2] Although for the last few years, the slow pulsar of the system is not visible and is believed to be beaming away from the Earth due to its spin-precession [16].

2. Precession in Double Neutron Star Binaries

Precessions (of both the spin and the orbit) of neutron stars in DNSs are very important. Following Barker and O'Connell [19], the rate of the change of the unit spin vector (s_a) of a spinning neutron star (*a*) in a binary can be written as:

$$\dot{s}_a = \vec{\Omega}_{sa} \times s_a \,, \tag{1}$$

where the angular spin-precession frequency $\vec{\Omega}_{sa}$ can be written as:

$$\vec{\Omega}_{sa} = \vec{\Omega}_{sPN_a} + \vec{\Omega}_{sLT_a} \,. \tag{2}$$

$\vec{\Omega}_{sPN_a}$ is the angular spin-precession frequency of the neutron star due to the space-time curvature around its companion and can be calculated within the 'Post-Newtonian' (PN) formalism. $\vec{\Omega}_{sLT_a}$ is the angular spin-precession frequency of that neutron star due to the Lense-Thirring effect of its spinning companion. We can further write:

$$\vec{\Omega}_{sPN_a} = A_{PN_a} \mathbf{k} \,, \tag{3}$$

where $\mathbf{k} = \vec{L}/|\vec{L}|$ is the unit vector along the orbital angular momentum, $\vec{L}$ is the orbital angular momentum, and:

$$\vec{\Omega}_{sLT_a} = A_{LT_a} [\mathbf{s}_{a+1} - 3(\mathbf{k}.\mathbf{s}_{a+1})\mathbf{k}] \,. \tag{4}$$

Here, if *a* is the neutron star under consideration, $a + 1$ is its companion. The amplitudes in Equations (3) and (4) are given by:

$$A_{PN_a} = \left(\frac{G}{c^3}\right)^{2/3} \frac{n^{5/3}}{(1-e^2)} \frac{M_{a+1}(4M_a + 3M_{a+1})}{2(M_a + M_{a+1})^{4/3}} \,, \tag{5}$$

$$A_{LT_a} = \frac{G}{c^3} \beta_{s\,a+1} \frac{n^2}{(1-e^2)^{3/2}} \frac{M_{a+1}^2}{2(M_a + M_{a+1})} \,, \tag{6}$$

where *G* is the gravitational constant, *c* is the speed of light in a vacuum and $n = 2\pi/P_b$ is the angular orbital frequency. M_a is the mass; I_a is the moment of inertia; P_{sa} is the spin period of the *a*-th neutron star; and

$$\beta_{s\,a+1} = \frac{cI_{a+1}}{GM_{a+1}^2} \cdot \frac{2\pi}{P_{s\,a+1}}, \quad \beta_{sa} = \frac{cI_a}{GM_a^2} \cdot \frac{2\pi}{P_{sa}}. \tag{7}$$

For pulsars in DNSs, the Lense-Thirring term ($\vec{\Omega}_{sLT_a}$) can be ignored. Even for the case of the slow pulsar (B) of the double pulsar, where the Lense-Thirring effect due to the spin of the fast pulsar (A) contributes to $\vec{\Omega}_{sB}$, I find $A_{PN_B} = 5.07481$ deg y^{-1} and $A_{LT_B} = 3.48865 \times 10^{-5}$ deg y^{-1} (using the values of the parameters given in Table 1). This leads to $\Omega_{sLT_B} \sim -6.97731 \times 10^{-5}$ deg y^{-1} using the fact that s_A is almost parallel to $\mathbf{k}$ [20,21], and hence, $\Omega_{sB} = 5.07474$ deg y^{-1}, which is close to the observed median value of $\Omega_{sB} = 4.77$ deg y^{-1} [22]. It is unlikely that companions of other pulsars would be much faster than PSR A, and hence, $\vec{\Omega}_{sLT_a}$ can be neglected. Therefore, Equation (1) becomes:

$$\dot{s}_a = A_{PN_a} \mathbf{k} \times \mathbf{s}_a = A_{PN_a} \sin \chi_a \mathbf{u} \,, \tag{8}$$

where χ_a is the angle between $\mathbf{k}$ and $\mathbf{s}_a$ and $\mathbf{u}$ is a unit vector perpendicular to the plane containing $\mathbf{k}$ and $\mathbf{s}_a$ in the direction given by the right-hand rule. Therefore, the measurement of the spin-precession of a pulsar helps estimate χ_a when other parameters involved in Equation (5) are already known.

Similarly, neglecting the tidal and the spin-quadrupole effects, as well as the spin-spin interaction, the orbital angular precession frequency can be written as:

$$\vec{\Omega}_b = \vec{\Omega}_{bPN} + \vec{\Omega}_{bLT} \,, \tag{9}$$

where $\vec{\Omega}_{bPN}$ and $\vec{\Omega}_{bLT}$ are contributions from the space-time curvature and the Lense-Thirring effect, respectively. Note that both members of the binary contribute to each term. $\vec{\Omega}_b$ leads to the precession of both the Laplace–Runge–Lenz vector $\vec{A}$, as well as $\vec{L}$, but none is directly observable. Damour and Schafer [23] first studied the manifestation of $\vec{\Omega}_b$ in terms of observable parameters by decomposing $\vec{\Omega}_b$ as:

$$\vec{\Omega}_b = \frac{d\varphi_a}{dt}\mathbf{h}_a + \frac{d\omega_a}{dt}\mathbf{k} + \frac{di}{dt}\mathbf{Y}_a, \tag{10}$$

where $\mathbf{h}_a$ is the unit vector along the line-of-sight, i.e., from the Earth to the a-th neutron star (the pulsar), $\mathbf{Y}_a = \frac{\mathbf{h}_a \times \mathbf{k}}{|\mathbf{h}_a \times \mathbf{k}|}$ is the unit vector along the line of the ascending node, φ_a is the longitude of the ascending node of the a-th neutron star and i is the angle between the orbit and the sky-plane. Damour and Schafer [23] has also given:

$$\dot{\varphi}_a = \frac{1}{\sin^2 i}[\vec{\Omega}_b \cdot \mathbf{h}_a - \cos i (\vec{\Omega}_b \cdot \mathbf{k})], \tag{11}$$

$$\dot{\omega}_a = \frac{1}{\sin^2 i}[\vec{\Omega}_b \cdot \mathbf{k} - \cos i (\vec{\Omega}_b \cdot \mathbf{h}_a)], \tag{12}$$

$$\frac{di}{dt} = \vec{\Omega}_b \cdot \mathbf{Y}_a. \tag{13}$$

As already mentioned, $\dot{\omega}_a$ is the parameter of interest as it has the potential to put constraints on the dense matter EoS. Therefore, I explore properties of $\dot{\omega}_a$ in the next and subsequent sections.

2.1. Periastron Advance

The observed rate of the periastron advance ($\dot{\omega}_{a,\text{obs}}$) of the a-th member of a relativistic binary can be written as:

$$\dot{\omega}_{a,\text{obs}} = \dot{\omega}_a + \dot{\omega}_{\text{Kop}_a} = \dot{\omega}_{\text{PN}_a} + \dot{\omega}_{\text{LT}_a} + \dot{\omega}_{\text{LT}_{a+1}} + \dot{\omega}_{\text{Kop}_a}, \tag{14}$$

where $\dot{\omega}_{\text{PN}_a}$ is due to the space-time curvature caused by both members of the binary, $\dot{\omega}_{\text{LT}_a}$ is due to the Lense-Thirring effect of the star a and $\dot{\omega}_{\text{LT}_{a+1}}$ is due to the Lense-Thirring effect of the star $a+1$. These three terms together come into the expressions of $\dot{\omega}_a$ (Equation (12)) caused by the precession of the orbit. $\dot{\omega}_{\text{Kop}_a}$ is a secular variation due to the gradual change of the apparent orientation of the orbit with respect to the line-of-sight due to the proper motion of the barycenter of the binary [24]. The expressions for different terms are:

$$\dot{\omega}_{\text{PN}_a} = \frac{3\beta_0^2 n}{1-e^2}[1+\beta_0^2 f_{0a}], \tag{15}$$

$$\dot{\omega}_{\text{LT}_a} + \dot{\omega}_{\text{LT}_{a+1}} = -\frac{3\beta_0^3 n}{1-e^2}(g_{sa}\beta_{sa} + g_{s\,a+1}\beta_{s\,a+1}), \tag{16}$$

$$\dot{\omega}_{\text{Kop}_a} = K \csc i\,(\mu_\alpha \cos\varphi_a + \mu_\delta \sin\varphi_a)\,,\ K = 502.65661\,, \tag{17}$$

where only the first and the second order terms are retained in the expression of $\dot{\omega}_{\text{PN}_a}$. μ_α and μ_δ are the proper motion of the barycenter of the binary (which is measured as the proper motion of the visible object) in the right ascension and the declination respectively, both expressed in units of milliarcseconds per year. All other parameters are in SI units. The parameters introduced in Equations (15)–(17) are as follow:

$$\beta_0 = \frac{(GMn)^{1/3}}{c}, \tag{18}$$

$$f_{0a} = \frac{1}{1-e^2}\left(\frac{39}{4}X_a^2 + \frac{27}{4}X_{a+1}^2 + 15X_a X_{a+1}\right) - \left(\frac{13}{4}X_a^2 + \frac{1}{4}X_{a+1}^2 + \frac{13}{3}X_a X_{a+1}\right), \tag{19}$$

where $X_a = M_a/M$, $M = M_a + M_{a+1}$ is the total mass of the system. β_{sa} is defined in Equation (7), and

$$\begin{aligned}
g_{sa} &= \frac{X_a(4X_a+3X_{a+1})}{6(1-e^2)^{1/2}\sin^2 i} \times \left[(3\sin^2 i - 1)\,\mathbf{k}\cdot\mathbf{s_a} + \cos i\,\mathbf{h}_a\cdot\mathbf{s_a}\right] \\
&= \frac{X_a(4X_a+3X_{a+1})}{6(1-e^2)^{1/2}\sin^2 i} \times \left[(3\sin^2 i - 1)\cos\chi_a + \cos i\,\cos\lambda_a\right]
\end{aligned} \quad (20)$$

where λ_a is the angle between $\mathbf{h}_a$ and $\mathbf{s_a}$ (see Figure 1). It is obvious from Equation (20) that the maximum value of g_{sa} occurs when $\mathbf{s}_a$ is parallel to the vector $(3\sin^2 i - 1)\,\mathbf{k} + \cos i\,\mathbf{h}_a$, giving:

$$g_{sa,\,max} = \left[3 + \frac{1}{\sin^2 i}\right]^{1/2} \frac{X_a(4X_a + 3X_{a+1})}{6(1-e^2)^{1/2}}. \quad (21)$$

If $\mathbf{s}_a \parallel \mathbf{k}$, i.e., $\chi_a = 0$ and $\lambda_a = i$:

$$g_{sa,\,\parallel} = \frac{X_a(4X_a + 3X_{a+1})}{3(1-e^2)^{1/2}}. \quad (22)$$

Using the expression of $g_{sa,\,\parallel}$ in Equation (16), I get $\dot{\omega}_{LT_a,\,\parallel}$, and using the expression of $g_{sa,\,max}$, I get $\dot{\omega}_{LT_a,\,max}$. The negative sign in the Lense-Thirring term (Equation (16)) implies the fact that actually $\dot{\omega}_{LT_a,\,max}$ is the minimum value. However, depending on the values of i, χ_p, and λ_p, g_{sa} (Equation (20)) can be negative making $\dot{\omega}_{LT_a}$ (Equation (16)) positive. Note that the Lense-Thirring effects from both the pulsar and the companion come in the total periastron advance rate (Equations (14) and (16)). However, if the companion is much slower than the pulsar, as in the case of the double pulsar, then the contribution from the companion can be neglected. Due to the non-detection of any pulsation, we do not know the values of the spin periods of the companions for other DNSs and cannot rule out the possibility of significant Lense-Thirring effect from those companions. However, most of the pulsars in DNSs are recycled (at least mildly), suggesting the fact that pulsars were born as neutron stars earlier than their companions. The companions, i.e., the second born neutron stars in DNSs, are expected to be slow; because even if neutron stars are born with spin periods in the range of a few tens of milliseconds up to a few hundred milliseconds, they quickly spin down to periods of a few seconds unless they gain angular momentum via mass accretion from their Roche lobe filling giant companions and become fast rotators. Such spin-up is possible only for the first born neutron star in a DNS. That is why in the present article, I ignore the Lense-Thirring effect from the companions of pulsars in DNSs.

In such a case, i.e., when $\dot{\omega}_{LT_{a+1}}$ is negligible, $\dot{\omega}_{LT_a}$ can be extracted by subtracting $\dot{\omega}_{PN_a} + \dot{\omega}_{Kop_a}$ from $\dot{\omega}_{a,obs}$ (Equation (14)). One can estimate the value of I_a from $\dot{\omega}_{LT_a}$ if all other relevant parameters like P_b, e, M_a, M_{a+1}, $\sin i$, χ_a, λ_a are known (Equations (7), (18)–(20)). Then, from the knowledge of I_a, M_a and P_{sa}, a new constraint on the EoS can be placed, as it is a well-known fact that the moment of inertia of a neutron star depends on its mass, spin-period and the EoS (see [25] and the references therein). For this purpose, DNSs with large values of $\dot{\omega}_{LT_a}$ (larger than the measurement uncertainty in $\dot{\omega}_{a,obs}$) are preferable. Large values of the orbital eccentricity e and small values of P_b and P_s make $\dot{\omega}_{LT_a}$ larger.

The procedure is actually a bit trickier than it seems at first. Usually, timing analysis of a binary pulsar reports the most accurate measurement of $\dot{\omega}_{a,obs}$ out of all PK parameters, and the masses of the pulsar and the companion are estimated by equating $\dot{\omega}_{a,obs}$ with $\dot{\omega}_{PN_a}$ and using the second most accurately-measured PK parameter. If $\dot{\omega}_{LT_a}$ is large, then this procedure would become erroneous. One should instead estimate masses using two PK parameters other than $\dot{\omega}_{a,obs}$, and these mass values should also be accurate enough to give a precise estimate of $\dot{\omega}_{PN_a}$. Accurate measurements of proper motion are also needed to evaluate $\dot{\omega}_{Kop_a}$. Only then, $\dot{\omega}_{PN_a} + \dot{\omega}_{Kop_a}$ can be subtracted from the observed $\dot{\omega}_{a,obs}$ to get the value of $\dot{\omega}_{LT_a}$. In short, one needs very precise measurements of at least three PK parameters, $\dot{\omega}_{a,obs}$ and two more, as well as the distance and the proper motion of the pulsar (the a-th neutron star).

Because of its small values of P_b and P_s, PSR J0737−3039A is expected to have a large value of $\dot{\omega}_{LT_a}$ in spite of its low eccentricity. Similarly, as PSR J1757−1854 has a slightly larger P_b, slightly smaller P_s and much larger e, it is also expected to have a large value of $\dot{\omega}_{LT_a}$. For this reason, in the next section, I explore $\dot{\omega}_{LT_a}$ for these two DNSs in detail. Note that, although, PSR J1946+2052 might be a useful system for this purpose, it is not possible to extend my study for this system due to lack of knowledge of masses of the pulsar and the companion.

One should also remember the dependence of $\dot{\omega}_{LT_a}$ (via g_{sa}) on the orientation of $\mathbf{s}_a$ with respect to $\mathbf{k}$ and $\mathbf{h}_a$ (Equation (20)). If I assume that the a-th neutron star is the pulsar, then all 'a's in the subscripts of the above equations will be replaced by 'p's, e.g., $\mathbf{s_p}$, $\mathbf{h_p}$, etc. Most of the time, the value of $g_{sp,\parallel}$ is estimated in the literature. However, at least for some pulsars, $\chi_p \neq 0$. In such cases, the values of χ_p and λ_p can be measured by analyzing the change of the pulse profile shape and the polarization. Conventionally, χ_p is denoted by δ, and λ_p is denoted by ζ. Note that, even if χ_p is a constant over time, λ_p would vary due to the spin precession and can be written as [26]:

$$\cos \lambda_p = -\cos \chi_p \cos i + \sin \chi_p \sin i \cos[\Omega_{sp}(t - T_{p0})]. \qquad (23)$$

where Ω_{sp} is the angular spin precession frequency (amplitude of $\vec{\Omega}_{sp}$ in Equation (2)), t is the epoch of the observation and T_{p0} is the epoch of the precession phase being zero. Note that i itself might change with time as given in Equation (13).

In Table 1, with other relevant parameters, I compile values of χ_p wherever available. I do not compile values of λ_p, although available in some cases, as it is a time-dependent parameter. From Figure 1, it is clear that the spin would precess in such a way that it would lie on the surface of a fiducial cone with the vertex at the pulsar and having the half-opening angle as χ_p. The constraint on λ_p is: (i) if $\chi_p < i$, then $\lambda_{p,max} = i + \chi_p$ and $\lambda_{p,min} = i - \chi_p$ (Figure 1a), (ii) if $\chi_p > i$, then $\lambda_{p,max} = \chi_p + i$ and $\lambda_{p,min} = \chi_p - i$ (Figure 1b).

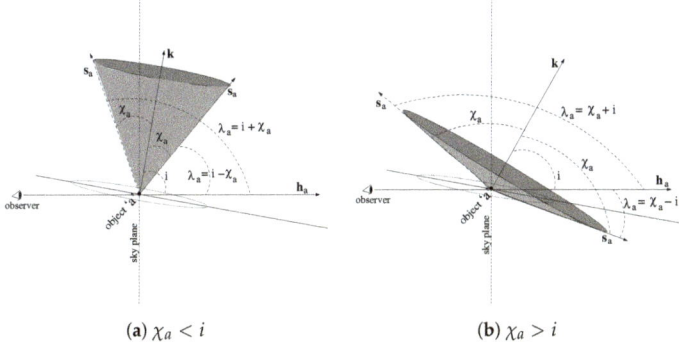

(a) $\chi_a < i$ (b) $\chi_a > i$

Figure 1. Orientation of different vectors relevant for the estimation of $\dot{\omega}_{LTa}$ where the subscript a refers to the object being observed, i.e., the pulsar (so the companion would be represented by the subscript $a + 1$). The vectors are as follow: $\mathbf{k}$ is the unit vector along the orbital angular momentum, $\mathbf{h}_a$ is the unit vector along the line-of-sight and $\mathbf{s}_a$ is the unit spin vector. The angles in the figures are: i is the inclination angle between the orbit of the pulsar and the sky-plane, as well as the angle between $\mathbf{k}$ and $\mathbf{h}_a$, χ_a is the angle between $\mathbf{k}$ and $\mathbf{s}_a$ and λ_a is the angle between $\mathbf{h}_a$ and $\mathbf{s}_a$. The left and right panels are for $\chi_a < i$ and $\chi_a > i$, respectively. In both panels, two positions of $\mathbf{s}_a$ are shown corresponding to maximum and minimum values of λ_a (see the text). The pulsar, i.e., the object a, is located at the vertex of the fiducial cone (gray in color) made by the precessing $\mathbf{s}_a$ around $\mathbf{k}$.

3. Results

For the purpose of demonstration, I compute the value of the moment of inertia (I_p) for the two most interesting pulsars, PSR J0737−3039A and PSR J1757−1854 using the Akmal-Pandharipande-Ravenhall (APR) equation of state [27] and the RNS code[3][28]. I find that, both of these pulsars have $I_p = 1.26 \times 10^{45}$ gm cm^2. I use this value in my calculations, remembering the fact that the true values of I_p would be different depending on the true EoS, and we are actually seeking an answer whether it would be possible to know the value of I_p by singling out $\dot{\omega}_{LT_p}$ from the total observed $\dot{\omega}_{p,obs}$.

I tabulate values of $\dot{\omega}_{PN_p}$, $\dot{\omega}_{LT_p,\|}$ and $\dot{\omega}_{LT_p, max}$ for all pulsars in DNSs in Table 2. I exclude candidate DNSs, i.e., PSR B1820−11 and PSR J1753−2240 from this calculation due to poor mass constraints. For other cases, I use the limiting mass values where actual values are unavailable. Only $\dot{\omega}_{LT_p,\|}$ can be calculated for the systems with undetermined/unpublished values of $\sin i$. As expected, PSR J0737−3039A shows the largest Lense-Thirring effect, followed by PSR J1757−1854. In fact, $|\dot{\omega}_{LT_p, max}|/|\dot{\omega}_{PN_p}|$ is slightly larger for PSR J1757−1854 (2.92×10^{-5}) than that for PSR J0737−3039A (2.81×10^{-5}). This fact makes it a very interesting system, and I investigate this system in detail. The discovery paper [17] also mentioned the large Lense-Thirring effect causing significant amount of di/dt (Equation (13)).

I find that, for PSR J1757−1854, $|\dot{\omega}_{LT_p, max}| = 3.03170 \times 10^{-4}$ deg y^{-1} is achieved for $\chi_p = 3°$, $\lambda_p = 99°$. However, as the value of χ_p for this system is not yet known (although Cameron et al. [17] obtained a constraint $\chi_p \sim 25°$ based on their simulation of the kick velocity), I vary χ_p in the range of $0°$–$60°$. It is obvious that $\dot{\omega}_{LT_p}$ depends more on χ_p than on λ_p as the absolute value of the factor (see Equation (20)) with $\cos \chi_p$, i.e., $3 \sin^2 i - 1$, is larger than that of the factor with $\cos \lambda_p$, i.e., $\cos i$. Furthermore, the measured value of $\sin i$ gives two values of i (as $\sin \theta = \sin(\pi - \theta)$), and I use the larger one to get a bigger range of λ_p. The minimum value of $|\dot{\omega}_{LT_p}| = 1.37597 \times 10^{-4}$ deg y^{-1} is achieved for $\chi_p = 60°$, $\lambda_p = 36°$. If I fix χ_p strictly as $25°$, then the maximum and minimum values of $|\dot{\omega}_{LT_p}|$ are 2.81108×10^{-4} deg y^{-1} and 2.67663×10^{-4} deg y^{-1}, for $\lambda_p = 121°$ and $\lambda_p = 71°$, respectively. In Figure 2, I show the variation of $\dot{\omega}_{LT_p}$ for χ_p in the range of $0°$–$60°$, and for each value of χ_p, λ_p varies in the range of $i - \chi_p$, $i + \chi_p$. On the other hand, as PSR J0737−3039A has $\chi_p < 6°$ [20], $\dot{\omega}_{LT_p}$ always remains close to $\dot{\omega}_{LT_p,\|}$. The maximum and minimum values of $|\dot{\omega}_{LT_p}|$ are 4.74488×10^{-4} deg y^{-1} at $\chi_p = 0.65$, $\lambda_p = 91.95°$; and 4.712931×10^{-4} deg y^{-1} at $\chi_p = 6.0°$, $\lambda_p = 85.31°$. To check the validity of my assumption that the Lense-Thirring effect from the companion could be ignored, I find that $\dot{\omega}_{LT_c, \|} = -3.792899 \times 10^{-6}$ deg y^{-1} using $I_c = 1.16 \times 10^{45}$ gm cm^2, where the subscript 'c' stands for the companion, i.e., the slow PSR J0737−3039B. Varying χ_c in the range of $130°$–$150°$ [20], I find that the minimum value of $|\dot{\omega}_{LT_c}|$ is 2.60052×10^{-6} deg y^{-1}. These results support my argument that the Lense-Thirring effect from the companion of a pulsar in a DNS would have a negligible contribution to the rate of the periastron advance of the binary.

As Table 2 shows, $|\dot{\omega}_{LT_p, max}|$ for PSR J0737−3039A is around 1.4-times smaller than the presently published accuracy, $\dot{\omega}_{p,obs} = 16.89947 \pm 0.00068$ deg y^{-1} where the uncertainty is twice the formal 1σ value obtained in the timing solution [9]. The uncertainties in other PK parameters are even poorer (Table 1 of [9]), so if we exclude $\dot{\omega}_{p,obs}$, the mass estimates would be less accurate and should not be used to calculate $\dot{\omega}_{PN_p}$. Therefore, to achieve the goal of estimating $\dot{\omega}_{LT_p}$, lowering only the uncertainty in $\dot{\omega}_{p,obs}$ would not be enough; one needs to improve the accuracy of at least two other parameters that can be used to get masses at least as precise as the ones already published. It is impossible to improve the accuracy of the Keplerian parameter R used by Kramer et al. [9] in combination with $\dot{\omega}_{p,obs}$ to report the values of the masses, because $R = M_p/M_c = x_c/x_p$ involves x_c, the projected semi-major axis of the companion (PSR B), which is not visible presently and even when it was visible, it was not a good timer like PSR A. Solving the expressions for PK parameters,

---

[3] RNS stands for 'Rapidly Rotating Neutron Star', a package to calculate different properties of rotating neutron stars, freely available at http://www.gravity.phys.uwm.edu/rns/.

I find that the combination of s and $\dot{P}_b$ is the best choice left when $\dot{\omega}_{p,\text{obs}}$ and R are excluded, i.e., gives the narrowest ranges of masses $M_p = 1.34 \pm 0.02\ M_\odot$, $M_c = 1.251 \pm 0.007\ M_\odot$ using the published values of the uncertainties. I also find the fact that, to get masses as accurate as the ones reported by Kramer et al. [9], the uncertainty in $\dot{P}_b$ should be reduced at least to 8.0×10^{-16} assuming that s and relevant Keplerian parameters (P_b, e, x_p) would also be improved by at least an order of magnitude. The published uncertainty in $\dot{P}_b$ is 1.7×10^{-14}. However, the question arises whether for such an improved measurement of $\dot{P}_b$, the dynamical contribution coming from the proper motion of the pulsar (Shklovskii effect) and the relative acceleration between the pulsar and the solar system barycenter could be ignored. Using the proper motion reported by Kramer et al. [9] and the distance (1.1 kpc) estimated by Deller et al. [29] using the Very-long-baseline interferometry (VLBI) parallax measurement, I find that $\dot{P}_{b,\text{dyn}} = 4.91 \times 10^{-17}$, where I have used a realistic model of the galactic potential [30]. The improvement required in the timing solution of PSR J0737−3039A is expected by 2030 [31,32] accompanied by an improvement in $\dot{\omega}_{p,\text{obs}}$. Fortunately, $\dot{\omega}_{\text{Kop}_p}$ is smaller than $|\dot{\omega}_{\text{LT}_p,\|}|$; using the proper motion reported in Kramer et al. [9], I find that it can be at most 1.18×10^{-6} deg y^{-1} (I could not calculate the actual value, as φ_p is unknown). However, one will still need better estimates of χ_p and λ_p to be able to extract the value of the moment of inertia of the pulsar from the $\dot{\omega}_{\text{LT}_p}$.

On the other hand, for PSR J1757−1854, both $|\dot{\omega}_{\text{LT}_p,\text{max}}|$ and $|\dot{\omega}_{\text{LT}_p,\|}|$ are about 1.5-times larger than the presently-published accuracy, $\dot{\omega}_{p,\text{obs}} = 10.3651 \pm 0.0002$ deg y^{-1}. However, even this system cannot be used at present to estimate $\dot{\omega}_{\text{LT}_p}$, not only because of the unknown values of χ_p and λ_p, but also because of the less accurate values of other PK parameters. If one uses γ and $\dot{P}_b$, then the uncertainties in these parameters should be improved by one and two orders of magnitudes respectively to get the uncertainty in masses of about $0.0008\ M_\odot$. This seems plausible, as the present accuracy in $\dot{P}_b$ is only 0.2×10^{-12} [17]. However, due to the lack of measurements of the proper motion and the parallax distance, I could not calculate $\dot{P}_{b,\text{dyn}}$ for this pulsar and only can give rough estimates based on the distance guessed from the dispersion measure using two different models of the galactic electron density, e.g., the NE2001 model [33,34] and the YMW16 model [35]. The contribution due to the relative acceleration is -2.11×10^{-15} for the NE2001 distance (7.4 kpc) and -1.65×10^{-14} for the YMW16 distance (19.6 kpc). Both of the values seem to be too large to be canceled out by the Shklovskii term. Measurements of the proper motion and the distance (using parallax) are essential to estimate $\dot{P}_{b,\text{dyn}}$. The measurement of the proper motion will also help estimate $\dot{\omega}_{\text{Kop}_p}$.

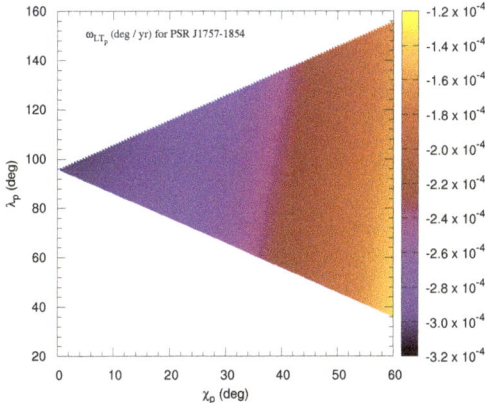

Figure 2. Variation of $\dot{\omega}_{\text{LT}_p}$ with χ_p (in degrees) along the horizontal axis and λ_p (in degrees) along the vertical axis for PSR J1757−1854. The color code represents values of $\dot{\omega}_{\text{LT}_p}$ in deg y^{-1}.

Table 1. Relevant observed parameters for known pulsars in DNSs. See the references for these parameters with more significant digits and uncertainties. The timing model (e.g., DDGR) is mentioned when more than one timing solution is available in the original reference.

DNS	Masses $M_\odot$	e	P_b Days	$\sin i$	P_{s1} ms	$\dot\omega$ deg y^{-1}	χ_p deg	Refs.
J0453+1559 (DDGR)	1.559 1.174	0.11251832	4.072468588	0.9671343	45.781816163093	0.0379412	—	R1
J0737−3039A J0737−3039B	1.3381 1.2489	0.0877775	0.10225156248	0.99974	22.70 2773.46	16.89947	<6.1 or <2.3 138	R2, Rχ_1
J1411+2551	≤1.62 ≥0.92	0.1699308	2.61585677939	—	62.452895517590	0.0768	—	R3
J1518+4904	≤1.17 ≥1.55	0.24948451	8.6340050964	≤0.73	40.934988908	0.0113725	—	R4
B1534+12 (DDGR)	1.3455 1.3330	0.27367740	0.420737298881	0.97496	37.9044411783	1.755795	27.0 ± 3.0 153.0 ± 3.0	R5
J1753−2240 †	— ≥0.4875	0.303582	13.6375668	—	95.1378086771	—	—	R6
J1756−2251	1.341 1.230	0.1805694	0.3196390143	0.93	28.4615890259983	2.58240	<5.9 (1σ value)	R7
J1757−1854	1.3384 1.3946	0.6058142	0.18353783587	0.9945	21.497231890027	10.3651	∼25 (theoretical)	R8
J1807−2500B † (g_1)	1.3655 1.2064	0.747033198	9.9566681588	0.9956	4.18617720284089	0.0183389	—	R9
J1811−1736	≤1.64 ≥0.93	0.828011	18.7791691	—	104.1819547968	0.0090	—	R10
B1820−11 †	—	0.794608	357.76199	—	279.828696565349	0.00007	—	R11

Table 1. Cont.

DNS	Masses $M_\odot$	e	P_b Days	$\sin i$	P_{s1} ms	$\dot\omega$ deg y^{-1}	χ_p deg	Refs.
J1829+2456	≤1.38 ≥1.22	0.13914	1.176028	–	41.00982358	0.28	–	R12
J1906+0746 (DDGR)	1.291 1.322	0.0852996	0.16599304686	0.690882411	144.07315538	7.5844	110^{+21}_{-55}	R13
B1913+16	1.4398 1.3886	0.6171334	0.322997448911	0.68 ‡	59.03000322	4.226598	22	R14, Rχ_1
J1930−1852	≤1.32 ≥1.30	0.39886340	45.0600007	–	185.52016047926	0.00078	–	R15
J1946+2052	–	0.063848	0.07848804	–	16.9601753230	25.6	–	R16
B2127+11C (g_2)	1.358 1.354	0.681395	0.33528204828	–	30.52929614864	4.4644	–	R17

† Candidate DNS; ‡ was never published, but here, I quote the value used by Weisberg & Huang [36]; pulsars denoted with g are in globular clusters; g_1 is in NGC 6544, and g_2 is in M15; References: R1: Martinez et al. [37], R2: Kramer et al. [9], R3: Martinez et al. [38], R4: Janssen et al. [39], R5: Fonseca, Stairs, Thorsett [40], R6: Keith et al. [41], R7: Ferdman et al. [42], R8: Cameron et al. [17], R9: Lynch et al. [43], R10 Corongiu et al. [44], R11: Hobbs et al. [45], R12: Champion et al. [46], R13: van Leeuwen et al. [47], R14: Weisberg, Nice, & Taylor [48], R15: Swiggum et al. [49], R16: Stovall et al. [18], R17: Jacoby et al. [50], Rχ_1 (for χ_p): Perera et al. [20].

Table 2. Values of $\dot{\omega}_{PN_p}$, $\dot{\omega}_{LT_p, \parallel}$ and $\dot{\omega}_{LT_p, max}$ for known pulsars in DNSs using $I_p = 1.26 \times 10^{45}$ gm cm^2 and neglecting the Lense-Thirring effect of the companion. To show the difference between $\dot{\omega}_{LT_p, \parallel}$ and $\dot{\omega}_{LT_p, max}$, I keep values up to five decimal places even for the cases where sin i is known with less accuracy.

DNS	$\dot{\omega}_{PN_p}$ (deg y^{-1})	$\dot{\omega}_{LT_p, \parallel}$ (deg y^{-1})	$\dot{\omega}_{LT_p, max}$ (deg y^{-1})
J0453+1559	0.03794	-1.30190×10^{-7}	-1.31310×10^{-7}
J0737−3039A	16.90312	-4.74458×10^{-4}	-4.74489×10^{-4}
J1411+2551	0.07681	-2.32510×10^{-7}	–
J1518+4904	0.01138	-4.48005×10^{-8}	–
B1534+12	1.75533	-1.84880×10^{-5}	-1.86078×10^{-5}
J1753−2240	–	–	–
J1756−2251	2.58363	-4.01930×10^{-5}	-4.09703×10^{-5}
J1757−1854	10.36772	-3.02752×10^{-4}	-3.03170×10^{-4}
J1807−2500B	0.01834	-8.98082×10^{-7}	-8.99076×10^{-7}
J1811−1736	0.00895	-1.45038×10^{-8}	–
B1820−11	–	–	–
J1829+2456	0.29284	-1.96704×10^{-6}	–
J1906+0746	7.58528	-2.91884×10^{-5}	-3.29423×10^{-5}
B1913+16	4.22760	-3.43984×10^{-5}	-3.90790×10^{-5}
J1930−1852	0.00079	-3.86975×10^{-10}	–
J1946+2052	–	–	–
B2127+11C	4.46458	-8.11241×10^{-5}	–

4. Summary and Conclusions

Measurement of the moment of inertia of a pulsar will provide yet another constraint on the dense matter equation of state. This issue has been discussed in the literature in the past (see [32,51,52] and the references therein), but unfortunately, the effect of the misalignment between the spin and the orbital angular momentum vectors has not been explored in detail mainly because these two vectors are almost parallel for PSR J0737−3039, the only suitable system existing until recently.

However, even a small misalignment angle would have significant consequences, because for a particular neutron star; the theoretical values of the moment of inertia using different EsoS are sometimes very close (depending on the stiffness of the EoS). One can see Figure 1 of Bejger et al. [53] for predicted values of the moment of inertia for PSR J0737−3039A using a sample of EsoS. The present article extensively studied the effects of such misalignment.

I have demonstrated the fact that the recently-discovered DNS, PSR J1757−1854, is almost as good as PSR J0737−3039A regarding the prospect of determining the moment of inertia of the pulsar from $\dot{\omega}_{LT_p}$. For PSR J0737−3039A, $\dot{\omega}_{LT_p}$ lies in the range of -4.757×10^{-4}−−4.713×10^{-4} deg y^{-1}, whereas for PSR J1757−1854, $\dot{\omega}_{LT_p}$ lies in the range of -1.376×10^{-4}−−3.032×10^{-4} deg y^{-1}. Future measurement of χ_p for PSR J1757−1854 will narrow down the range of $\dot{\omega}_{LT_p}$.

For both of the systems, significant improvements in the timing solution are needed. Moreover, the proper motion and the parallax measurements for PSR J1757−1854 are crucial. Additionally, it is essential to measure χ_p and λ_p for PSR J1757−1854; hence, long-term studies of polarization properties and profile variation would be very useful.

Note that, although all of the calculations presented in this article are within the framework devised by Barker & O'Connell [19] and Damour & Schäfer [23], alternative formalisms to incorporate the Lense-Thirring effect in $\dot{\omega}_{p,obs}$ exist. One example is Iorio [54], but for systems with $i \sim 90°$, $\chi_p \sim 0°$, these two formalisms take similar forms. Precisely, Iorio [54] reports $\dot{\omega}_{LT_p, \parallel} = -3.7 \times 10^{-4}$ deg y^{-1} for PSR J0737−3039A using $I_p = 1.00 \times 10^{45}$ gm cm^2, which is close to what I get, i.e., $\dot{\omega}_{LT_p, \parallel} = -3.766 \times 10^{-4}$ deg y^{-1} if I use the same value of I_p instead of 1.26×10^{45} gm cm^2 used in this article. A fundamentally different idea, i.e., to incorporate the Lense-Thirring effect into the delays in the pulse arrival times, has been proposed recently [55].

If this can be implemented in future algorithms for pulsar timing, the effect of the Lense-Thirring effect could be measured directly.

Finally, if the measurement accuracy of $\dot{\omega}_{p,\text{obs}}$ improves by a few orders of magnitudes, one will need to subtract the Kopeikin term (the secular variation due to the proper motion). The measurement of the proper motion of PSR J1757−1854 will help us calculate this term. Moreover, in order to use $\dot{P}_b$ to estimate masses accurately, one needs to subtract the contributions from the proper motion and the relative acceleration between the pulsar and the solar system barycenter. A package to perform these tasks based on the realistic potential of the Milky Way has been recently developed [30], and gradual improvement in the model of the galactic potential is expected with the help of Gaia data.

Acknowledgments: The author thanks the organizers of 'Compact stars in the QCD phase diagram VI' (especially David Blashcke) for their excellent hospitality, all three reviewers for extensive suggestions on the earlier version of the manuscript and ANI Technologies Pvt. Ltd. and Uber Technologies Inc. for helping to survive in Chennai.

Conflicts of Interest: The authors declare no conflict of interest.

References

1. Van Straten, W.; Bailes, M.; Britton, M.; Kulkarni, S.R.; Anderson, S.B.; Manchester, R.N.; Sarkissian, J. A test of general relativity from the three-dimensional orbital geometry of a binary pulsar. *Nature* **2001**, *412*, 158–160.
2. Deller, A.T.; Boyles, J.; Lorimer, D.R.; Kaspi, V.M.; McLaughlin, M.A.; Ransom, S.; Stairs, I.H.; Stovall, K. VLBI Astrometry of PSR J2222−0137: A Pulsar Distance Measured to 0.4% Accuracy. *Astrophys. J.* **2013**, *770*, 145.
3. Deller, A.T.; Vigeland, S.J.; Kaplan, D.L.; Goss, W.M.; Brisken, W.F.; Chatterjee, S.; Cordes, J.M.; Janssen, G.H.; Lazio, T.J.W.; Petrov, L.; et al. Microarcsecond VLBI Pulsar Astrometry with PSRπ. I. Two Binary Millisecond Pulsars with White Dwarf Companions. *Astrophys. J.* **2017**, *828*, 8.
4. Lorimer, D.; Kramer, M. *Handbook of Pulsar Astronomy*; Cambridge Observing Handbooks for Research Astronomers; Cambridge University Press: Cambridge, UK, 2005; pp. 32–58.
5. Freire, P.C.C.; Wex, N. The orthometric parametrization of the Shapiro delay and an improved test of general relativity with binary pulsars. *Mon. Not. R. Astron. Soc.* **2010**, *409*, 199–212.
6. Kramer, M.; Stairs, I.H.; Manchester, R.N. Strong-field tests of gravity with the double pulsar. *Annalen der Physik* **2006**, *15*, 34–42.
7. Stairs, I. Testing General Relativity with Pulsar Timing. *Living Rev. Relat.* **2003**, *6*, 5.
8. Kramer, M. Pulsars as probes of gravity and fundamental physics. *Int. J. Mod. Phys.* **2016**, *25*, 1630029–1630061.
9. Kramer, M.; Stairs, I.H.; Manchester, R.N. Tests of General Relativity from Timing the Double Pulsar. *Science* **2006**, *314*, 97–102.
10. Demorest, P.B.; Pennucci, T.; Ransom, S.M.; Roberts, M.S.E.; Hessels, J.W.T. A two-solar-mass neutron star measured using Shapiro delay. *Nature* **2010**, *467*, 1081–1083.
11. Antoniadis, J.; Freire, P.C.C.; Wex, N.; Tauris, T.M.; Lynch, R.S. A Massive Pulsar in a Compact Relativistic Binary. *Science* **2013**, *340*, 1233232.
12. Abbott, B.P. GW170817: Observation of Gravitational Waves from a Binary Neutron Star Inspiral. *Phys. Rev. Lett.* **2017**, *119*, 161101.
13. Annala, E.; Gorda, T.; Kurkela, A.; Vuorinen, A. Gravitational-wave constraints on the neutron-star-matter Equation of State. *arXiv* **2017**, arXiv:1711.02644.
14. Ayriyan, A.; Bastian, N.U.; Blaschke, D.; Grigorian, H.; Maslov, K.; Voskresensky, D.N. How robust is a third family of compact stars against pasta phase effects? *arXiv* **2017**, arXiv:1711.03926.
15. Kurkela, A.; Romatschke, P.; Vuorinen, A. Cold quark matter. *Phys. Rev. D* **2010**, *81*, 105021.
16. Perera, B.B.P.; McLaughlin, M.A.; Kramer, M.; Stairs, I.H.; Ferdman, R.D.; Freire, P.C.C.; Possenti, A.; Breton, R.P.; Manchester, R.N.; Burgay, M.; et al. The Evolution of PSR J0737−3039B and a Model for Relativistic Spin Precession. *Astrophys. J.* **2010**, *721*, 1193–1205.
17. Cameron, A.D.; Champion, D.J.; Kramer, M.; Bailes, M.; Barr, E.D. The High Time Resolution Universe Pulsar Survey—XIII. PSR J1757−1854, the most accelerated binary pulsar. **2018**, *475*, L57–L61.

18. Stovall, K.; Freire, P.C.C.; Chatterjee, S.; Demorest, P.B.; Lorimer, D.R. PALFA Discovery of a Highly Relativistic Double Neutron Star Binary. *Astrophys. J.* **2018**, *231*, 453.
19. Barker, B.M.; O'Connell, R.F. The gravitational interaction: Spin, rotation, and quantum effects—A review. *Gen. Relat. Gravit.* **1979**, *11*, 149–175.
20. Perera, B.B.P.; Kim, C.; McLaughlin, M.A.; Ferdman, R.D.; Kramer, M.; Stairs, I.H.; Freire, P.C.C.; Possenti, A. Realistic Modeling of the Pulse Profile of PSR J0737−3039A. *Astrophys J.* **2014**, *787*, 51.
21. Pol, N.; McLaughlin, M.; Kramer, M.; Stairs, I.; Perera, B.B.P.; Possenti, A. A direct measurement of sense of rotation of PSR J0737−3039A. *Astrophys. J.* **2018**, *853*, 73.
22. Breton, R.P.; Kaspi, V.M.; Kramer, M. Relativistic Spin Precession in the Double Pulsar. *Science* **2008**, *321*, 104–107.
23. Damour, T.; Schafer, G. Higher-order relativistic periastron advances and binary pulsars. *Nuovo Cim. B* **1988**, *101*, 127–176.
24. Kopeikin, S.M. Proper Motion of Binary Pulsars as a Source of Secular Variations of Orbital Parameters. *Astrophys. J.* **1996**, *467*, L93.
25. Bagchi, M. Rotational parameters of strange stars in comparison with neutron stars. *New Astron.* **2010**, *15*, 126–134.
26. Ferdman, R.D.; Stairs, I.H.; Kramer, M.; Breton, R.P.; McLaughlin, M.A.; Freire, P.C.C.; Possenti, A.; Stappers, B.W.; Kaspi, V.M.; Manchester, R.N.; et al. The Double Pulsar: Evidence for Neutron Star Formation without an Iron Core-collapse Supernova. *Astrophys. J.* **2013**, *767*, 85.
27. Akmal, A.; Pandharipande, V.R.; Ravenhall, D.G. Equation of state of nucleon matter and neutron star structure. *Phys. Rev. C* **1998**, *58*, 1804–1828.
28. Nozawa, T.; Stergioulas, N.; Gourgoulhon, E.; Eriguchi, Y. Construction of highly accurate models of rotating neutron stars—Comparison of three different numerical schemes. *Astron. Astrophys. Suppl. Ser.* **1998**, *132*, 431–454.
29. Deller, A.T.; Bailes, M.; Tingay, S.J. Implications of a VLBI Distance to the Double Pulsar J0737−3039A/B. *Science* **2009**, *323*, 1327–1329.
30. Pathak, D.; Bagchi, M. GalDynPsr: A package to estimate dynamical contributions in the rate of change of the period of radio pulsars. *arXiv* **2017**, arXiv:1712.06590.
31. Kramer, M.; Wex, N. TOPICAL REVIEW: The double pulsar system: A unique laboratory for gravity. *Class. Quantum Gravity* **2009**, *26*, 073001.
32. Kehl, M.S.; Wex, N.; Kramer, M.; Liu, K. Future measurements of the Lense-Thirring effect in the Double Pulsar. In Proceedings of the Fourteenth Marcel Grossmann Meeting, Rome, Italy, 12–18 July 2015.
33. Cordes, J.M.; Lazio, T.J.W. NE2001.I. A New Model for the Galactic Distribution of Free Electrons and its Fluctuations. *arXiv* **2001**, arXiv:astro-ph/0207156.
34. Cordes, J.M.; Lazio, T.J.W. NE2001. II. Using Radio Propagation Data to Construct a Model for the Galactic Distribution of Free Electrons. *arXiv* **2003**, arXiv:astro-ph/0301598.
35. Yao, J.M.; Manchester, R.N.; Wang, N. A New Electron-density Model for Estimation of Pulsar and FRB Distances. *Astrophys. J.* **2017**, *835*, 29.
36. Weisberg, J.M.; Huang, Y. Relativistic measurements from timing the binary pulsar PSR B1913+16. *Astrophys. J.* **2016**, *829*, 55.
37. Martinez, J.G.; Stovall, K.; Freire, P.C.C.; Deneva, J.S.; Jenet, F.A.; McLaughlin, M.A.; Bagchi, M.; Bates, S.D.; Ridolfi, A. Pulsar J0453+1559: A Double Neutron Star System with a Large Mass Asymmetry. *Astrophys. J.* **2015**, *812*, 143.
38. Martinez, J.G.; Stovall, K.; Freire, P.C.C.; Deneva, J.S.; Tauris, T.M.; Ridolfi, A.; Wex, N.; Jenet, F.A.; McLaughlin, M.A.; Bagchi, M. Pulsar J1411+2551: A Low-mass Double Neutron Star System. *Astrophys. J.* **2017**, *851*, L29.
39. Janssen, G.H.; Stappers, B.W.; Kramer, M.; Nice, D.J.; Jessner, A.; Cognard, I.M.B.; Purver, M.B. Multi-telescope timing of PSR J1518+4904. *Astron. Astrophys.* **2008**, *490*, 753–761.
40. Fonseca, E.; Stairs, I.H.; Thorsett, S.E. A Comprehensive Study of Relativistic Gravity Using PSR B1534+12. *Astrophys. J.* **2014**, *787*, 82.
41. Keith, M.J.; Kramer, M.; Lyne, A.G.; Eatough, R.P.; Stairs, I.H.; Possenti, A.; Camilo, F.; Manchester, R.N. PSR J1753−2240: A mildly recycled pulsar in an eccentric binary system. *Mon. Not. R. Astron. Soc.* **2009**, *393*, 623–627.

42. Ferdman, R.D.; Stairs, I.H.; Kramer, M.; Janssen, G.H.; Bassa, C.G.; Stappers, B.W.; Demorest, P.B.; Cognard, I.; Desvignes, G.; Theureau, G.; et al. PSR J1756−2251: A pulsar with a low-mass neutron star companion. *Mon. Not. R. Astron. Soc.* **2014**, *443*, 2183–2196.
43. Lynch, R.S.; Freire, P.C.C.; Ransom, S.M.; Jacoby, B.A. The Timing of Nine Globular Cluster Pulsars. *Astrophys. J.* **2012**, *745*, 109.
44. Corongiu, A.; Kramer, M.; Stappers, B.W.; Lyne, A.G.; Jessner, A.; Possenti, A.; D'Amico, N.; Löhmer, O. The binary pulsar PSR J1811-1736: Evidence of a low amplitude supernova kick. *Astron. Astrophys.* **2007**, *462*, 703–709.
45. Hobbs, G.; Lyne, A.G.; Kramer, M.; Martin, C.E.; Jordan, C. Long-term timing observations of 374 pulsars. *Mon. Not. R. Astron. Soc.* **2004**, *353*, 1311–1344.
46. Champion, D.J.; Lorimer, D.R.; McLaughlin, M.A.; Cordes, J.M.; Arzoumanian, Z.; Weisberg, J.M.; Taylor, J.H. PSR J1829+2456: A relativistic binary pulsar. *Mon. Not. R. Astron. Soc.* **2004**, *350*, L61–L65.
47. Van Leeuwen, J.; Kasian, L.; Stairs, I.H.; Lorimer, D.R.; Camilo, F. The Binary Companion of Young, Relativistic Pulsar J1906+0746. *Astrophys. J.* **2015**, *798*, 118.
48. Weisberg, J.M.; Nice, D.J.; Taylor, J.H. Timing Measurements of the Relativistic Binary Pulsar PSR B1913+16. *Astrophys. J.* **2010**, *722*, 1030–1034.
49. Swiggum, J.K.; Rosen, R.; McLaughlin, M.A.; Lorimer, D.R.; Heatherly, S. PSR J1930−1852: A Pulsar in the Widest Known Orbit around Another Neutron Star. *Astrophys. J.* **2015**, *805*, 156.
50. Jacoby, B.A.; Cameron, P.B.; Jenet, F.A.; Anderson, S.B.; Murty, R.N.; Kulkarni, S.R. Measurement of Orbital Decay in the Double Neutron Star Binary PSR B2127+11C. *Astrophys. J.* **2006**, *644*, L113–L116.
51. Iorio, L. Prospects for measuring the moment of inertia of pulsar J0737−3039A. *New Astron.* **2009**, *1*, 40–43.
52. O'Connell, R.F. Proposed New Test of Spin Effects in General Relativity. *Phys. Rev. Lett.* **2004**, *93*, 081103.
53. Bejger, M.; Bulik, T.; Haensel, P. Constraints on the dense matter equation of state from the measurements of PSR J0737−3039A moment of inertia and PSR J0751+1807 mass. *Mon. Not. R. Astron. Soc.* **2005**, *364*, 635–639.
54. Iorio, L. General relativistic spin-orbit and spin-spin effects on the motion of rotating particles in an external gravitational field. *Gen. Relat. Gravit.* **2012**, *44*, 719–736.
55. Iorio, L. Post-Keplerian perturbations of the orbital time shift in binary pulsars: an analytical formulation with applications to the galactic center. *Eur. Phys. J. C* **2017**, *77*, 439.

© 2018 by the author. Licensee MDPI, Basel, Switzerland. This article is an open access article distributed under the terms and conditions of the Creative Commons Attribution (CC BY) license (http://creativecommons.org/licenses/by/4.0/).

Article

On Cooling of Neutron Stars with a Stiff Equation of State Including Hyperons

Hovik Grigorian [1,2], Evgeni E. Kolomeitsev [3], Konstantin A. Maslov [4,5] and Dmitry N. Voskresensky [4,5,*]

1. Laboratory for Information Technologies, Joint Institute for Nuclear Research, Dubna RU-141980, Russia; hovikgrigorian@gmail.com
2. Faculty of Physics, Yerevan State University, Alek Manyukyan 1, Yerevan 0025, Armenia
3. Depertment of Physics, Matej Bel University, Tajovskeho 40, SK-97401 Banska Bystrica, Slovakia; E.Kolomeitsev@gsi.de
4. Bogoliubov Laboratory for Theoretical Physics, Joint Institute for Nuclear Research, Dubna RU-141980, Russia; const.maslov@gmail.com
5. Department of Theoretical Nuclear Physics, National Research Nuclear University (MEPhI), Kashirskoe shosse 31, Moscow RU-115409, Russia
* Correspondence: D.Voskresensky@gsi.de

Received: 29 November 2017; Accepted: 15 January 2018; Published: 8 February 2018

Abstract: Exploiting a stiff equation of state of the relativistic mean-field model MKVORHϕ with σ-scaled hadron effective masses and couplings, including hyperons, we demonstrate that the existing neutron-star cooling data can be appropriately described within "the nuclear medium cooling scenario" under the assumption that different sources have different masses.

Keywords: neutron stars; equation of state; in-medium effects; neutrino

1. Introduction

The equation of state (EoS) of the neutron-star matter should be stiff, cf. [1,2], in order to describe measured masses of the heaviest known pulsars PSR J1614-2230 (of mass $M = 1.928 \pm 0.017 M_\odot$) [3] and PSR J0348+0432 (of mass $M = 2.01 \pm 0.04 M_\odot$) [4]. The presence of hyperons in neutron stars leads to a softening of the EoS which results in a decrease of the maximum neutron-star mass below the measured values of masses for PSR J1614-2230 and PSR J0348+0432 pulsars, if one exploits ordinary relativistic mean-field (RMF) models (hyperon puzzle [5,6]). However, within RMF, EoSs with σ-scaled hadron effective masses and coupling constants, the maximum neutron-star mass remains above $2M_\odot$ even when hyperons are included [7,8]. Additionally, other important constraints on the equation of state, e.g., the flow constraint from heavy-ion collisions [9,10] are fulfilled. We demonstrate how a satisfactory explanation of all existing observational data for the temperature–age relation is reached within the "nuclear medium cooling" scenario [11], now with the RMF EoS MKVORHϕ with σ-scaled hadron effective masses and coupling constants, including hyperons [7,8].

2. Equation of State and Pairing Gaps

The EoS of the cold hadronic matter should:

- satisfy experimental information on properties of dilute nuclear matter;
- fulfil empirical constraints on global characteristics of atomic nuclei;
- satisfy constraints on the pressure of the nuclear matter from the description of particle transverse and elliptic flows and the K^+ production in heavy-ion collisions, cf. [9,10];
- allow for the heaviest known pulsars, i.e., PSR J1614-2230 (of mass $M = 1.928 \pm 0.017 M_\odot$) [3] and PSR J0348+0432 (of mass $M = 2.01 \pm 0.04 M_\odot$) [4];

- allow for an adequate description of the compact star cooling [11], most probably without direct Urca (DU) neutrino processes in the majority of the known pulsars detected in soft X rays [12];
- yield a mass–radius relation comparable with the empirical constraints including recent gravitation wave LIGO-Virgo detection [13];
- when extended to non-zero temperature T (for $T < T_c$ where T_c is the critical temperature of the deconfinement), appropriately describe supernova explosions, proto-neutron stars, and heavy-ion collision data, etc.

The most difficult task is to simultaneously satisfy the flow of heavy-ion collision and the maximum neutron-star mass constraints. To fulfil the flow constraints [9,10], a rather soft EoS of isospin-symmetric matter (ISM) is required, whereas the EoS of the beta-equilibrium matter (BEM) should be stiff in order to predict the maximum mass of a neutron star to be higher than the measured mass $M = 2.01 \pm 0.04 M_\odot$ [4] of the pulsar PSR J0348+0432, this mass being the heaviest among the known pulsars.

In standard RMF models, hyperons may already appear in neutron-star cores for $n \gtrsim (2-3)n_0$, which results in a decrease of the maximum neutron-star mass below the observed limit. Within the RMF models with the σ field-dependent hadron effective masses and coupling constants, the hyperon puzzle is resolved, see [7,8]. Here, we use the MKVOR-based models from these works. Most other constraints on the EoS, including the flow constraints, are also appropriately satisfied. In Figure 1, we demonstrate the neutron-star mass as a function of the central density for the MKVOR model without hyperons and for the MKVORHϕ model which includes hyperons, cf. Figures 20 and 25 in [8]. For the MKVOR model, the maximum neutron-star mass reaches $2.33M_\odot$ and the DU reaction is allowed for $M > 2.14M_\odot$. For the MKVORHϕ model, the maximum neutron-star mass is $2.22M_\odot$. The DU reactions on Λ hyperons $\Lambda \to p + e + \bar{\nu}$, $p + e \to \Lambda + \nu$, become allowed for $M > 1.43M_\odot$. The DU reactions with participation of Ξ^-, $\Xi^- \to \Lambda + e + \bar{\nu}$ and $\Lambda + e \to \Xi^- + \bar{\nu}$ become allowed for $M > 1.65M_\odot$. The neutrino emissivity in these processes is typically smaller than that in the standard DU processes on nucleons due to a smaller coupling constant for the hyperons (cf. 0.0394 factor for the DU process $\Lambda \to p + e + \bar{\nu}$ and 0.0175 for $\Xi^- \to \Lambda + e + \bar{\nu}$ compared to 1 for the DU process on nucleons). Besides, we should bear in mind that the pairing suppression R-factors for the DU processes on nucleons and hyperons are different and in our model, for the EoS in the region in which there are hyperons, the R-factor for DU processes on nucleons is larger than that on hyperons.

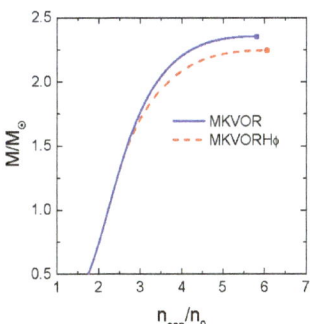

Figure 1. Neutron-star masses versus the central density for the MKVOR model without the inclusion of hyperons and for the MKVORHϕ model with hyperons included.

We adopt here all cooling inputs such as the neutrino emissivities, specific heat, crust properties, etc., from our earlier works performed on the basis of the HHJ equation of state (EoS) [11,14,15], a stiffer HDD EoS [16] and even stiffer DD2 and DD2vex EoSs [17] for the hadronic matter. These works exploit the nuclear medium cooling scenario, where the most efficient processes are the medium-modified

Urca (MMU) processes, $nn \to npe\bar{\nu}$ and $np \to ppe\bar{\nu}$, medium-modified nucleon bremstrahlung (MNB) processes $nn \to nn\nu\bar{\nu}$, $np \to np\nu\bar{\nu}$, $pp \to pp\nu\bar{\nu}$, and the pair-breaking-formation (PBF) processes $n \to n\nu\bar{\nu}$ and $p \to p\nu\bar{\nu}$. The latter processes are allowed only in supefluid matter.

The results are rather insensitive to the value of the nn pairing gap since the $^1S_0$ neutron pairing does not spread in the interior region of the neutron star. We use the same values as we have used in our previous works, e.g., see Figure 5 in [11] for details. Within our scenario, we continue to exploit the assumption that the value of the $^3P_2$ nn pairing gap is tiny and its actual value does not affect the calculations of the neutrino emissivity [14]. For calculation of the proton pairing gaps, we use the same models as in [17] but now we exploit EoSs of the MKVORHϕ model. The corresponding gaps are shown on the left panel in Figure 2.

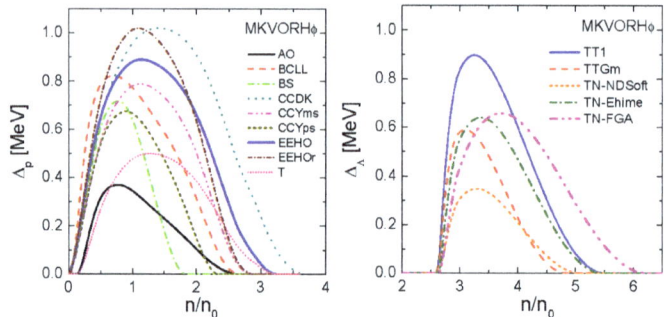

Figure 2. Pairing gaps for protons (left panel) and Λ hyperons (right panel) as functions of baryon density for the MKVORHϕ EoS including hyperons. Proton gaps are evaluated using the same models as in [17] and the Λ hyperon gaps are from [18,19].

With the increase of the density in the MKVORHϕ model, the Λ hyperons are the first to appear at the density $n_c^{(\Lambda)} = 2.63n_0$, and then the Ξ^- hyperons appear at $n_c^{(\Xi^-)} = 2.93n_0$. We take the values of the Λ gaps from the calculations [18,19]. The model TT1 uses the ND-soft model by the Nijmegen group for bare $\Lambda\Lambda$ interaction and model TTGm uses the results of G-matrix calculations by Lanskoy and Yamamoto [20] at density $2.5n_0$. The other three models include three nucleon forces and TNI6u forces for several $\Lambda\Lambda$ pairing potentials: ND-Soft, Ehime and FG-A. In the right panel, we show the Λ hyperon pairing gaps, which we exploit in this work. Ξ^- are considered unpaired.

The quantity

$$-G_\pi^{R-1}(\mu_\pi, k, n) = \omega^{*2}(k) = k^2 + m_\pi^2 - \mu_\pi^2 + \mathrm{Re}\Sigma_\pi(\mu_\pi, k, n) \tag{1}$$

in dense neutron-star matter (for $n \gtrsim n_0$) has a minimum for $k = k_m \simeq p_F$, where p_F is the neutron Fermi momentum. For π^0, the minimum occurs for $\mu_\pi = 0$. The value $\omega^{*2}(k_m)$ has the meaning of the squared effective pion gap. It enters the NN interaction amplitude and the emissivity of the MMU and MNB processes, instead of the quantity $m_\pi^2 + p_F^2$ entering the calculation of the NN interaction amplitude and the emissivity of the modified Urca (MU) and nucleon bremstrahlung (NB) processes in the minimal cooling scheme. The inequality $\omega^{*2}(k_m) < m_\pi^2 + p_F^2$ demonstrates the effect of the pion softening. Of key importance is the fact that here we use the very same density dependence of the effective pion gap $\omega^*(n)$ as in our previous works, e.g., see Figure 2 of [17]. To be specific, we assume a saturation of the pion softening and absence of the pion condensation for $n > n_c$. We plot this pion gap in Figure 3.

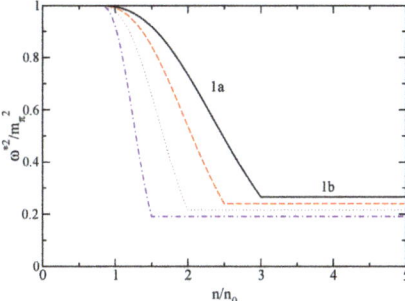

Figure 3. Density dependence of the squared effective pion gap used in the given work. We assume that the pion softening effect saturates above a critical density, the value of which we vary from 1.5 to $3n_0$.

3. Results

In the left panel in Figure 4, we show the cooling history of neutron stars calculated using the EoS of the MKVOR model without inclusion of hyperons. The demonstrated calculations employ the proton gap following the EEHOr model shown in Figure 2, and the dotted curve in Figure 3 was used for the effective pion gap.

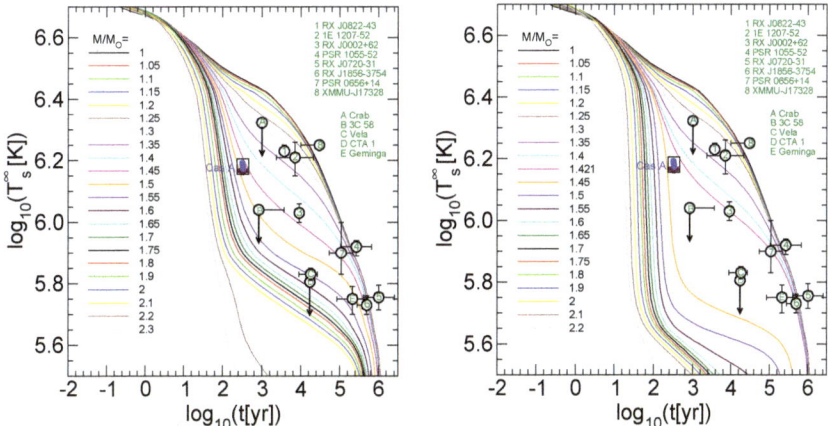

Figure 4. Redshifted surface temperature as a function of the neutron-star age for various neutron-star masses and choice of the EoS. Left panel: MKVOR model without the inclusion of hyperons. Right panel: MKVORHϕ model with hyperons included with the gaps following from the TN-FGA parameter choice. Proton gaps for both calculations, without and with hyperons, are taken following the EEHOr model.

In the right panel in Figure 4, we show the cooling history of neutron stars calculated using the EoS of the MKVORHϕ model with the inclusion of hyperons. As shown in the left panel, the proton gap is given by the EEHOr model and the effective pion gap is given by the dotted curve in Figure 3. Hyperons are taken following the TN-FGA parameter choice. We see a rather appropriate description of the data. With the given model, the DU reactions on hyperons are responsible for the cooling of the intermediate and rapid coolers (objects Cas A, B, 3, C, D, E).

With the pion gaps given by the solid and dashed curves and with proton gaps following the EEHO, EEHOr, CCDK, CCYms, and T curves, we may also rather appropriately describe the cooling history of neutron stars within our scenario. Finally, we should mention that the effect of hyperons on the cooling of neutron stars could be diminished, if for some reasons the hyperon pairing gaps had larger values compared to those shown in Figure 2 and would spread to a higher density. These results will be shown in our subsequent publication.

4. Conclusions

In this study, we have demonstrated that the presently known cooling data can be appropriately described within our nuclear medium cooling scenario under the assumption that different sources have different masses, provided we use the EoS of the MKVORHϕ model (with hyperons included) at appropriately selected proton and hyperon pairing gaps.

Acknowledgments: The research was supported by the Ministry of Education and Science of the Russian Federation within the state assignment, project No 3.6062.2017/BY. The work was also supported by Slovak grant VEGA-1/0348/18. We also acknowledge the support of the Russian Science Foundation, project No 17-12-01427.

Author Contributions: All authors contributed equally to the formulation of the model, numerical analysis and writing of the paper.

Conflicts of Interest: The authors declare no conflict of interest.

References

1. Gandolfi, S.; Gezerlis, A.; Carlson, J. Neutron Matter from Low to High Density. *Ann. Rev. Nucl. Part. Sci.* **2015**, *65*, 303–328.
2. Lattimer, J.; Prakash, M. The Equation of State of Hot, Dense Matter and Neutron Stars. *Phys. Rep.* **2016**, *621*, 127–164.
3. Fonseca, E.; Pennucci, T.T.; Ellis, J.A.; Stairs, I.H.; Nice, D.J.; Ransom, S.M.; Demorest, P.B.; Arzoumanian, Z.; Crowter, K.; Dolch, T.; et al. The NANOGrav nine-year data set: Mass and geometric measurements of binary millisecond Pulsars. *Astrophys. J.* **2016**, *832*, 167.
4. Antoniadis, J.; Freire, P.C.C.; Wex, N.; Tauris, T.M.; Lynch, R.S.; van Kerkwijk, M.H.; Kramer, M.; Bassa, C.; Dhillon, V.S.; Driebe, T.; et al. A massive pulsar in a compact relativistic binary. *Science* **2013**, *340*, 448A.
5. Schaffner-Bielich, J. Hypernuclear physics for neutron stars. *Nucl. Phys. A* **2008**, *804*, 309–321.
6. Djapo, H.; Schaefer, B.; Wambach, J. On the appearance of hyperons in neutron stars. *Phys. Rev. C* **2010**, *81*, 035803.
7. Maslov, K.A.; Kolomeitsev, E.E.; Voskresensky, D.N. Solution of the hyperon puzzle within a relativistic mean-field model. *Phys. Lett. B* **2015**, *748*, 369–375.
8. Maslov, K.A.; Kolomeitsev, E.E.; Voskresensky, D.N. Relativistic mean-field models with scaled hadron masses and couplings: Hyperons and maximum neutron star mass. *Nucl. Phys. A* **2016**, *950*, 64–109.
9. Danielewicz, P.; Lacey, R.; Lynch, W. Determination of the equation of state of dense matter. *Science* **2002**, *298*, 1592–1596.
10. Lynch, W.; Tsang, M.; Zhang, Y.; Danielewicz, P.; Famiano, M.; Li, Z.; Steiner, A. Probing the symmetry energy with heavy ions. *Prog. Part. Nucl. Phys.* **2009**, *62*, 427–432.
11. Blaschke, D.; Grigorian, H.; Voskresensky, D.N. Cooling of neutron stars: Hadronic model. *Astron. Astrophys.* **2004**, *424*, 979–992.
12. Klahn, T.; Blaschke, D.; Typel, S.; van Dalen, E.N.E.; Faessler, A.; Fuchs, C.; Gaitanos, T.; Grigorian, H.; Ho, A.; Kolomeitsev, E.E.; et al. Constraints on the high-density nuclear equation of state from the phenomenology of compact stars and heavy-ion collisions. *Phys. Rev.* **2006**, *74*, 035802.
13. Abbott, B.P. LIGO Scientific and Virgo Collaborations. GW170817: Observation of gravitational waves from a binary neutron star inspiral. *Phys. Rev. Lett.* **2017**, *119*, 161101.
14. Grigorian, H.; Voskresensky, D.N. Medium effects in cooling of neutron stars and 3P(2) neutron gap. *Astron. Astrophys.* **2005**, *444*, 913–929.
15. Blaschke, D.; Grigorian, H.; Voskresensky, D.N.; Weber, F. On the cooling of the neutron star in Cassiopeia A. *Phys. Rev. C* **2012**, *85*, 022802.
16. Blaschke, D.; Grigorian, H.; Voskresensky, D.N. Nuclear medium cooling scenario in the light of new Cas A cooling data and the $2M_\odot$ pulsar mass measurements. *Phys. Rev. C* **2013**, *88*, 065805.

17. Grigorian, H.; Voskresensky, D.N.; Blaschke, D. Influence of the stiffness of the equation of state and in-medium effects on the cooling of compact stars. *Eur. Phys. J. A* **2016**, *52*, 67.
18. Takatsuka, T.; Tamagaki, R. Λ hyperon superfluidity in neutron star cores. *Nucl. Phys. A* **2000**, *670*, 222c.
19. Takatsuka, T.; Nishizaki, S.; Yamamoto, Y.; Tamagaki, R. Occurrence of hyperon superfluidity in neutron star cores. *Prog. Theor. Phys.* **2006**, *115*, 355–379.
20. Lanskoy, D.E.; Yamamoto, Y. Skyrme-Hartree-Fock treatment of Λ and ΛΛ hypernuclei with G-matrix motivated interactions. *Phys. Rev. C* **1997**, *55*, 2330.

© 2018 by the authors. Licensee MDPI, Basel, Switzerland. This article is an open access article distributed under the terms and conditions of the Creative Commons Attribution (CC BY) license (http://creativecommons.org/licenses/by/4.0/).

Conference Report
Cracking Strange Stars by Torsional Oscillations

Francesco Tonelli [1,2,†] and Massimo Mannarelli [1,*,†]

1 Laboratori Nazionali del Gran Sasso, Istituto Nazionale di Fisica Nucleare, I-67100 L'Aquila, Italy; francesco.tonelli@lngs.infn.it
2 Dipartimento di Scienze Fisiche e Chimiche, Universitá di L'Aquila, I-67010 L'Aquila, Italy
* Correspondence: massimo@lngs.infn.it
† These authors contributed equally to this work.

Received: 8 December 2017; Accepted: 29 January 2018; Published: 17 February 2018

Abstract: Strange stars are one of the possible compact stellar objects formed in the core collapse of supernovae. These hypothetical stars are made by deconfined quark matter and are selfbound. In our study, we focus on the torsional oscillations of a non bare strange star, i.e., a strange star with a thin crust made of standard nuclear matter. We construct a theoretical model assuming that the inner parts of the star are in two different phases, namely the color flavour locked phase and the crystalline colour superconducting phase. Since the latter phase is rigid, with a large shear modulus, it corresponds to a first stellar crust. Above this crust a second small crust made by standard nuclear matter is suspended thanks to a strong electromagnetic dipolar moment. We focus on the electromagnetically coupled oscillations of the two stellar crusts. Notably, we find that if a small fraction of the energy of a glitch event like a typical Vela glitch is conveyed in torsional oscillations, the small nuclear crust will likely break. This is due to the fact that in this model the maximum stress, due to torsional oscillations, is likely located near the star surface.

Keywords: neutron stars; star oscillations

1. Introduction

The properties of hadronic matter at densities higher than the nuclear saturation density are under intense theoretical and experimental inspection [1,2]. The high temperature regime is studied in relativistic heavy ion experiments [1], leading to the production and identification of the quark gluon plasma. The low temperature regime ($T \lesssim 1$ MeV) is relevant in the physics of compact stellar objects (CSOs), originating from the collapse of a supernova. The CSOs can be divided into three classes: neutron stars, hybrid stars and strange stars. Neutron stars are the widely studied class of CSOs and are mainly made of nucleons, electrons and muons (to ensure the charge neutrality of the star). If deconfined quarks are present in the core, we are in the presence of the second class of CSOs, the hybrid stars. Strange stars are instead almost completely made of deconfined quark matter [3,4]. The astronomical observations indicate that the mass of CSOs is between 1.2 $M_\odot$ and 2 $M_\odot$, where $M_\odot$ is the solar mass. The estimated radius is of the order of ten kilometers. Unfortunately using these observed values it is not possible to determine the nature of the CSOs because strange stars and hybrid stars can *masquerade* as standard neutron stars [5].

The existence of strange stars is based on the hypothesis of Bodmer [6] and Witten [7] that standard nuclei are in a metastable state. According to this hypothesis, the real ground state of hadronic matter is a configuration that corresponds to an hypothetical short range free-energy minimum of the strong interaction. This is a *collapsed state* of matter and we can imagine a strange star as a huge collapsed state of hadrons. The interaction that binds the star is the strong interaction, i.e., the star is self bound, with gravity playing a role only for very massive stars. So strange stars have no lower limit on mass and can be arbitrarily small.

Unfortunately, assuming that strange matter is the ground state of hadronic matter does not unambiguously define the properties of the system, even at the densities reachable in the CSOs. The essential point is that Quantum Chromodynamics (QCD) is not in the perturbative regime, so it is not under quantitative control. Therefore, we have to use some approximate scheme. Analysis with various methods indicates that strange matter should be in a superconducting phase [8], i.e., a phase in which quarks form Cooper pairs and break the SU(3) color gauge symmetry. In the inner part of the star, at high density, the color flavour locked (CFL) phase should be favored [9]. In the CFL phase u, d, s quarks of all colors pair coherently in a BCS-like state, maximizing the free-energy gain. However, it is conceivable that CFL is not the unique color superconducting phase realized in strange stars. In fact, at lower densities, the chemical potential of the strange quark becomes comparable with its mass, so coherent pairing cannot happen. For that reason a different superconducting phase can be favored; one possibility is the crystalline color superconducting (CCSC) phase [10,11]. One important feature is that the CCSC phase is characterized by a periodic modulation of the diquark pairing. The periodic modulation of the pairing implies that the CCSC phase is mechanically rigid and it turns out that it has an extremely large shear modulus [11,12], which is a key ingredient for torsional oscillations [13,14]. Indeed, the existence of a phase with a large shear modulus suggests that torsional oscillations of large amplitude can be sustained by this structure. Torsional oscillations of strange stars with a CCSC crust have been first analyzed in [15], while in [16] the coupled oscillations of the CCSC crust and of the ionic crust have been studied. In the present contribution to the proceedings of the CSQCD VI conference we report on the latter study.

2. The Model

2.1. Background Configuration

The star composition we have hypothesized in [16] is shown in Figure 1.

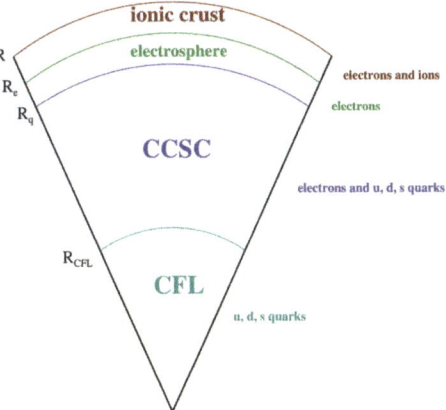

Figure 1. Schematic picture (not in scale) of the stellar model we propose. The mass of the structure is around 1.4 $M_\odot$, and the total radius is $R \simeq 9.2$ km. The electrosphere has a thickness of few hundred fm, while the ionic crust has a thickness of about 200 m [16].

The inner part of the strange star, called the *quarksphere*, is populated by deconfined quarks (u, d and s). In the core, at the highest density, the CFL phase is realized and above it, at a radius R_{CFL}, there is a transition to the CCSC phase. The actual radius R_{CFL} is unknown, so we will consider it as a parameter of the model. In the CCSC phase, due to the lack of strange quarks, electrons are present to guarantee the charge neutrality. Above the quarksphere, there is a very small layer (about a hundred fermi thick) populated only by electrons, forming the so-called *Electrosphere*. This is possible because quarks are confined within the quarksphere of radius R_q by the strong interaction, but electrons do not feel the strong interaction and can therefore get outside the quarksphere. They are only bound by the electrostatic interaction on a range of hundreds of fm. On the top of this structure there is a small standard ionic crust, which is electromagnetically suspended due to the positive charge present at the surface of the quarksphere [3].

To determine the mass and the radius of our structure, we solve the Tolman Oppenheimer Volkov (TOV) equations that are a generalization of the hydrostatic equilibrium in non rotating spherical metric. For that reason we have to choose an Equation of State (EoS) that can describe our system. Since QCD is not perturbative, it is not possible to determine the actual EoS. The strange star temperature is much lower than the typical scale of QCD, thus it could be considered equal to zero. This means that we can consider our system as a Fermi liquid at zero temperature. To take into account the strong interaction we use a Taylor expansion of the grand potential as a function of the average baryonic chemical potential, μ, as proposed in [5]:

$$\Omega_{QM} = -\frac{3}{4\pi^2} a_4 \mu^4 + \frac{3}{4\pi^2} a_2 \mu^2 + B_{eff} \tag{1}$$

where a_4, a_2 and B_{eff} are independent of μ. We use the set of parameters $a_4 = 0.7$, $a_2 = (200 \text{ MeV})^2$ and $B_{eff} = (165 \text{ MeV})^4$ discussed in [15].

We assume that the previous EoS describes the whole quarksphere and is matched with an EoS valid for the standard ionic crust at R_q, corresponding to a pressure of the quarksphere equal to the pressure corresponding to the neutron drip in the standard ionic crusts. In fact if the density exceeds the neutron drip density, neutrons in the ionic crust are ripped off from the nuclei and fall down into the quarksphere, where they are eventually converted in deconfined light quarks. For the ionic crust we assume that it consists of a Coulomb crystal embedded in a degenerate electron gas, and we use the data reported in [17].

Setting the central density to $\rho_c = 1.5 \times 10^{15}$ (g/cm^3) we obtain a star of 1.4 $M_\odot$ with a radius of $R \simeq 9.2$ km. The ionic crust is about 200 m thick, and so the radius at which we have the transition between quark matter and standard nuclear matter is $R_q \simeq 9$ km.

2.2. Torsional Oscillation

The two considered crusts have a nonvanishing shear modulus and so non radial modes can be excited. We briefly review the equations governing the non radial modes. Defining as $\vec{u}$ the displacement vector, the non radial modes obey to $\nabla \cdot \vec{u} = 0$, with a vanishing radial component, that is $u_r = 0$. Furthermore, assuming that $u = e^{i\sigma t} \vec{\zeta}(r)$, in the newtonian limit and considering the fluid description, according to [13] we can write the Euler's equation in spherical coordinates as follows

$$\sigma^2 W_i(r) = v_i^2 \left[-\frac{d\log v_i}{dr} \left(\frac{dW_i}{dr} - \frac{W_i}{r} \right) - \frac{1}{r}\frac{d}{dr}\left(r^2 \frac{dW_i}{dr}\right) + \frac{l(l+1)}{r} W_i \right], \tag{2}$$

where $i = 1, 2$ characterizes the two crusts, v_i are the corresponding shear moduli and W is a function of r defined as

$$\zeta_\vartheta = W \frac{1}{\sin\vartheta} \frac{\partial Y_{lm}}{\partial \phi}, \quad \zeta_\phi = -W \frac{\partial Y_{lm}}{\partial \vartheta}. \tag{3}$$

To solve this equation in both the two crusts we have to know the shear modulus of the CCSC crust, of the ionic crust and set the appropriate boundary conditions.

The rigidity of the CCSC crust has been studied in [12]. The shear modulus calculated for the CCSC matter is:

$$v \simeq v_0 \left(\frac{\Delta}{10\text{MeV}}\right)^2 \left(\frac{\mu_q}{400\text{MeV}}\right)^2, \tag{4}$$

where Δ is the pairing energy of the condensate, μ_q is the quark chemical potential and the reference value is:

$$\nu_0 = 2.47 \frac{\text{MeV}}{\text{fm}^3}. \tag{5}$$

This is the value that we will use in our study, because we expect that $\Delta \sim 10$ MeV and the quark chemical potential is roughly constant within the CCSC phase, see [15] for a discussion. The actual value of the CCSC shear modulus can be different from the reference value also because the procedure for its evaluation relies on a number of approximations, see [11]. However, the most important aspect for our discussion is that the shear modulus of the CCSC matter is much larger than the shear modulus of standard nuclear matter. Note also that the effect of the pairing gap in the EoS is effectively included in the a_2 coefficient of Equation (1), see [5] and the discussion in [15].

The shear modulus of the ionic crust strongly depends on the particular crystalline structure of the crust and on the plane of application. Actually, we do not know the crystalline structure of the ionic crust, but we can calculate an effective shear modulus [18] as:

$$\nu_{\text{eff}} = c \frac{n_N(r)(Z(r)e)^2}{a_N(r)}, \tag{6}$$

where $n_N(r)$ is the density of nucleons as a function of the radial coordinate r, $Z(r)$ is the number of protons in the nuclei, and $a_N(r) = (3/(4\pi n_N))^{(1/3)}$ is the average inter-ion spacing. The constant c has been evaluated in [18,19] and it is $c \simeq 0.1$.

The boundary conditions that we impose are the followings

$$\left(\frac{dW_1}{dr} - \frac{W_1}{r} \right) \bigg|_{R=R_{\text{CFL}}} = 0 \tag{7}$$

$$\nu_1 \left(\frac{dW_1}{dr} - \frac{W_1}{r} \right) \bigg|_{R=R_q} = \nu_2 \left(\frac{dW_2}{dr} - \frac{W_2}{r} \right) \bigg|_{R=R_q} \tag{8}$$

$$W_1(R_q) = W_2(R_q) \tag{9}$$

$$\left(\frac{dW_2}{dr} - \frac{W_2}{r} \right) \bigg|_{R=R_{\text{ocean}}} = 0, \tag{10}$$

where $R_{\text{ocean}} \simeq 9.15$ km and the ocean is the region in which the density is less than 10^7 g/cm^3. The first, the second and the last equations are no-traction conditions, while the third one is a no-slip condition. The no-traction condition means that there is no force acting between two adjacent layers. The no-slip condition means that the displacement at the interface of two layers is the same. See [16] for more details on these boundary conditions.

3. Results

In our study we focus on the $l = 1$ modes that we call $_1t_n$, where n indicates the number of nodes, corresponding to a twist of the two crusts. We numerically solve the equation for the non radial modes, considering the density inside the CCSC crust as a constant (the estimated error made with this approximation is less than 10%) and a realistic radial density dependence in the ionic crust. Since the transition between the CFL and CCSC phase depends on the unknown pressure difference between the two phases, we define $R_{\text{CFL}} = aR_q$ with a a parameter that varies between 0 and 1, and we study the problem varying a.

In Figure 2 we show the obtained frequencies of the $_1t_1$, $_1t_2$, $_1t_3$ modes as a function of a. As we can easily see, the typical frequencies are of the order of 10 kHz and we can identify two different behaviors. There is one kind of oscillations, dependent on the parameter a, associated to a non radial coupled crusts oscillation (CCSO), in which both crusts are sensibly displaced. A second kind of oscillations is almost independent of a and is associated to a sensible displacement of only the ionic crust (ICO). To show the amplitude of the oscillation we assume that the energy of the order of the one released in a Vela-like glitch ($E \sim 5 \times 10^{42}$ erg) is conveyed to the $_1t_n$ modes.

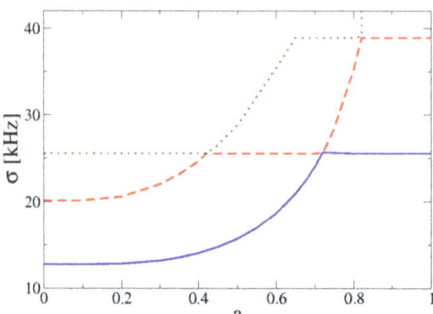

Figure 2. Frequencies of the modes $_1t_1$ (solid blue), $_1t_2$ (dashed red) and $_1t_3$ (dotted green) as a function of $a = R_q/R_{CFL}$. As it is possible to see, we have two kind of features. There are oscillations whose frequencies depends on a and oscillations with frequencies independent of a [16].

In Figure 3 we show the amplitude of $_1t_1$ oscillations in two different cases. In the left panel we choose $a = 0.4$ while for the right panel we choose $a = 0.8$. For $a = 0.4$ the energy is divided by the CCSC crust and the ionic crust, corresponding to a CCSO-type mode. The maximum amplitude of the oscillation is of the order of 20 cm. In the right panel we consider the $a = 0.8$ case. In this case all the energy is conveyed on the ionic crust, corresponding to a ICO-type mode, and the amplitude of the oscillation is of the order of the km.

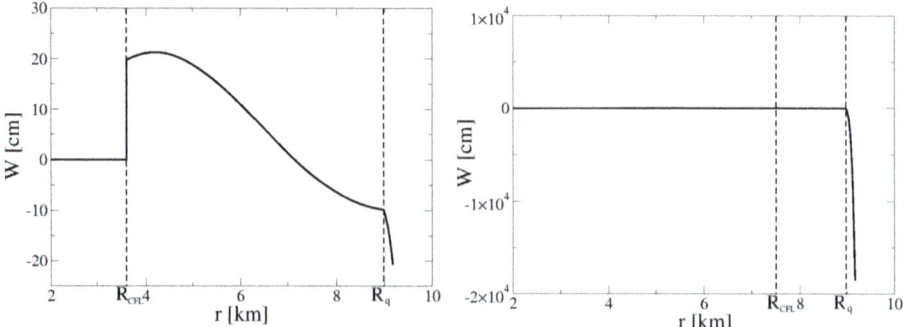

Figure 3. Radial dependence of the amplitude of the $_1t_1$ mode oscillation inside our strange star model. **Left:** Oscillation of the coupled crusts oscillation (CCSO)-type (see text) obtained choosing $a = 0.8$. The amplitude of the oscillation is of the order of 20 cm. **Right:** Oscillation of the ISO-type (see text) obtained choosing $a = 0.4$. In the nuclear crust the amplitude can reach values of the order of few km [16].

To better understand our results we study the deformation of the solid crusts due to the torsional oscillation. A measure of the deformation is the shear strain. For our study, due to symmetry of the problem, we restrict our analysis to the radial component of the strain, that is defined as:

$$|s| = \left| \frac{dW}{dr} - \frac{W}{r} \right|. \tag{11}$$

For the shear strain in the case $a = 0.8$ we obtain a maximum near the surface of the star. This is a particularly relevant result because in previous analysis, such as [13], the maximum strain is far inside the inner crust and it is impossible to break the crust. For the ionic crust, despite the high uncertainties, the maximum strain should be between 10^{-4} and 10^{-2} [20] but in perfect crystals values of the order of 10^{-1} could be appropriate [21]. Since in our model we obtain maximum strains larger than these values, of the order of unity, this probably means that in our model it is possible to break the ionic crust during a glitch or in any other event releasing a comparable amount of energy.

4. Conclusions

We have considered a model of a nonbare strange star comprising a quarksphere of superconducting quark matter surmounted by a standard nuclear matter crust [16]. The quarksphere is in two different phases: the CFL phase and the rigid CCSC phase. The CCSC crust and the ionic crust are separated by an electron layer a few hundred fm thick. We have solved the TOV equations using a simple parameterization of the EoS of quark matter in function of the baryon chemical potential matched with a realistic EoS for the description of the ionic crust.

Both the CCSC and the ionic crusts are rigid, so electromagnetically coupled torsional oscillations are possible. We have found two types of oscillations, the first involves the two crusts with comparable amplitude (CCS0-type), and the second confined in the ionic crust (ICO-type). If the CCSC is thinner than around 2 km, ICOs are the only relevant oscillations.

We have studied in detail the $l = 1$ torsional modes, obtaining frequencies of the order of 10 kHz. These modes correspond to oscillatory twists of the crust and do not conserve angular momentum. For that reason we have assumed that they are triggered by a pulsar glitch and following this idea we have assumed that the energy of a Vela-like glitch is conveyed to the strange star crust. If the CCSC crust is sufficiently thin, then ISOs are triggered and it is possible to break the ionic crust. Indeed, computing the strain as a function of the radial coordinate, we find that its maximum is located a few tens of meters below the stellar surface and is of the order of unity. This probably means that within our model it is possible to break the ionic crust during a glitch or any other stellar event conveying a comparable amount of energy to the torsional oscillations.

Possible observables related to torsional oscillations are giant Gamma Ray Burst (see [22]) and Quasi Periodic Oscillations (see [23–27]).

Author Contributions: These authors contributed equally to this work.

Conflicts of Interest: The authors declare no conflict of interest.

References

1. Shuryak, E. Strongly coupled quark-gluon plasma in heavy ion collisions. *Rev. Mod. Phys.* **2017**, *89*, 035001.
2. Alford, M.G.; Schmitt, A.; Rajagopal, K.; Schafer, T. Color superconductivity in dense quark matter. *Rev. Mod. Phys.* **2008**, *80*, 1455–1515.
3. Alcock, C.; Farhi, E.; Olinto, A. Strange stars. *Astrophys. J.* **1986**, *310*, 261.
4. Haensel, P.; Zdunik, J.L.; Schaeffer, R. Strange quark stars. *Astron. Astrophys.* **1986**, *160*, 121–128.
5. Alford, M.; Braby, M.; Paris, M.W.; Reddy, S. Hybrid stars that masquerade as neutron stars. *Astrophys. J.* **2005**, *629*, 969.
6. Bodmer, A. Collapsed nuclei. *Phys. Rev. D* **1971**, *4*, 1601–1606.
7. Witten, E. Cosmic separation of phases. *Phys. Rev. D* **1984**, *30*, 272–285.
8. Rajagopal, K.; Wilczek, F. The Condensed Matter Physics of QCD. In *At the frontier of particle physics. Handbook of QCD*; Shifman, M., Ed.; World Scientific: Hackensack, NJ, USA, 2000.
9. Alford, M.; Rajagopal, K.; Wilczek, F. The Minimal CFL-Nuclear Interface. *Nucl. Phys. B* **2000**, *537*, 443–458.
10. Alford, M.; Bowers, J.A.; Rajagopal, K. Crystalline Color Superconductivity. *Phys. Rev. D* **2001**, *63*, 074016.
11. Anglani, R.; Casalbuoni, R.; Ciminale, M.; Ippolito, N.; Gatto, R.; Mannarelli, M.; Ruggieri, M. Crystalline color superconductors. *Rev. Mod. Phys.* **2014**, *86*, 509–561.

12. Mannarelli, M.; Rajagopal, K.; Sharma, R. The rigidity of crystalline color superconducting quark matter. *Phys. Rev. D* **2007**, *76*, 074026.
13. McDermott, P.N.; van Horne, H.M.; Hansen, C.J. Nonradial oscillations of neutron stars. *Astrophys. J.* **1988**, *325*, 725–748.
14. Samuelsson, L.; Andersson, N. Neutron Star Asteroseismology. Axial Crust Oscillations in the Cowling Approximation. *Mon. Not. R. Astron. Soc.* **2007**, *374*, 256–268.
15. Mannarelli, M.; Pagliaroli, G.; Parisi, A.; Pilo, L. Electromagnetic signals from bare strange stars. *Phys. Rev. D* **2014**, *89*, 103014.
16. Mannarelli, M.; Pagliaroli, G.; Parisi, A.; Pilo, L.; Tonelli, F. Torsional oscillation of non bare strange stars. *Astrophys. J.* **2015**, *815*, 81.
17. Haensel, P.; Pichon, B. Experimental nuclear masses and the ground state of cold dense matter. *Astron. Astrophys.* **1994**, *283*, 313–318.
18. Strohmayer, T.; van Horn, H.M.; Ogata, S.; Iyetomi, H.; Ichimaru, S. The shear modulus of the neutron star crust and nonradial oscillations of neutron stars. *Astrophys. J.* **1991**, *375*, 679–686.
19. Hoffman, K.; Heil, J. Mechanical properties of non-accreting neutron star crusts. *Mon. Not. R. Astron. Soc.* **2012**, *426*, 2404–2412.
20. Horowitz, C.J.; Kadau, K. The Breaking Strain of Neutron Star Crust and Gravitational Waves. *Phys. Rev. Lett.* **2009**, *102*, 191102.
21. Kittel, C. *Introduction to Solid State Physics*; Wiley: New York, NY, USA, 1976.
22. Thompson, C.; Duncan, R.C. The Giant Flare of 1998 August 27 from SGR 1900+14: II. Radiative Mechanism and Physical Constraints on the Source. *Astrophys. J.* **2001**, *561*, 980.
23. Israel, G.; Belloni, T.; Stella, L.; Rephaeli, Y.; Gruber, D.E.; Casella, P.; Dall'Osso, S.; Rea, N.; Persic, M.; Rothschild, R.E. Discovery of Rapid X-ray Oscillations in the Tail of the SGR 1806-20 Hyperflare. *Astrophys. J. Lett.* **2005**, *628*, L53–L56.
24. Watts, A.L.; Strohmayer, T.E. Detection with RHESSI of high frequency X-ray oscillations in the tail of the 2004 hyperflare from SGR 1806-20. *Astrophys. J. Lett.* **2005**, *632*, L111.
25. Glampedakis, K.; Samuelson, L.; Anderson, N. Elastic or magnetic? A toy model for global magnetar oscillations with implications for QPOs during flares. *Mon. Not. R. Astron. Soc. Lett.* **2006**, *371*, L74–L77.
26. Watts, A.L.; Reddy, S. Magnetar oscillations pose challenges for strange stars. *Mon. Not. R. Astron. Soc.* **2007**, *379*, L63–L66.
27. Watts, A.L. Neutron starquakes and the dynamic crust. *arXiv* **2011**, arXiv:1111.0514.

© 2018 by the authors. Licensee MDPI, Basel, Switzerland. This article is an open access article distributed under the terms and conditions of the Creative Commons Attribution (CC BY) license (http://creativecommons.org/licenses/by/4.0/).

Article

Many Aspects of Magnetic Fields in Neutron Stars

Rodrigo Negreiros [1],*, Cristian Bernal [2], Veronica Dexheimer [3] and Orlenys Troconis [1]

1. Instituto de Física, Universidade Federal Fluminense, Praia Vermelha 24210-346, Brazil; troconisorlenys@gmail.com
2. Instituto de Matemática, Estatísca e Física, Universidade Federal de Rio Grande, de Rio Grande 96200-970, Brazil; cgbernal@furg.br
3. Department of Physics, Kent State University, Kent, OH 44240, USA; vdexheim@kent.edu
* Correspondence: rnegreiros@id.uff.br

Received: 11 December 2017; Accepted: 7 February 2018; Published: 26 February 2018

Abstract: In this work, we explore different aspects in which strong magnetic fields play a role in the composition, structure and evolution of neutron stars. More specifically, we discuss (i) how strong magnetic fields change the equation of state of dense matter, alter its composition, and create anisotropies, (ii) how they change the structure of neutron stars (such mass and radius) and the formalism necessary to calculate those changes, and (iii) how they can affect neutron stars' evolution. In particular, we focus on how a time-dependent magnetic field modifies the cooling of a special group known as X-ray dim neutron stars.

Keywords: neutron stars; stellar magnetic field; stellar structure; stellar evolution

1. Introduction

Since the detection of a soft gamma repeater in 1979 and an anomalous X-ray pulsar in 1981, people became interested in neutrons stars that could be powered by their strong magnetic field. In 1992 and 1993, Duncan and Thompson proposed the magnetar model [1,2] and, since then, approximately 30 Soft-Gamma Ray Repeaters (SGRs) and Anomalous X-Ray Pulsars (AXP's) have been observed [3] with thousands of them being expected to exist in our galaxy. Today, these two kinds of objects are understood as being one class of objects, magnetars [4–6]. In recent years, several measurements have shed new light on the strength of magnetic fields on the surface and in the interior of neutron stars. While measurements using anharmonic precession of star spin down have estimated surface magnetic fields to be on the magnitude of 10^{15} G for the sources 1E 1048.1−5937 and 1E 2259+586 [7], data for slow phase modulations in star hard X-ray pulsations (interpreted as free precession) suggest internal magnetic fields to be on the magnitude of 10^{16} G for the source 4U 0142+61 [8]. Together, these estimates have motivated a large amount of research on the issue of how magnetic fields modify the physics of neutron stars.

In addition to the objects discussed above, magnetic fields are a key aspect in the evolution of neutron stars. Particularly interesting are the objects known as X-ray dim neutron stars (XDINS). These objects show bright soft X-ray emission, exhibit blackbody-like spectra, and have high temperatures with respect to their spin-down age. In fact, one of the most puzzling aspects of the XDINS is that their spin-down properties (period and period derivative) indicate old ages (the so-called spin-down age), which is unexpected, considering their high observed temperatures.

Motivated by the scenarios discussed above, in this work we will discuss the most prominent aspects of magnetic field for the composition, structure and evolution of neutron stars. Initially we will focus on the microscopic aspects and, in particular, how magnetic fields may change the equation of state of dense matter through Landau quantization, which generates pressure anisotropies, and affect the particle composition. After that, we will concentrate on the macroscopic aspects, in particular on the question of how strong magnetic fields change the structure of neutron stars (such mass and radius), we will also briefly discuss the formalism which is necessary to take care of the anisotropies generated by the magnetic field. Finally, we are going to discuss how magnetic fields may affect neutron stars' temporal evolution, particularly, we expand on the idea that the spin-down age may mask the true age of the neutron star if its magnetic field experiences a non-canonical behavior, namely if it is evolving with time. We note here that in this work we will address these three fronts (microscopic, macroscopic and evolutionary) in a up to some extent independent manner, given the significant challenge that is to treat all of them on the same footing. Our ultimate goal is to build a framework in which such treatment may be viable, thus in here we lay the ground work for such investigation and present our perspective to performing such self-consistent overarching study.

2. Microscopic Aspects

The inclusion of an external magnetic field in one direction generates the quantization of energy levels in the directions perpendicular to the magnetic field, what is referred to as Landau levels. These levels are related to the particles' angular momentum quantum number, but also charge and spin. The levels are degenerated, except for the zeroth level. This formalism was originally derived for the non-relativistic case by Landau [9]. In the case of finite temperature, Landau levels are summed from zero to infinity but, in the special case of zero temperature, there is a maximum Landau level beyond which the particles' Fermi momenta in the direction of the field become negative. As a consequence, the 3-dimensional integrals for the thermodynamical quantities become 1-dimensional but summed over all the possible levels. On one hand, very strong magnetic fields restrict the summations to the lower Landau levels, whereas very week fields require summations over many levels, becoming at some point continuous. In this case, the integrals become once more 3-dimensional. This formalism, when applied to neutron-star matter with homogeneous magnetic fields [10–14] generates softer equations of state, as the enhancement of charged particles also turns the system more isospin symmetric. Of course, a possible suppression of hyperons by the magnetic field has exactly the opposite effect on the equation of state (as hyperons turn the equation of state softer) and phase transitions to quark matter can have different effects [15].

In addition, the energy momentum tensor becomes anisotropic, as the pressure in the directions parallel and perpendicular to the magnetic field become distinct. This issue first raised by Canuto [16,17] happens due to an extra contribution from the fermions, which move in their quantized orbits perpendicular to the external magnetic field. This issue was recently revisited in recent publications [18–20]. Another effect created by the magnetic field concerns its interaction with the fermions' spin, the anomalous magnetic moment [18,21]. In this case, the magnetic field turns the equation of state of hadronic matter stiffer, due to a modification of the amount of particles with different spins (polarization) [22,23]. For quarks, the efficiency of anomalous magnetic moment has been estimated to be very small [24,25].

As it is going to be shown in the next section, the effects discussed above concerning stiffness and anisotropy of the equation of state and particle population do not have a substantial effect on macroscopic properties of neutrons stars, such as stellar masses [26,27]. On the other hand strong magnetic fields decrease substantially the central density of neutron star. In this case, the particle population inside a neutron star can be substantially modified, as shown in Figure 1. As explained in detail in Ref. [27], very strong magnetic fields (but still realistic) can suppress not only a phase transition to quark matter, but also the appearance of hyperons (bottom panels of Figure 1). But, interestingly, as the magnetic field strength throughout the star decays over time (more details about magnetic field temporal evolution in Section 4), an increase in central density or chemical potential allows the appearance of hyperons and then quarks, as shown in the top panels of Figure 1. These calculations were performed by solving numerically the Einstein-Maxwell equations at fixed baryon mass $M_B = 2.2$ $M_\odot$ using the chiral mean field (CMF) model [28].

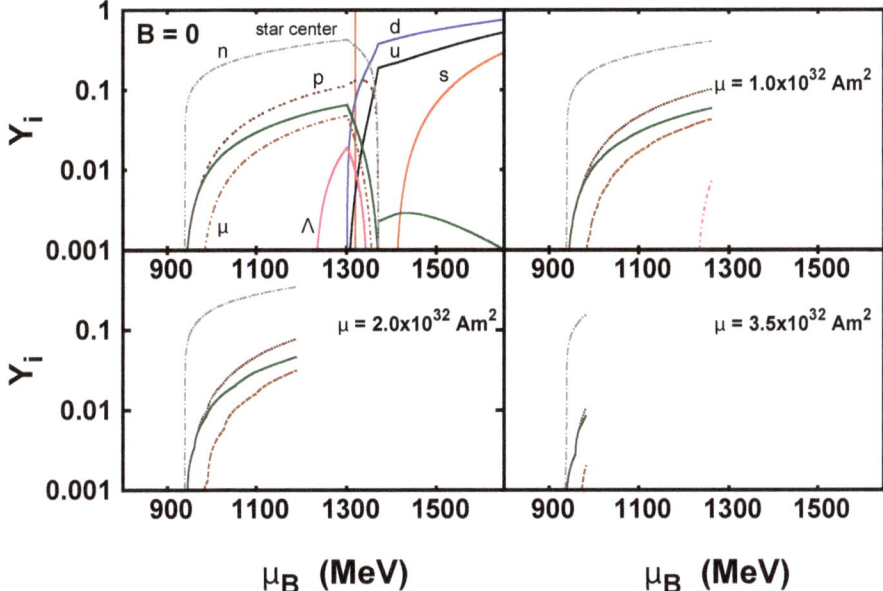

Figure 1. Particle population as a function of the baryon chemical potential calculated using the CMF model shown for different dipole magnetic moments μ. All figures represent equilibrium sequences obtained from the numerical solution of he Einstein-Maxwell equations at fixed baryon mass. For the non-magnetized case, the vertical red curve represents the chemical potential reached at the center of the star, namely, 1320 MeV.

Another relevant question is how much does the magnetic field vary inside neutron stars. As already pointed out by Menezes et al. in Ref. [29], ad hoc formulas for magnetic field profiles in neutron stars do not fulfill Maxwell's equations (more specifically, Gauss' law) and, therefore, are incorrect. The Ref. [30] showed that strong magnetic fields inside magnetic neutron stars increase quadratically with baryon chemical potential in the polar direction and not by more than one order of magnitude when solving the Einstein-Maxwell's equations. This study was performed independently for several different equations of state but did not provide profiles for the magnetic field strength in the stellar equatorial direction, as those are much more complicated.

3. Macroscopic Aspects

Following the microscopic discussion we now address how the magnetic field may alter the macroscopic properties of a neutron star. For that we must fully describe magnetic field in a general relativistic environment as well as the deviation from spherical symmetry that are associated with it. Since the seminal paper by Oppenheimer and Snyder [31], most of the work dedicated to the problem of general relativistic sources, deal with spherically symmetric fluid distribution. In reality, the study of self gravitating compact objects one usually assumes that small deviations from spherical symmetry are likely to take place. Such small deviations are not appropriate for stars with strong magnetic fields where a full axially symmetric treatment is necessary to properly describe the system. The population statistics of SGRs suggest that more than 10% of neutron stars are born as magnetars, with their magnetic field decaying as they age [32–34]. Hence it seems likely that some mechanism is capable of generating large magnetic fields in nascent neutron stars.

In this section we will describe such highly magnetized neutron stars as a perfect fluid coupled with a dipolar magnetic field. First, we will discuss the theoretical aspects relevant to the magnetic field role on the structure of neutron stars, more specifically, we will discuss conservation equations, which will be written in terms of three quantities - whose physical interpretation will be given. Furthermore, we will also discuss calculations of highly magnetized neutron stars, whose structure have been calculated numerically. We will discuss how the presence of a strong (albeit realistic) magnetic field affects global, and potentially observational properties of neutron stars such as gravitational mass and radius.

3.1. Formal Aspects of the Magnetic Field on the Structure of Neutron Stars

Now we briefly discuss the formal aspects of the magnetic field in the stellar structure and gravitational equations in the context of Einstein's general relativity. We begin by considering a bound, static and axially symmetric source. The line element may be written in cylindrical coordinates as

$$ds^2 = -A^2(dx^0)^2 + B^2[(dx^1)^2 + (dx^2)^2] + D^2(dx^3)^2, \tag{1}$$

where we identify $x^0 = t$, $x^1 = \rho$, $x^2 = z$, $x^3 = \phi$ and A, B, D are positive functions of the coordinates ρ and z. In the Weyl spherical coordinates, the line element (1) is

$$ds^2 = -A^2(dt)^2 + B^2[(dr)^2 + r^2(d\theta)^2] + D^2(d\phi)^2, \tag{2}$$

where $\rho = r\sin\theta$ and $z = r\cos\theta$. We denote the coordinates as $x^\mu = (t, r, \theta, \phi)$, and $A(r,\theta), B(r,\theta), D(r,\theta)$ are three independent functions.

The sources of curvature in Einstein's general relativity is represented by the energy-momentum tensor. For a magnetized neutron star, we describe the system as a perfect fluid coupled to a poloidal magnetic field. The perfect fluid assumption simplifies the mathematical treatment dramatically, however, there has also been research considering spherically symmetric dissipative and anisotropic fluid distribution and some of them with analytical solutions (see for instance [35,36]. As mentioned at the beginning of this section, highly magnetized neutron stars should be modeled using an axially symmetric metric tensor which increases the complexity of the problem considerably.

The motivation behind the assumption of a poloidal magnetic field is that such assumption is compatible with the circularity of the space-time [37]. It is important to note, however, that non-negligible toroidal magnetic fields are likely to exist in neutron stars, making the study considerably more complicated. The study of toroidal magnetic fields, in addition to poloidal ones is beyond the scope of this work.

Following the scenario discussed above, the energy-momentum tensor for the system is written as that of a perfect-fluid in addition to the energy-momentum tensor of the electromagnetic field,

$$T_{\mu\nu} = T^{PF}_{\mu\nu} + T^{EM}_{\mu\nu}. \tag{3}$$

The perfect fluid (PF) contribution is

$$T^{PF}_{\mu\nu} = (\rho + P)u_\mu u_\nu + P g_{\mu\nu}, \tag{4}$$

where ρ and P are, respectively, the rest-frame energy density and pressure, u^μ is the fluid 4-velocity with $u^\mu u_\mu = -1$. The electromagnetic part (EM) in (3) is

$$T^{EM}_{\mu\nu} = \frac{1}{4\pi}\left(F_\mu{}^\alpha F_{\nu\alpha} - \frac{1}{4}g_{\mu\nu}F^{\alpha\beta}F_{\alpha\beta}\right), \tag{5}$$

where the Maxwell tensor $F_{\mu\nu}$ is defined in terms of the electromagnetic 4-potential A_μ as

$$F_{\mu\nu} = A_{\nu,\mu} - A_{\mu,\nu}. \tag{6}$$

We are interested in describing a distribution without free-charge and with only poloidal magnetic field, thus the electromagnetic 4-potential is written as

$$A_\mu = (0, 0, 0, A_\phi(r, \theta)). \quad (7)$$

The matrix form of $F_{\mu\nu}$ is written as

$$F_{\mu\nu} = \begin{pmatrix} 0 & 0 & 0 & 0 \\ 0 & 0 & 0 & \frac{\partial A_\phi}{\partial r} \\ 0 & 0 & 0 & \frac{\partial A_\phi}{\partial \theta} \\ 0 & -\frac{\partial A_\phi}{\partial r} & -\frac{\partial A_\phi}{\partial \theta} & 0 \end{pmatrix}, \quad (8)$$

and the electromagnetic energy-momentum tensor is

$$T^{EM\mu}{}_\nu = \begin{pmatrix} T^{EM0}{}_0 & 0 & 0 & 0 \\ 0 & T^{EM1}{}_1 & T^{EM1}{}_2 & 0 \\ 0 & \frac{1}{r^2}T^{EM1}{}_2 & -T^{EM1}{}_1 & 0 \\ 0 & 0 & 0 & -T^{EM0}{}_0 \end{pmatrix}, \quad (9)$$

where the non-vanishin components, in terms of the electromagnetic 4-potential, are given by

$$T^{EM0}{}_0 = -\frac{1}{8\pi} g^{\phi\phi} \left[g^{rr} \left(\frac{\partial A_\phi}{\partial r}\right)^2 + g^{\theta\theta} \left(\frac{\partial A_\phi}{\partial \theta}\right)^2 \right], \quad (10)$$

$$T^{EM1}{}_1 = \frac{1}{8\pi} g^{\phi\phi} \left[g^{rr} \left(\frac{\partial A_\phi}{\partial r}\right)^2 - g^{\theta\theta} \left(\frac{\partial A_\phi}{\partial \theta}\right)^2 \right], \quad (11)$$

$$T^{EM1}{}_2 = \frac{1}{4\pi} g^{rr} g^{\phi\phi} \left(\frac{\partial A_\phi}{\partial r}\right) \left(\frac{\partial A_\phi}{\partial \theta}\right). \quad (12)$$

Now, inspired by Equation (10) we define the following electromagnetic quantities

$$B_\theta = \sqrt{g^{rr}} \left(\frac{\partial A_\phi}{\partial r}\right), \quad (13)$$

$$B_r = \sqrt{g^{\theta\theta}} \left(\frac{\partial A_\phi}{\partial \theta}\right), \quad (14)$$

It is important to realize that these components are not exactly the components measured by the Eulerian observer, but rather convenient definitions of electromagnetic functions that allow us to write the components of T^{EM} in a more intuitive manner, as

$$T^{EM0}{}_0 = -\frac{1}{8\pi} g^{\phi\phi} \left(B_r^2 + B_\theta^2\right), \quad (15)$$

$$T^{EM1}{}_1 = -\frac{1}{8\pi} g^{\phi\phi} \left(B_r^2 - B_\theta^2\right), \quad (16)$$

$$T^{EM1}{}_2 = \frac{1}{8\pi} 2 g^{\phi\phi} \sqrt{\frac{g^{rr}}{g^{\theta\theta}}} B_r B_\theta. \quad (17)$$

Here, if we want to fully comprehend the physical meaning of the components of the electromagnetic energy-momentum tensor, we must draw a parallel with its flat-space counter-part, given (in S.I. units) as [38]

$$T^{EM\mu\nu} = \begin{pmatrix} \frac{1}{2}(\epsilon_0 E^2 + \frac{1}{\mu_0}B^2) & S_x/c & S_y/c & S_z/c \\ S_x/c & -\sigma_{xx} & -\sigma_{yy} & -\sigma_{zz} \\ S_y/c & -\sigma_{yx} & -\sigma_{yy} & -\sigma_{yz} \\ S_z/c & -\sigma_{zx} & -\sigma_{zy} & -\sigma_{zz} \end{pmatrix}, \quad (18)$$

where $\vec{S} = \frac{1}{\mu_0}\vec{E}\times\vec{B}$ is the Poynting vector and the components σ_{ij} are given by

$$\sigma_{ij} = \epsilon_0 E_i E_j + \frac{1}{\mu_0}B_i B_j - \frac{1}{2}\left(\epsilon_0 E^2 + \frac{1}{\mu_0}B^2\right)\delta_{ij}. \quad (19)$$

The first term in Equation (18) is easily identified as the electromagnetic energy density, the other terms in the diagonal, i.e., σ_{xx}, σ_{yy}, σ_{zz} can be read as the electromagnetic pressure and the terms σ_{ij} for $i \neq j$ represents shear stress.

Inspired in the electromagnetic energy-momentum tensor for flat space-time, we define the following quantities

$$W \equiv \frac{1}{8\pi}g^{\phi\phi}\left(B_r^2 + B_\theta^2\right), \quad (20)$$

$$\Pi \equiv \frac{1}{8\pi}g^{\phi\phi}\left(B_r^2 - B_\theta^2\right), \quad (21)$$

$$\sigma \equiv \frac{1}{8\pi}2g^{\phi\phi}B_r B_\theta. \quad (22)$$

With these definitions, the matrix form of the electromagnetic energy-momentum tensor looks like

$$T^{EM\mu}{}_\nu = \begin{pmatrix} -W & 0 & 0 & 0 \\ 0 & -\Pi & r\sigma & 0 \\ 0 & \frac{1}{r}\sigma & \Pi & 0 \\ 0 & 0 & 0 & W \end{pmatrix}. \quad (23)$$

From Equation (23) we can extract the following properties for T^{EM}: it is symmetric, traceless and the component T^{EM00} is positive definite, which are the expected properties of an electromagnetic energy-momentum tensor. One must note that Equation (23) correspond to the mixed components of the electromagnetic energy-momentum tensor, whereas the properties just defined are related to the contra-variant components.

Combining Equations (4) and (23), the matrix form of the energy-momentum tensor describing a perfect fluid coupled with a poloidal magnetic field for the line element (2) is

$$T^{\mu\nu} = \begin{pmatrix} \frac{1}{A^2}(\rho + W) & 0 & 0 & 0 \\ 0 & \frac{1}{B^2}(P - \Pi) & \frac{1}{rB^2}\sigma & 0 \\ 0 & \frac{1}{rB^2}\sigma & \frac{1}{(Br)^2}(P + \Pi) & 0 \\ 0 & 0 & 0 & \frac{1}{D^2}(P + W) \end{pmatrix}. \quad (24)$$

The first term, i.e., T^{00} in Equation (24) represents the total energy density of the system which comes from the perfect fluid distribution and the electromagnetic field, through the quantity W; the others diagonal terms correspond to the pressure and as we can see the quantities Π and W, which depend on the electromagnetic four potential, and compose the total pressure of the system. Finally, the off-diagonal terms depend only on the electromagnetic four potential and represents the shear stress of the system σ.

With the goal of providing a physical interpretation to the quantities W, Π and σ we now derive the conservation equations for a perfect fluid coupled with a poloidal magnetic field and compare these equations with those obtained in [39] where no electromagnetic contribution was considered.

The non-vanishing components of the conservation equations $T^{\mu\nu}{}_{;\nu} = 0$ for the energy-momentum tensor (24) are

For $\mu = 0$

$$\dot{\rho} + \dot{W} = 0 \tag{25}$$

where the dot denotes derivative with respect to t. Equation (25) is a consequence of the staticity.

The other non-vanishing components are

$\mu = 1$

$$(P - \Pi)' + \frac{A'}{A}(\rho + W + P - \Pi) - \frac{B'}{B} 2\Pi - \frac{D'}{D}(W + \Pi) + \tag{26}$$
$$+ \frac{1}{r}\left[\sigma_{,\theta} + \left(\frac{A_{,\theta}}{A} + 2\frac{B_{,\theta}}{B} + \frac{D_{,\theta}}{D}\right)\sigma - 2\Pi\right] = 0,$$

$\mu = 2$

$$(P + \Pi)_{,\theta} + \frac{A_{,\theta}}{A}(\rho + W + P + \Pi) + \frac{B_{,\theta}}{B} 2\Pi - \frac{D_{,\theta}}{D}(W - \Pi) + \tag{27}$$
$$+ r\left[\sigma' + \left(\frac{A'}{A} + 2\frac{B'}{B} + \frac{D'}{D}\right)\sigma\right] + 2\sigma = 0,$$

where f' and $f_{,\theta}$ means derivative with respect the coordinates r and θ, respectively. Equations (26) and (27) are the hydrostatic equilibrium equations. In the special case of vanishing magnetic field and istropic fluid, these equations correspond to the Tolman–Oppenheimer–Volkoff Equations [40–42].

Herrera et al. [39], studied axially symmetric, static bound sources. The matter content considered for the authors in a local Minkowski coordinates (τ, x, y, z) is given by

$$\hat{T}_{\alpha\beta} = \begin{pmatrix} \mu & 0 & 0 & 0 \\ 0 & P_{xx} & P_{xy} & 0 \\ 0 & P_{yx} & P_{yy} & 0 \\ 0 & 0 & 0 & P_{zz} \end{pmatrix}, \tag{28}$$

where $\mu, P_{xx}, P_{yy}, P_{zz}, P_{xy} = P_{yx}$ denote the energy density and different stresses, respectively, measured by a local Minkowski observer. In a spacetime described by (2), the canonical form of the energy-momentum tensor is

$$T_{\alpha\beta} = (\mu + P)V_\alpha V_\beta + P g_{\alpha\beta} + \Pi_{\alpha\beta}, \tag{29}$$

with

$$\Pi_{\alpha\beta} = (P_{xx} - P_{zz})\left(K_\alpha K_\beta - \frac{h_{\alpha\beta}}{3}\right)$$
$$+ (P_{yy} - P_{zz})\left(L_\alpha L_\beta - \frac{h_{\alpha\beta}}{3}\right) + 2P_{xy}K_{(\alpha}L_{\beta)}, \tag{30}$$

$$P = \frac{P_{xx} + P_{yy} + P_{zz}}{3}, \quad h_{\alpha\beta} = g_{\alpha\beta} + V_\alpha V_\beta, \tag{31}$$

where
$$V_\alpha = (-A, 0, 0, 0), \quad K_\alpha = (0, B, 0, 0), \quad L_\alpha = (0, 0, Br, 0) \tag{32}$$

The conservation equations calculated by [39] are

$$P'_{xx} + \tfrac{A'}{A}(\mu + P_{xx}) + \tfrac{B'}{B}(P_{xx} - P_{yy}) + \tfrac{D'}{D}(P_{xx} - P_{zz}) + \tfrac{1}{r}\left[P_{xy,\theta} + \left(\tfrac{A_\theta}{A} + 2\tfrac{B_\theta}{B} + \tfrac{D_\theta}{D}\right) P_{xy} + P_{xx} - P_{yy}\right] = 0, \tag{33}$$

$$P_{yy,\theta} + \tfrac{A_\theta}{A}(\mu + P_{yy}) + \tfrac{B_\theta}{B}(P_{yy} - P_{xx}) + \tfrac{D_\theta}{D}(P_{yy} - P_{zz}) + r\left[P'_{xy} + \left(\tfrac{A'}{A} + 2\tfrac{B'}{B} + \tfrac{D'}{D}\right) P_{xy}\right] + 2P_{xy} = 0. \tag{34}$$

Comparing Equations (26) and (27), which describe a perfect fluid coupled with a poloidal magnetic field, with the hydrostatic Equations (33) and (34), calculated in [39], which describe an anisotropic fluid (without electromagnetic contribution), we can read the quantities $\rho + W$ as the total energy density of our distribution. In fact, the definition of W given by Equation (20) remind us of the typical definition of the electromagnetic energy density. The quantity 2Π can be read as the anisotropy of the distribution, and it is a direct consequence of the poloidal magnetic field. The quantity σ given by (22) can be identified as the shear stress experienced by the fluid. The quantities $W + \Pi$ and $W - \Pi$ can be read as an anisotropy defined with respect to z-axis. In conclusion, if we apply the Bondi approach [43] then a locally Minkowskian observer measures, for the perfect fluid coupled with a poloidal magnetic field, $\rho + W$ as the total energy density, 2Π as the anisotropy caused by the different components of the magnetic field and σ as the shear stress experienced by the fluid.

3.2. Global Structural Properties

In the previous section we have addressed the formal aspects of the magnetic field in a general relativistic, axis-symmetric environment. Such studies are useful in aiding us in understanding how the presence of a poloidal magnetic field alters the geometry of star and to properly identify relevant quantities such as anisotropy, shear stress and energy density. One must, however, resort to numerical calculations if one wants to quantitatively describe such alterations in the structure of neutron stars. With that in mind we now move forward to discuss numerical results of the effects of magnetic fields on macroscopic stellar properties, obtained from using the LORENE (Langage Objet pour la RElativité NumériquE) code [37,44]. In this axisymmetric formalism, a global poloidal magnetic field is generated though a global current and the field strength generated depends on the stellar radius, angle θ (with respect to symmetry axis), and dipole magnetic moment μ, being different for each equation of state. Note that different approaches agree that the maximum central magnetic field inside neutron stars cannot be larger than a couple of times 10^{18} G [37,44–47].

As numerical calculations show, there are substantial changes in masses and radii for neutron stars that possess strong magnetic fields, but the change comes mainly form the pure magnetic field contribution to the energy momentum tensor, which is highly anisotropic, as discussed in the previous section. Figure 2 shows the gravitational stellar mass and central enthalpy for families calculated with a fixed dipole magnetic moment. The different lines for each color show (1) effects only from the pure magnetic field contribution, (2) effects from pressure anisotropy and pure the magnetic field contribution, and (3) the effects from Landau levels, pressure anisotropy, and the pure magnetic field contribution . Note that fixed baryon masses (in the case of isolated stars) would imply moving up but at the same time left in the figure, in a way that central enthalpy decreases with magnetic field strength. For more details on the value for the mass increase in magnetic stars (which is of the order of percents in the case of fixed baryonic mass), see for example [27,48]. Note that it has been shown that such decrease in central enthalpy/chemical potential/density is equation of state and stiffness dependent [48].

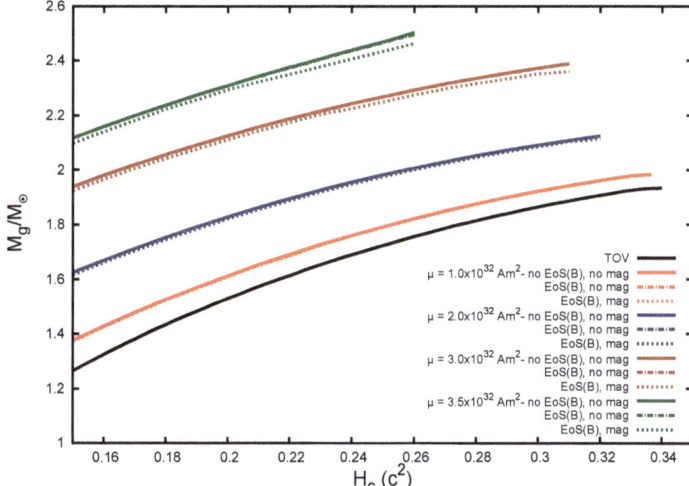

Figure 2. Equilibrium sequence obtained from the numerical solution of he Einstein-Maxwell equations for the CMF model equation of state shown for different dipole magnetic moments including different magnetic field effects.

In this formalism, the stellar radii are largely modified, as poloidal magnetic fields generate oblate stars, with increased/decreased values of about tens or percent in the equatorial/polar direction, respectively. For extreme magnetic fields, for instance, the ratio $r_{polar}/r_{equatorial}$ can reach values near to 0.5 in the more extreme configurations (near the critical magnetic field after which the star starts to attain a torus geometry). One must note that such extreme configuration and the critical values of the magnetic field depend on the equation of state properties.

Note however that different stellar shapes are generated when toroidal magnetic fields are considered [49]. Although it is understood that the long-term stability of magnetic stars requires poloidal and toroidal components, those configurations are much more expansive to model numerically; see Ref. [30].

4. Evolution Aspects

Finally, we now investigate evolutionary aspects of the magnetic field. One must notice that in previous sections the aspects of the magnetic fields on neutron stars were studied under the assumption that such fields are static (or quasi-static). This does not need to be the case, and, in fact, observations show that the magnetic field in neutron stars is evolving. In this section we propose a phenomenological model for the evolution of magnetic fields that may affect indirectly (by masking the true age of the star) and directly (by generating heat) the thermal evolution of magnetized neutron star. We note, however, that we perform such studies independently, that is, ignoring micro and macroscopic effects. We stress that a complete, self-consistent study must take all of this into account, which is our ultimate goal for future investigations. For now, we focus in the objects known as XDINS. These are a few objects first detected by the ROSAT All-Sky Survey that are radio quiet but exhibit bright emission in the soft X-Ray band [50]. These objects do not exhibit non-thermal hard emission and their spectra are blackbody-like [51]. Furthermore, there is no known association with any supernova remnant, which could indicate that these objects cannot be too young [51]. Considerable information regarding the spin properties of XDINS has been collected along the years; see Table 1 for a summary of them.

Table 1. Observational data for Isolated X-ray Neutron Stars (XDINS).

XDIN Name	Period (s) P (s)	Period Derivative $\dot{P}$ (s s^{-1})	Blackbody Temperature T_{bb} (K)	References	Spin-Down Age (Years) t (yr)
RX J0720.4-3125	8.39	6.98×10^{-14}	9.7×10^{5}	[52,53]	1.9×10^{6}
RX J0806.4-4123	11.37	5.5×10^{-14}	1.17×10^{6}	[53,54]	3.3×10^{6}
RX J1308.6+2127	10.31	1.12×10^{-13}	1.09×10^{6}	[53,55]	1.5×10^{6}
RX J1856.5-3754	7.055	2.97×10^{-14}	7.30×10^{5}	[56,57]	3.8×10^{6}
RX J2143.0+0654	9.437	4.14×10^{-14}	1.24×10^{6}	[58,59]	3.6×10^{6}
RX J0420.0-5022	3.45	2.8×10^{-14}	6.0×10^{5}	[58,59]	2.0×10^{6}
RX J1605.3+3249	3.39	1.6×10^{-12}	1.15×10^{6}	[60,61]	3.4×10^{4}

As shown in Table 1, the temperature obtained from their blackbody-like spectra (in the range of 0.6–1.4 $\times$ 10^{6} K) is normally associated with young neutron stars, however their long periods and absence of radio emission (both of which are normally observed in young neutron star) may indicate that these objects are not too young. Furthermore age estimations based on their spin-down properties, namely the spin-down age $\tau_{SD} \equiv P/(2\dot{P})$, indicate very old ages. Evidently, the spin-down properties may not be a good indicator for the true age of the neutron stars [53], since it makes a series of strong assumptions, in particular with respect to the magnetic field that is assumed to be constant throughout the evolution. On the other hand, if the temperature of these objects is to be explained solely due to their young age, they would have to been born with very high magnetic fields or unusually long periods, as to explain how they can attain such long periods during their thermal evolution life [51]. This presents a challenge: how to explain such objects that show temperature of young neutron stars and rotation properties of old ones, when evidence indicates that they can be neither? We attempt to answer this question by computing the neutron star thermal evolution combined with a magnetic field evolution model. We show that the combination of their age, masked by the magnetic field evolution, plus the thermal emission, and in some cases possibly with some moderate internal heating source, could indicate the true age of these pulsars to lie in the intermediary range, i.e. between young and old ages. This result is in agreement with recent measurements of the proper motion of three XDINS, that indicate the age of at least of these stars approximate to be 4 $\times$ 10^{5} years.

For that purpose we consider the thermal evolution of these objects, by making use of state of the art cooling calculations [62–64]. We employ the most recent thermal emission and cooling mechanisms available, that have also been used to explain the thermal properties of the cooling neutron star in Cassiopeia A [62,65–67]. In addition, we also consider a phenomenological model for the magnetic field evolution, first considering the magnetic field decay that could be used to explain the heat deposit in the neutron star crust [51], and also the magnetic field emergence (growth) after an initial hyper-critical accretion, followed by a final magnetic field decay (see Refs. [68–71], and references therein). We show how the thermal/spin properties of the XDINS may be explained in such scenario with and without the need of a heating mechanism.

4.1. Magnetic-Thermal Evolution

Here we aim to reconcile the observed properties of XDINS (in particular their thermal behavior) with the current trends on the theory of neutron star cooling. As mentioned before, if one uses the current models for neutron star thermal evolution and the observed properties of XDINS, namely T_{bb} and the spin-down age $P/(2\dot{P})$, one obtains a clear disagreement, as shown in Figure 3.

For the cooling calculation of Figure 3 we used a neutron star whose microscopic composition and equation of state (EoS) is given by the APR model [72]. The neutron star mass for this particular calculation is 1.55 $M_\odot$. Furthermore neutron and proton pairing are assumed to take place in the star such that agreement with Cassiopeia A is possible [65–67]. Figure 3 shows that the XDINS seem to be much warmer than what is predicted for their ages. This would indicate that, either there is a powerful heating mechanism (unbeknownst to us) that is keeping these objects warm at old ages, or, these objects are much younger than their apparent age. There is, of course, the possibility of both.

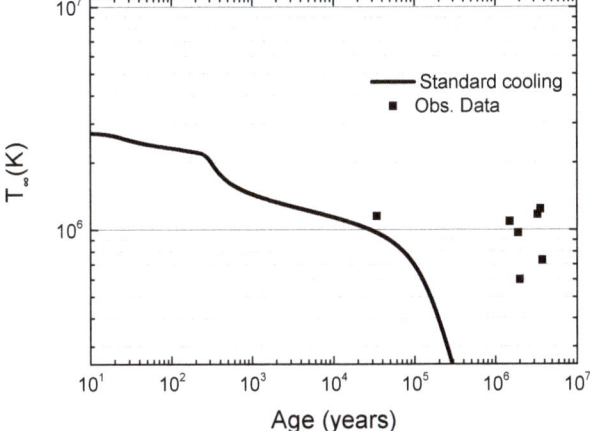

Figure 3. Temperature evolution of a 1.5 $M_\odot$ neutron star. The squares represent the observed temperature of the XDINS at their respective spin-down age.

As discussed above, the spin-down age is a far cry from a reliable measurement of the true age of a neutron star. In the few cases in which both the spin-down age and the kinematic age have been estimated, they were usually off by thousands of years [73]. Even for XDINS, the few cases in which it was possible to estimate the kinematic age, it has been found that the neutron star must be younger than what their spin-down age indicates (see [53,74], and references therein). A possible way in which the spin-down age of an object would be different than the actual age of the object is if the magnetic field of the star evolves over time. Several mechanisms have been proposed to explain the variation of the magnetic field, e.g., ohmic dissipation and ambipolar diffusion (both responsible for the decrease of the magnetic field) [53]. There has also been proposed that the magnetic field may be initially buried in the neutron star crust, due to a stage of hyper-critical accretion [71,75,76], followed by an increase over time [70].

Here, rather than assuming a particular model for the magnetic field evolution, we choose a phenomenological approach, that gives us the freedom to investigate different relaxation and evolution times for the magnetic field, as to investigate under which conditions the magnetic field evolution may lead to an agreement with the observed spin-down age and thermal properties of XDINS. This study does not exclude the possibility of heating also taking place; actually, the agreement with observed data can only be improved if heating is present.

We start considering the possibility of the object being born with a magnetic field that undergoes a decay over time. We adopt a function for the magnetic field decay in the form of a smoothed top-hat with the parameter τ_D indicating the relaxation time. We study different relaxation times for the magnetic field decay, as indicated in Figure 4.

Applying the magnetic breaking model for such magnetic field evolution, one can find the relevant rotation quantities, such as the rotation period and period time derivatives, which allow us to calculate the estimated spin-down age, $\tau_c = P/(2\dot{P})$, and contrast it to the real age of the object. This is shown in Figure 5. We must stress that the spin-down age $\tau_c = P/(2\dot{P})$ is only valid for the canonical case where the magnetic field is constant, which, evidently is not the case here. We have, however, chosen to use the quantity $\tau_c = P/(2\dot{P})$ as a parameter, even though we are not considering a constant magnetic field, since this is how the spin-down age is estimated based on observational data (such as that in Table 1). Therefore, by doing that, we make sure we are comparing the same quantities. This is, in fact, an important point of this work, that we must look at the quantity $\tau_c = P/(2\dot{P})$ as an observational parameter, rather than a true reflection of the age of the star.

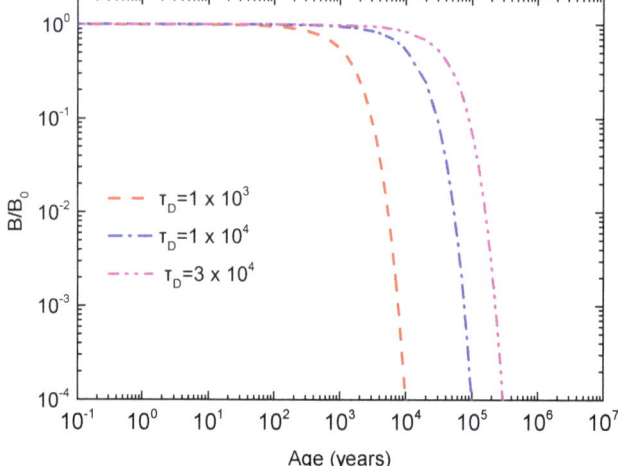

Figure 4. Different magnetic field evolution functions used to represent a neutron star undergoing magnetic field decay. τ_D represents the magnetic field relaxation time.

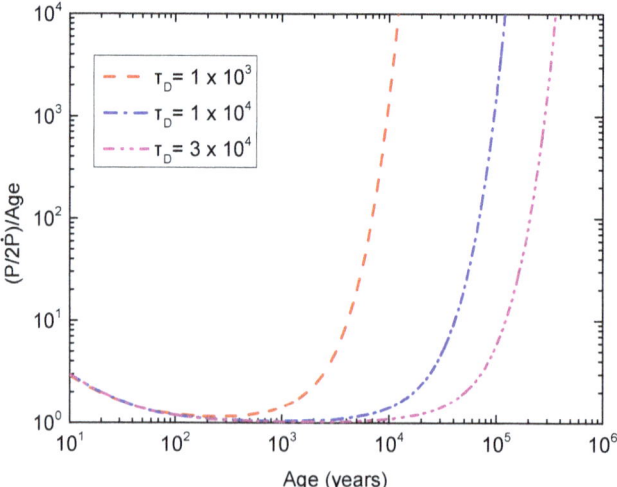

Figure 5. Estimated spin-down age $P/(2\dot{P})$ over the true age of the pulsar as estimated for the magnetic field evolutions of Figure 4.

As one can see in Figure 5, in the initial stages of evolution, the spin-down age differs very little from the true age. At later stages of evolution, we have a deviation from the canonical case, with the spin-down age indicating an age much older than the actual age of the star. This result shows, as discussed before, that comparing the observed temperature of neutron stars with their observed spin-down age may erroneously indicate than an object retains high temperature at later ages. We show this by calculating the temperature evolution of the same neutron star used in Figure 3, undergoing the magnetic field evolution depicted in Figure 4. These results are shown in Figure 6.

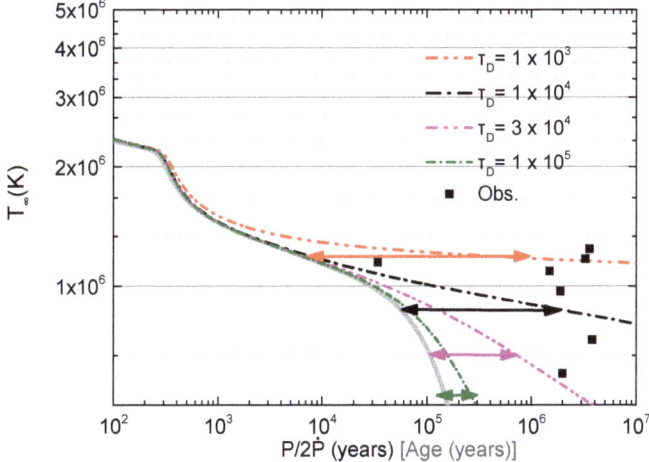

Figure 6. Thermal evolution of a 1.55 $M_\odot$ neutron star undergoing magnetic field decay. The solid washed gray line represents the temperature as a function of the real age of the object. For the other curves the x-axis represent the spin-down age observed. We also plot a few arrows depicting the discrepancy between the true and spin-down age.

As indicated by Figure 6, by using the spin-down age as a measurement of the age of the object, one may believe that a neutron star is much older than reality. One also sees that for stronger magnetic field decay (those that take place more quickly, with lower relaxation times), the deviation from the real age of the object becomes more pronounced. For instance, as can be seen from Figure 6, a star with surface temperature of 10^6 K may appear to be 10^6 years old as measured by the spin-down age while, if we assume that the magnetic field is decaying with $\tau_D = 10^{3-4}$ years, the star has a real age of approximately 5×10^4 years (see red arrow in Figure 6). The interesting fact of this result is that, if the magnetic field is indeed undergoing a decay process, we may use the observed temperature to estimate the age of the star, or, conversely, if we know the age of the star from other measurements, for instance kinematic estimates, we can then determine how the magnetic field of the object is evolving.

Based on the results just discussed, we can conclude that if the only reason of these objects appearing to be warm at old ages is due to magnetic field decay, the warmer the temperature of the object the stronger the magnetic field decay needs to be. This is due to the fact that high temperatures are associated with younger stars (if no heating is assumed, as is the case for now), thus only with a strong magnetic field decay one can explain the high deviation from the real age of the object. For lower observed temperatures, a moderate magnetic field decay is enough, since a lower temperature does not require very young ages. Our conclusion is that for objects with the highest temperatures it is unlikely that only the magnetic field decay and the consequent deviation from real age are enough to explain their observed properties. For those cases, a heating source may be necessary. For the high temperature objects in Table 1, if no heating is present, the real age of the star would be $\sim 10^3$ years, and the magnetic field would have to strongly decay in a matter of 10^3 years as well. This seems to be an unlikely scenario. As for the objects with lower temperatures, one may not need to resort to heating, since a moderate magnetic field decay with relaxation times $\tau_D = 10^{3-5}$ years would be enough to explain their temperatures, and the real age would be around 10^{4-5} years. It is important to notice that measurements of the proper motion of three of the magnificent seven have yielded to kinematic ages lower than the spin-down ones [53]; Furthermore, as mentioned before, the new distance calculated for RXJ1856.5–3754 [74], implies a kinematic age of 5×10^4 years, in agreement with the age we estimate for the moderately warm XDINS.

4.2. Complex Magnetic Field Evolution

In addition to the above study of the magnetic field decay scenario, we proceed now to consider as well a somewhat more complex situation. As it has been proposed in Refs. [68,71,77], after a hyper-accretion phase at the early stages of evolution, the magnetic field of a neutron star may be buried into the crust of the object. In the stellar core-collapse scenario (including catastrophic collapse of an iron core, generating a shock wave, and the dramatic ejection of the star envelope), the residual compact remnant could be a neutron star immersed in a dense environment. Such just born neutron star may be exposed to an hyper-accretion phase from fallback material a few seconds after the supernova event [78]. When the ejected matter collides with denser external layers, it is bifurcated and a reverse shock is formed. In that case, some part of ejected material continues on its way to the outside where the hydrodynamics effects are not very important, while another part of the material fallback onto the surface of the proto-neutron star. This reverse shock allows to deposit large amount of matter onto the proto-neutron star surface forming a new crust and hiding there any initial configuration of the magnetic field of the star, as it was shown in Ref. [71]. In that case, the magnetic field may be buried. After the hyper-accretion phase ends and the residual envelope of the newborn neutron star vanishes, the magnetic field can suffer a growth phase, emerging from the crust and appearing as a delayed-off pulsar, as it was pointed out in [68,70]. When the magnetic field is buried on the stellar surface, it can be amplified by compression due to the strong accretion and by a turbulent dynamo. In this scenario, the magnetic field could grow to higher values than those estimated for XDINS and then suffer the familiar decay stage as we studied above.

We now investigate such scenario by considering, as before, a phenomenological evolution for the magnetic field that takes into account the emergence and decay phase. We consider a two smoothed top-hat analytic function that take into account the early growth and the later decay. We consider all growth to have the same relaxation time, as indicated in Figure 7, and only vary the relaxation time of the magnetic field decay (τ_D). Such magnetic field evolution is represented in Figure 7.

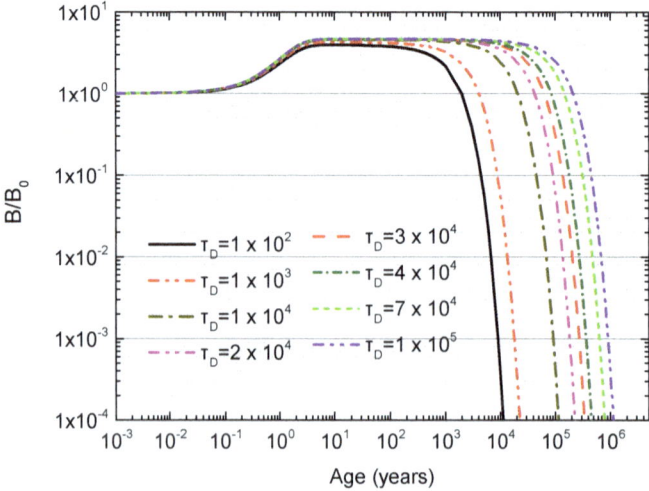

Figure 7. Magnetic field evolution representing the emerging after a hyper-accretion phase followed by the decay of the field. τ_D represents the relaxation time of the decay phase.

We show in Figure 8 the spin-down age over the true age of the star of a neutron star whose magnetic field undergoes such an evolution. This result shows us that initially the spin-down age is slightly higher than the true age (reflecting the growth of the magnetic field), followed by the a behavior equal to the canonical case, since once again the magnetic field is constant. Later, at the onset of the magnetic field decay, we see the same deviation from the canonical case as the previously studied case, with the spin-down age indicating objects older than their real age. As expected, since at this epoch the behavior is the same as in the decay-only scenario, we see that for stronger (faster) magnetic decays we obtain stronger deviations from the canonical case.

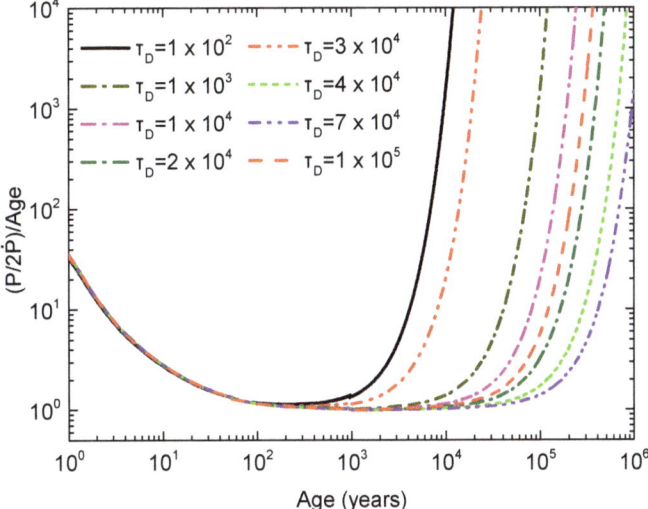

Figure 8. Spin-down age ($P/2\dot{P}$) over the true age of the star as a function of the real age for a neutron star whose magnetic field emerges, after the burial due to a hyper-accretion phase, and then it is followed by a decay.

We now calculate the thermal evolution of the same neutron stars that we have followed in this investigation, except now considering such magnetic field evolution. This, as it was the case before, allows us to keep track of their real and spin-down age and, by using the observed temperature and spin-down parameters (particularly P and $\dot{P}$), we can probe which magnetic field evolution would be appropriate to explain such observed quantities without resorting to any heating mechanisms. The thermal evolution calculations are shown in Figure 9.

As the case we studied before, we see that the magnetic field decay may lead one to believe that a neutron star is much older than reality (see black arrow in Figure 9). The emergence of the magnetic field has little impact on the spin-down age at later stages, thus we cannot use the thermal evolution to determine whether or not such emergence occurs. The same conclusions for the decay-only case hold here: the best scenario to describe the observed properties of the XDINS (without resorting to any heating mechanism) would be a magnetic field decay with relaxation time of the order of $\tau_D = 10^{3-5}$ years. This is true to the objects that are moderately warm, those that are hotter would need a more violent magnetic field decay, which would also indicate that they are actually much younger. As mentioned before, we believe that this is an unlikely scenario, due to the lack of evidence that such objects are much younger than 10^4 years.

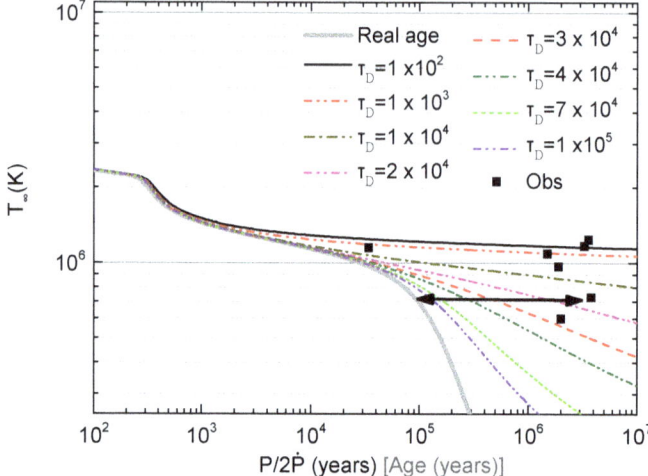

Figure 9. Thermal evolution of neutron stars whose magnetic field evolve according to Figure 7. With the exception of the washed gray curve, which shows the real age, the curves represent the temperature as observed by an observer at infinity as a function of the spin-down age ($P/2\dot{P}$). We also indicate by the arrow an example of the discrepancy between the true and spin-age (to avoid cluttering the image we only show one arrow).

Finally, we briefly consider the presence of heating, as to demonstrate that our model does not invalidate such scenario and that, in fact, it benefits from it. As mentioned before, we believe that the warmer objects of Table 1 would be better explained by a combination of heating and magnetic field decay. We have seen that the latter alone would indicate that these objects are undergoing a rather fast magnetic field decay (of the order of ~100–1000 years) and that they would need to be very young (as to explain their high temperature). On the other hand, if their temperature is to be explained solely by heating, a considerable amount of continuous heating would be needed to keep such objects so warm at such old ages. We thus consider a heating source that is uniformly distributed along the crust of the neutron star, with an intensity of $H \approx 10^{16}$ erg s^{-1} cm^{-3}, that is active for ~10^4 years. The magnetic field evolution is the same as we have used in the previous calculations.

As it is shown in Figure 10, the presence of a heating increases the temperature of the neutron star while the heating source is active. With this, one can now explain the high temperature observed in a few XDINS as a result of a magnetic field decay with $\tau_D \sim 10^4$ years combined with a heating acting with a relaxation time also of ~10^4 years. In this scenario, as indicated in Figure 10, the true age of the object would be ~10^5 years, rather than ~10^{3-4} years, as it would be in the absence of heating. Evidently, one could argue that a hypothetical source of heat could be active for 10^{6-7} years, thus solely explaining the high temperature of these stars. We cannot rule out this possibility; however, we believe that the combination of moderate heating plus a magnetic field evolution that masks the true age of the star is a more realistic scenario, since the explanation based only on the existence of a heating source would need a very large internal energy-power ($\dot{E}_H \sim 10^{34}$ erg s^{-1}) active for very long times of the order of 10^{6-7} years.

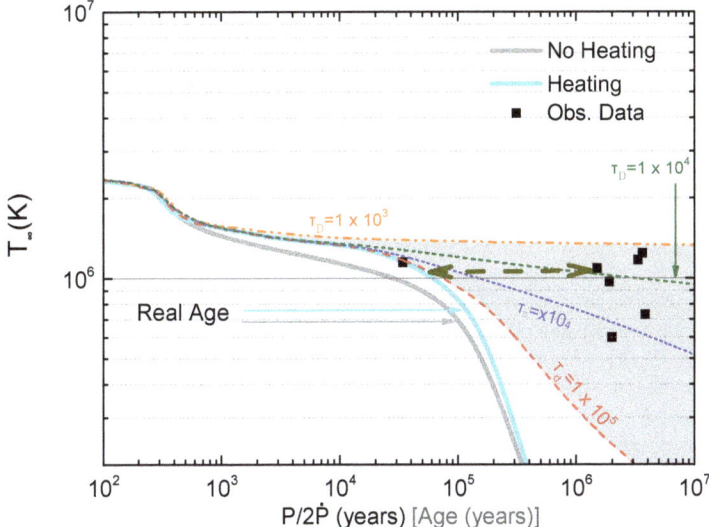

Figure 10. Same as Figure 9 but with a heating source of $H \approx 10^{16}$ erg s^{-1} cm^{-3} uniformly applied to the crust for the first 10^4 years. Curves accompanied by the indicated τ_D represent the temperature as a function of the spin-down age ($P/2\dot{P}$). The shaded area represents where the curves for magnetic field evolutions with $\tau_D = 10^{3-5}$ years lie. We also indicate, for the sake of example, the difference between the spin-down age and real age (in the presence of heating) for the XDINNS RXJ130862127.

Our results indicate that the magnetic field decay works well to explain the moderately warm XDINS (with $T < 10^6$ K). For the higher temperature XDINS, such scenario seems far fetched since it would require a rather fast magnetic field decay ($\tau_D \sim 10^{2-3}$ years) and a stellar true age of 10^{3-4} years. We believe that for such objects a moderate source of heating might be necessary. To illustrate that, we considered a heating source of $H \approx 10^{16}$ erg s^{-1} cm^{-3} acting in approximately the same timescale of the magnetic field decay, i.e., 10^4 years. We find that with such heating, the magnetic field decay shows a more palatable situation for the warmer XDINS, that may, under this scenario, be explained with a magnetic field decay with $\tau_D = 10^4$ years, leading to a true age of 10^{4-5} years. We stress, once again, that we cannot rule out the presence of intense heating that acts on the neutron star for 10^{6-7} years, making the magnetic field decay unnecessary. We believe, however, that the combination of magnetic field decay with a moderate heating source is a more realistic scenario, specially in view of the recent results regarding the kinematic age of a few XDINS [53,74] that indicate that such objects may be younger than the spin-down age indicates, with a true age $\sim 4 \times 10^5$ years.

One must note that other authors have investigated the evolution of magnetic fields taking a less phenomenological approach, based on numerical simulations; see, for instance, [79,80]. Such approach is certainly interesting, although different of our purpose in the research we present here, where rather than investigate a particular evolution model we chose to investigate a whole set of them, which, when compared to observed data, allow us to best choose the most likely behavior of the magnetic field evolution. Evidently we have no microscopic information, and we must eventually look for a microscopic model that agrees with the conclusions reached by our investigation, and for that one must resort to simulations as those found in [79,80]. Although beyond of the scope of this work, we have briefly compared our results with those of [79,80], and found that they are, for the most part compatible. In these studies, the authors presented numerical results of the magneto-thermal evolution of isolated neutron stars, including the most relevant physical parameters. They found that certain thermodynamic mechanisms in the inner crust of the neutron star are responsible for the high dissipation of the magnetic field on time-scales 100 kyr, which is compatible with our phenomenological approach that allow us to explore interesting characteristics of such evolution without the need for intensive numerical work. We also note that the authors in [81] have shown that including Hall effect in their numerical simulations leads to a saturation of the magnetic field evolution within 100 kyr, allowing a slow and sustained decay. The conclusion is that the most energetic behavior occurs while the star is young (high magnetic fields involved). This issue support our results that a phase of growth of the magnetic field was subsequent to the hypercritical regime of the newborn neutron star.

Finally we stress that most data suggests an unified model for neutron stars whose key parameter is the magnetic field [82,83]. However, the magneto-thermal model does not explain how neutron stars may be born with such high magnetic fields. If the hidden magnetic field scenario is correct, there exists the possibility that the magnetic field can be amplified (by two or three orders of magnitude) due a turbulent dynamo effects inside the neutron star crust in a timescale of months, and then subsequently undergo a phase of growth appearing as a delayed switch-on pulsar, several years after the supernova event, which may explain the highest magnetic fields associated with magnetars. Following that, the neutron star continues its canonical evolution until the thermo-magnetic effects can be relevant and cause the magnetic field to decay at the aforementioned time-scales. This phase of amplification of the magnetic field in the crust requires an intensive numerical treatment that is outside the scope of this work. But we aim to explore this possibility in a future work, as well as coupling such evolution to microscopic and macroscopic studies.

5. Conclusions

In this work we studied many ways in which a magnetic field may affect neutron stars, starting with the microscopic composition, going through the stellar structure, and culminating with evolutionary aspects. As we have shown, the proper account of the magnetic field in these different realms is no ordinary task. It is, however, imperative to do so. As we have seen, the magnetic field plays an important role in all regimes of the neutron star, altering its composition, affecting its structure and geometry, and being a key factor for the neutron star evolution.

For the composition, our study shows that the magnetic field may suppress the appearance of hyperons and the quark-matter phase transition, leading to a substantially different particle composition when compared to a neutron stars with small or vanishing magnetic fields.

In the macroscopic realm, we have discussed the formal aspects under which a dipolar magnetic field needs to be considered under the light of general relativity. We have shown that the presence of a dipolar magnetic field causes the fluid to become anisotropic and we identified the shear stress caused by it. We have also numerically calculated the structure of highly magnetized neutron stars and have shown that high magnetic fields can lead to more massive objects.

Finally we have shown that a better comprehension of the magnetic field evolution in neutron stars is fundamental for being able to accurately estimate the ages of XDINS. We propose to use XDINS as a way of probing the magnetic field evolution, by making use of the observed spin-down properties (particularly the spin-down age) and thermal properties. By making use of kinematic observations capable of making better estimates of the star's true age, we can use the temperature to determine the most likely scenario for the magnetic field evolution and/or presence of heating. We have found that for moderately warm XDINS there is no need of heating, and a magnetic field evolution with relaxation times of $\tau_D \sim 10^{4-5}$ years is enough to explain their thermal properties, making their true age $\sim 10^5$ years. As for the warmer XDINS, we have found that a moderate heating is necessary, in addition to magnetic field decay with $\tau_D \sim 10^4$ years, leading to a true age for these objects $\sim 10^{4-5}$ years.

As thoroughly discussed above we recall that as much as we strived to describe the different ways in which a magnetic field may affect neutron stars, we have certainly not covered all the possibilities. Furthermore, all of the studies we presented were done independently of each other, whereas a more appropriate (and challenging) treatment would consider all aspects on the same framework. That is our goal for future investigations. Considering that now we have a more comprehensive understanding of how the magnetic field affects these three realms (microscopic, macroscopic and evolutionary) we hope to perform self-consistent studies of all of them. That means that the equation of state (microscopic) would be affected by the macroscopic distribution of the magnetic field in the star, as well as the evolution of such fields. Evidently the macroscopic structure as well as the evolution of the object also depend on the microscopic equation of state, making such treatment highly non-linear and challenging, although still warranted if one aims to obtain a more complete description of neutron stars.

Acknowledgments: V.D. acknowledges support from NewCompStar COST Action MP1304. R.N. and O.T. acknowledges financial support from CAPES and CNPq. R.N acknowledges that this work is a part of the project INCT-FNA Proc. No. 464898/2014-5.

Author Contributions: All authors contributed equally to the formulation of the model, numerical analysis and writing of the paper.

Conflicts of Interest: The authors declare no conflict of interest.

Reference

1. Duncan, R.C.; Thompson, C. Formation of very strongly magnetized neutron stars—Implications for gamma-ray bursts. *Astrophys. J.* **1992**, *392*, L9–L13.
2. Thompson, C.; Duncan, R.C. Neutron star dynamos and the origins of pulsar magnetism. *Astrophys. J.* **1993**, *408*, 194–217.
3. Olausen, S.A.; Kaspi, V.M. The McGill Magnetar Catalog. *Astrophys. J. Suppl.* **2014**, *212*, 6.
4. Thompson, C.; Duncan, R.C. The soft gamma repeaters as very strongly magnetized neutron stars—I. Radiative mechanism for outbursts. *Mon. Not. R. Astron. Soc.* **1995**, *275*, 255–300.
5. Thompson, C.; Duncan, R.C. The Soft Gamma Repeaters as Very Strongly Magnetized Neutron Stars. II. Quiescent Neutrino, X-ray, and Alfven Wave Emission. *Astrophys. J.* **1996**, *473*, 322.
6. Kaspi, V.M.; Kramer, M. Radio Pulsars: The Neutron Star Population Fundamental Physics. *ArXiv* **2016**, arXiv:astro-ph.HE/1602.07738.
7. Melatos, A. Bumpy spindown of anomalous X-ray pulsars: The link with magnetars. *Astrophys. J.* **1999**, *519*, L77.
8. Makishima, K.; Enoto, T.; Hiraga, J.S.; Nakano, T.; Nakazawa, K.; Sakurai, S.; Sasano, M.; Murakami, H. Possible Evidence for Free Precession of a Strongly Magnetized Neutron Star in the Magnetar 4U 0142+61. *Phys. Rev. Lett.* **2014**, *112*, 171102.
9. Landau, L.D.; Lifshitz, E.M. *Quantum Mechanics: Non-Relativistic Theory*; Elsevier: Amsterdam, The Netherlands, 1977.
10. Lai, D.; Shapiro, S.L. Cold equation of state in a strong magnetic field—Effects of inverse beta-decay. *Astrophys. J.* **1991**, *383*, 745–751.
11. Chakrabarty, S. Quark matter in strong magnetic field. *Phys. Rev. D* **1996**, *54*, 1306–1316.
12. Chakrabarty, S.; Bandyopadhyay, D.; Pal, S. Dense nuclear matter in a strong magnetic field. *Phys. Rev. Lett.* **1997**, *78*, 2898–2901.
13. Yuan, Y.F.; Zhang, J.L. The Effects of Interior Magnetic Fields on the Properties of Neutron Stars in the Relativistic Mean-Field Theory. *Astrophys. J.* **1999**, *525*, 950–958.
14. Broderick, A.; Prakash, M.; Lattimer, J.M. The Equation of state of neutron star matter in strong magnetic fields. *Astrophys. J.* **2000**, *537*, 351.
15. Dexheimer, V.; Negreiros, R.; Schramm, S. Hybrid stars in a strong magnetic field. *Eur. Phys. J. A* **2012**, *48*, 189.
16. Canuto, V.; Chiu, H.Y. Quantum theory of an electron gas in intense magnetic fields. *Phys. Rev.* **1968**, *173*, 1210–1219.
17. Canuto, V.; Chiu, H.Y. Thermodynamic properties of a magnetized fermi gas. *Phys. Rev.* **1968**, *173*, 1220–1228.
18. Strickland, M.; Dexheimer, V.; Menezes, D.P. Bulk Properties of a Fermi Gas in a Magnetic Field. *Phys. Rev. D* **2012**, *86*, 125032.
19. Dexheimer, V.; Menezes, D.P.; Strickland, M. The influence of strong magnetic fields on proto-quark stars. *J. Phys. G Nucl. Part. Phys.* **2014**, *41*, 015203.

20. Andersen, J.O.; Naylor, W.R.; Tranberg, A. Phase diagram of QCD in a magnetic field. *Rev. Mod. Phys.* **2016**, *88*, 025001.
21. Broderick, A.E.; Prakash, M.; Lattimer, J.M. Effects of strong magnetic fields in strange baryonic matter. *Phys. Lett. B* **2002**, *531*, 167–174.
22. Bordbar, G.H.; Rezaei, Z. Magnetized hot neutron matter: Lowest order constrained variational calculations. *Phys. Lett. B* **2013**, *718*, 1125–1131.
23. Dexheimer, V.; Negreiros, R.; Schramm, S.; Hempel, M. Deconfinement to Quark Matter in Neutron Stars—The Influence of Strong Magnetic Fields. *AIP Conf. Proc.* **2013**, *1520*, 264–269.
24. Weinberg, S. Why do quarks behave like bare Dirac particles? *Phys. Rev. Lett.* **1990**, *65*, 1181–1183.
25. Ferrer, E.J.; de la Incera, V.; Manreza Paret, D.; Pérez Martínez, A.; Sanchez, A. Insignificance of the anomalous magnetic moment of charged fermions for the equation of state of a magnetized and dense medium. *Phys. Rev. D* **2015**, *91*, 085041.
26. Chatterjee, D.; Elghozi, T.; Novak, J.; Oertel, M. Consistent neutron star models with magnetic field dependent equations of state. *Mon. Not. Roy. Astron. Soc.* **2015**, *447*, 3785–3796.
27. Franzon, B.; Dexheimer, V.; Schramm, S. A self-consistent study of magnetic field effects on hybrid stars. *Mon. Not. Roy. Astron. Soc.* **2016**, *456*, 2937–2945.
28. Dexheimer, V.; Negreiros, R.; Schramm, S. Hybrid Stars in a Strong Magnetic Field. *Eur. Phys. J. A* **2012**, *48*, 189.
29. Alloy, M.D.; Menezes, D.P. Maxwell equation violation by density dependent magnetic fields in neutron stars. *Int. J. Mod. Phys. Conf. Ser.* **2017**, *45*, 1760031.
30. Dexheimer, V.; Franzon, B.; Gomes, R.O.; Farias, R.L.S.; Avancini, S.S.; Schramm, S. What is the magnetic field distribution for the equation of state of magnetized neutron stars? *Phys. Lett. B* **2017**, *773*, 487–491.
31. Oppenheimer, J.R.; Snyder, H. On Continued Gravitational Contraction. *Phys. Rev.* **1939**, *56*, 455–459.
32. Hurley, K.; Dingus, B.L.; Mukherjee, R.; Sreekumar, P.; Kouveliotou, C.; Meegan, C.; Fishman, G.J.; Band, D.; Ford, L.; Bertsch, D.; et al. Detection of a γ-ray burst of very long duration and very high energy. *Nature* **1994**, *372*, 652–654.
33. Galama, T.J.; Vreeswijk, P.M.; van Paradijs, J.; Kouveliotou, C.; Augusteijn, T.; Böhnhardt, H.; Brewer, J.P.; Doublier, V.; Gonzalez, J.F.; Leibundgut, B.; et al. An unusual supernova in the error box of the γ-ray burst of 25 April 1998. *Nature* **1998**, *395*, 670–672.
34. Popov, S.B.; Prokhorov, M.E. Progenitors with enhanced rotation and the origin of magnetars. *Mon. Not. R. Astron. Soc.* **2006**, *367*, 732–736.
35. Herrera, L.; di Prisco, A.; Martin, J.; Ospino, J.; Santos, N.O.; Troconis, O. Spherically symmetric dissipative anisotropic fluids: A general study. *Phys. Rev. D* **2004**, *69*, 084026.
36. Herrera, L.; di Prisco, A.; Fuenmayor, E.; Troconis, O. Dynamics of Viscous Dissipative Gravitational Collapse: A Full Causal Approach. *Int. J. Mod. Phys. D* **2009**, *18*, 129–145.
37. Bonazzola, S.; Gourgoulhon, E.; Salgado, M.; Marck, J.A. Axisymmetric rotating relativistic bodies: A new numerical approach for 'exact' solutions. *Astron. Astrophys.* **1993**, *278*, 421–443.
38. Misner, C.; Thorne, K.S.; Wheeler, J.A. *Gravitation*; Princeton University Press: Princeton, NJ, USA, 1973.
39. Herrera, L.; Di Prisco, A.; Ibáñez, J.; Ospino, J. Axially symmetric static sources: A general framework and some analytical solutions. *Phys. Rev. D* **2013**, *87*, 024014.
40. Tolman, R.C. Effect of Inhomogeneity on Cosmological Models. *Proc. Natl. Acad. Sci. USA* **1934**, *20*, 169–176.
41. Oppenheimer, J.R.; Volkoff, G.M. On Massive Neutron Cores. *Phys. Rev.* **1939**, *55*, 374–381.
42. Tolman, R.C. Static Solutions of Einstein's Field Equations for Spheres of Fluid. *Phys. Rev.* **1939**, *55*, 364–373.
43. Bondi, H. The contraction of gravitating spheres. *Proc. R. Soc. A* **1964**, *281*, 39–48.
44. Bocquet, M.; Bonazzola, S.; Gourgoulhon, E.; Novak, J. Rotating neutron star models with magnetic field. *Astron. Astrophys.* **1995**, *301*, 757–775.
45. Cardall, C.Y.; Prakash, M.; Lattimer, J.M. Effects of strong magnetic fields on neutron star structure. *Astrophys. J.* **2001**, *554*, 322–339.
46. Frieben, J.; Rezzolla, L. Equilibrium models of relativistic stars with a toroidal magnetic field. *Mon. Not. Roy. Astron. Soc.* **2012**, *427*, 3406–3426.
47. Pili, A.G.; Bucciantini, N.; Del Zanna, L. Axisymmetric equilibrium models for magnetized neutron stars in General Relativity under the Conformally Flat Condition. *Mon. Not. Roy. Astron. Soc.* **2014**, *439*, 3541–3563.

48. Gomes, R.O.; Franzon, B.; Dexheimer, V.; Schramm, S. Many-body forces in magnetic neutron stars. *Astrophys. J.* **2017**, *850*, 20.
49. Pili, A.G.; Bucciantini, N.; Del Zanna, L. General relativistic models for rotating magnetized neutron stars in conformally flat space-time. *Mon. Not. Roy. Astron. Soc.* **2017**, *470*, 2469–2493.
50. Kaplan, D.L.; Bassa, C.; Wang, Z.; Cumming, A.; Kaspi, V.M. Nearby, Thermally Emitting Neutron Stars. *AIP Conf. Proc.* **2008**, *983*, 331–339.
51. Haberl, F. AXPs and X-ray dim neutron stars: Recent XMM-Newton and Chandra results. In Proceedings of the 34th COSPAR Scientific Assembly, Houston, TX, USA, 10–19 October 2002.
52. Kaplan, D.L.; van Kerkwijk, M.H. A Coherent Timing Solution for the Nearby Isolated Neutron Star RX J0720.4-3125. *Astrophys. J. Lett.* **2005**, *628*, L45–L48.
53. Gill, R.; Heyl, J.S. Statistical ages and the cooling rate of X-ray dim isolated neutron stars. *Mon. Not. R. Astron. Soc.* **2013**, *435*, 3243–3250.
54. Kaplan, D.L.; van Kerkwijk, M.H. Constraining the Spin-down of the Nearby Isolated Neutron Star RX J0806.4-4123, and Implications for the Population of Nearby Neutron Stars. *Astrophys. J.* **2009**, *705*, 798–808.
55. Kaplan, D.L.; van Kerkwijk, M.H. A Coherent Timing Solution for the Nearby Isolated Neutron Star RX J1308.6+2127/RBS 1223. *Astrophys. J. Lett.* **2005**, *635*, L65–L68.
56. Walter, F.M.; Wolk, S.J.; Neuhäuser, R. Discovery of a nearby isolated neutron star. *Nature* **1996**, *379*, 233–235.
57. Van Kerkwijk, M.H.; Kaplan, D.L. Timing the Nearby Isolated Neutron Star RX J1856.5-3754. *Astrophys. J. Lett.* **2008**, *673*, L163–L166.
58. Zane, S.; Cropper, M.; Turolla, R.; Zampieri, L.; Chieregato, M.; Drake, J.J.; Treves, A. XMM-Newton Detection of Pulsations and a Spectral Feature in the X-ray Emission of the Isolated Neutron Star 1RXS J214303.7+065419/RBS 1774. *Astrophys. J.* **2005**, *627*, 397–403.
59. Kaplan, D.L.; van Kerkwijk, M.H. Constraining the Spin-Down of the Nearby Isolated Neutron Star RX J2143.0+0654. *Astrophys. J. Lett.* **2009**, *692*, L62–L66.
60. Van Kerkwijk, M.H.; Kaplan, D.L.; Durant, M.; Kulkarni, S.R.; Paerels, F. A Strong, Broad Absorption Feature in the X-ray Spectrum of the Nearby Neutron Star RX J1605.3+3249. *Astrophys. J.* **2004**, *608*, 432–443.
61. Pires, A.M.; Haberl, F.; Zavlin, V.E.; Motch, C.; Zane, S.; Hohle, M.M. XMM-Newton reveals a candidate period for the spin of the "Magnificent Seven" neutron star RX J1605.3+3249. *Astron. Astrophys.* **2014**, *563*, A50.
62. Negreiros, R.; Schramm, S.; Weber, F. Impact of Rotation-Driven Particle Repopulation on the Thermal Evolution of Pulsars *Phys. Lett. B* **2011**, *718*, 1–5.
63. Negreiros, R.; Schramm, S.; Weber, F. Thermal evolution of neutron stars in two dimensions. *Phys. Rev. D* **2012**, *85*, 104019.
64. De Carvalho, S.M.; Negreiros, R.; Orsaria, M.; Contrera, G.A.; Weber, F.; Spinella, W. Thermal evolution of hybrid stars within the framework of a nonlocal Nambu-Jona-Lasinio model. *Phys. Rev. C* **2015**, *92*, 035810.
65. Shternin, P.S.; Yakovlev, D.G.; Heinke, C.O.; Ho, W.C.G.; Patnaude, D.J. Cooling neutron star in the Cassiopeia A supernova remnant: evidence for superfluidity in the core. *Mon. Not. R. Astron. Soc. Lett.* **2011**, *412*, L108–L112.
66. Yakovlev, D.G.; Ho, W.C.G.; Shternin, P.S.; Heinke, C.O.; Potekhin, A.Y. Cooling rates of neutron stars and the young neutron star in the Cassiopeia A supernova remnant. *Mon. Not. R. Astron. Soc.* **2011**, *411*, 1977–1988.
67. Page, D.; Prakash, M.; Lattimer, J.M.; Steiner, A.W. Rapid Cooling of the Neutron Star in Cassiopeia A Triggered by Neutron Superfluidity in Dense Matter. *Phys. Rev. Lett.* **2011**, *106*, 081101.
68. Muslimov, A.; Page, D. Delayed switch-on of pulsars. *Astrophys. J. Lett.* **1995**, *440*, L77–L80.
69. Geppert, U.; Page, D.; Zannias, T. Submergence and re-diffusion of the neutron star magnetic field after the supernova. *Astron. Astrophys.* **1999**, *345*, 847–854.
70. Viganò, D.; Pons, J.A. Central compact objects and the hidden magnetic field scenario. *Mon. Not. R. Astron. Soc.* **2012**, *425*, 2487–2492.
71. Bernal, C.G.; Page, D.; Lee, W.H. Hypercritical Accretion onto a Newborn Neutron Star and Magnetic Field Submergence. *Astrophys. J.* **2013**, *770*, 106.
72. Heiselberg, H.; Hjorth-Jensen, M. Phase Transitions in Rotating Neutron Stars. *Phys. Rev. Lett.* **1998**, *80*, 5485–5488.
73. Page, D.; Lattimer, J.M.; Prakash, M.; Steiner, A.W. Minimal Cooling of Neutron Stars: A New Paradigm. *Astrophys. J. Suppl. Ser.* **2004**, *155*, 623–650.

74. Kaplan, D.L.; van Kerkwijk, M.H.; Anderson, J. The Parallax and Proper Motion of RX J1856.53754 Revisited. *Astrophys. J.* **2002**, *571*, 447–457.
75. Chevalier, R.A. Neutron star accretion in a supernova. *Astrophys. J.* **1989**, *346*, 847–859.
76. Bernal, C.G.; Lee, W.H.; Page, D. Hypercritical accretion onto a magnetized neutron star surface: A numerical approach. *Revista Mexicana de Astronomía y Astrofísica* **2010**, *46*, 309–322.
77. Muslimov, A.; Page, D. Magnetic and Spin History of Very Young Pulsars. *Astrophys. J.* **1996**, *458*, 347.
78. Brown, G.E.; Weingartner, J.C. Accretion onto and radiation from the compact object formed in SN 1987A. *Astrophys. J.* **1994**, *436*, 843–847.
79. Pons, J.A.; Miralles, J.A.; Geppert, U. Magneto-thermal evolution of neutron stars. *Astron. Astrophys.* **2009**, *496*, 207–216.
80. Viganò, D.; Rea, N.; Pons, J.A.; Perna, R.; Aguilera, D.N.; Miralles, J.A. Unifying the observational diversity of isolated neutron stars via magneto-thermal evolution models. *Mon. Not. R. Astron. Soc.* **2013**, *434*, 123–141.
81. Gourgouliatos, K.N.; Wood, T.S.; Hollerbach, R. Magnetic field evolution in magnetar crusts through three-dimensional simulations. *Proc. Natl. Acad. Sci. USA* **2016**, *113*, 3944–3949.
82. Kaspi, V.M. Grand unification of neutron stars. *Proc. Natl. Acad. Sci. USA* **2010**, *107*, 7147–7152.
83. Harding, A.K. The neutron star zoo. *Front. Phys.* **2013**, *8*, 679–692.

© 2018 by the authors. Licensee MDPI, Basel, Switzerland. This article is an open access article distributed under the terms and conditions of the Creative Commons Attribution (CC BY) license (http://creativecommons.org/licenses/by/4.0/).

Article
Rotating Quark Stars in General Relativity

Enping Zhou [1,2,*,†], Antonios Tsokaros [2,4], Luciano Rezzolla [2,3], Renxin Xu [1,5] and Kōji Uryū [6]

1. State Key Laboratory of Nuclear Science and Technology and School of Physics, Peking University, Beijing 100871, China; r.x.xu@pku.edu.cn
2. Institute for Theoretical Physics, Frankfurt am Main 60438, Germany; tsokaros@illinois.edu (A.T.); rezzolla@th.physik.uni-frankfurt.de (L.R.)
3. Frankfurt Institute of Advanced Studies, 60438 Frankfurt am Main, Germany
4. Department of Physics, University of Illinois at Urbana-Champaign, Urbana, IL 61801, USA
5. Kavli Institute for Astronomy and Astrophysics, Peking University, Beijing 100871, China
6. Department of Physics, University of the Ryukyus, Senbaru, Nishihara, Okinawa 903-0213, Japan; uryu@sci.u-ryukyu.ac.jp
* Correspondence: zhouenpingz715@sina.com
† Current address: State Key Laboratory of Nuclear Science and Technology and School of Physics, Peking University, Beijing 100871, China

Received: 4 January 2018; Accepted: 26 February 2018; Published: 5 March 2018

Abstract: We have built quasi-equilibrium models for uniformly rotating quark stars in general relativity. The conformal flatness approximation is employed and the Compact Object CALculator (COCAL) code is extended to treat rotating stars with surface density discontinuity. In addition to the widely used MIT bag model, we have considered a strangeon star equation of state (EoS), suggested by Lai and Xu, that is based on quark clustering and results in a stiff EoS. We have investigated the maximum mass of uniformly rotating axisymmetric quark stars. We have also built triaxially deformed solutions for extremely fast rotating quark stars and studied the possible gravitational wave emission from such configurations.

Keywords: pulsars; quark stars; general relativity

1. Introduction

The gravitational-wave (GW) event GW170817 and the associated electromagnetic emission observations [1,2] from a binary neutron star (BNS) merger has announced the birth of a multi-messenger observation era. Apart from enriching our knowledge on origins of short gamma-ray bursts [3] and heavy elements in the universe [4,5], it provides an effective way for us to constrain the equation of state (EoS) of neutron stars (NSs). In addition to BNS systems , rapidly rotating compact stars are also important candidates of GW sources [6], which could be detected by ground-based GW observatories [7–11]. Further, the properties of both uniformly and differentially rotating stars is tightly related to the evolution of the post-merger product during a BNS merger, for example, whether or not there will be a prompt collapse to a black hole. Hence, studying the properties of rotating compact stars has long been important and of great interest.

Following the first study on the equilibrium models of uniformly rotating, incompressible fluid stars in a Newtonian gravity scheme [12], various works have been done with more realistic EoSs and general relativity [13,14]. Among those studies, quasi-universal relationship has been found for both uniformly rotating and differentially rotating NSs [15–20]. Quasi-equilibrium figures of triaxially rotating NSs have also been created and studied in full general relativity [21,22].

However, it is worth noting that the EoS of compact stars is still a matter of lively debate as it originates from complicated problems in non-perturbative quantum choromodynamics (QCD). In addition to the conventional NS model, strange quark stars (QSs) are also suggested [23,24], after it was conjectured that strange quark matter (SQM) consists of up, down, and strange quarks that could be absolutely stable [25,26]. Additionally, the small tidal deformability of QSs passes the test of GW170817 [27], which requires that a dimensionless tidal deformability of a 1.4 solar mass star is smaller than 800. A more detailed analysis based on the probability distribution of each star in the binary system also indicates that the strangeon star model is consistent with the observation GW170817 together with other EoSs such as APR4. Possible models are also suggested to explain the electromagnetic counterparts (c.f. [27,28]).

Following this possibility, we here use the Compact Object CALculator code, COCAL, to build general-relativistic rotating QS solution sequences using different EoS models. COCAL is a code to calculate general-relativistic equilibrium and quasi-equilibrium solutions for binary compact stars (black hole and NSs) as well as rotating NSs [21,22,29–32]. The EoS part of COCAL is modified to treat quark stars that have a surface density discontinuity. With the modified code, we have built a uniformly rotating axisymmetric and triaxial sequence for quark stars.

2. Results

2.1. Maximum Mass of Axisymmetric Rotating Quark Stars

The maximum mass of a static spherical compact star and an axisymmetric rotating compact star depends on the EoS and is also closely related to the post-merger phase of a BNS merger. The total mass of the binary system could be obtained according to the GW observation. By comparing the total mass of the system with the maximum mass of a rotating star, it can be interpreted whether the post-merger product is a long-lived supramassive NS or short-lived hypermassive NS.

Various nuclear EoS models have been applied to build both uniform and differentially rotating NSs. It has been found that the maximum mass of uniformly rotating NSs, compared with the TOV maximum mass, depends very weakly on EoSs [15]. More specifically, regardless of the EoS model, the star could support approximately 20% more mass by uniform rotation. Another universal relationship is also discussed by [16] for differentially rotating NSs. Such relations have been invoked to interpret the observation of GW170817, and constraints on the maximum mass of NSs have been set accordingly [33]. In order to see whether this relationship still holds for rotating QSs, we have built axisymmetric rotating QS sequences for both the MIT bag model and the strangeon star model [34]. We first build a TOV solution sequence for both EoSs with 24 successive central densities. From each of those TOV solutions, we construct a rotating QS solution sequence by fixing the central density and decrease the axis ratio R_z/R_x. The axis ratio parameter determines the rotation of the star and preserves axisymmetry at the same time. Those rotating QS solutions terminate at the mass shedding limit. In this way, we manage to explore the parameter space for rotating QSs for various central densities and angular velocities.

Once we have all the solutions ready, we can obtain the TOV maximum mass (M_{TOV}) and angular momentum at the mass shedding limit (J_{kep}) as well as the maximum mass for a certain angular momentum (M_{crit}). The relationship between normalized mass (M_{crit}/M_{TOV}) and angular momentum (J/J_{kep}) can therefore be re-investigated for rotating QSs. The result is shown in Figure 1. As can be seen, the universal relationship for NSs no longer holds for QSs. Moreover, even for rotating QSs with different EoSs, the relation is quite different.

Although we couldn't extend the universal relationship or find a new one for rotating QSs, it does provide a potential to distinguish between NS and QS models from a BNS merger event. In particular, quark stars could be more massive when supported by uniform rotation compared with neutron stars (40% compared with 20%). Consequently, a post-merger phase might be longer before collapsing to a black hole if a QS is formed during the merger.

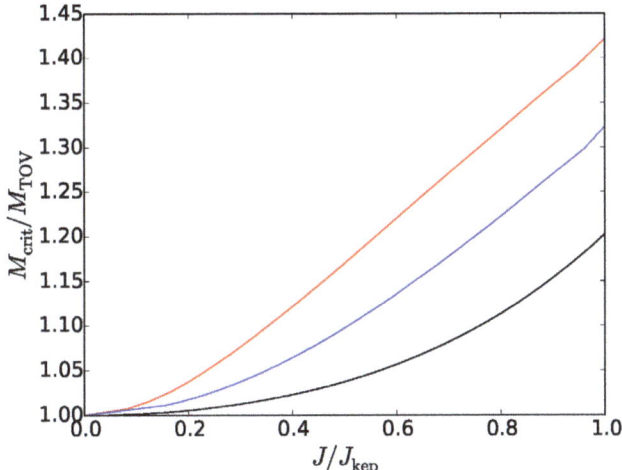

Figure 1. Relationship between normalized critical mass and the normalized angular momentum for rotating eutron stars (NSs) and quark stars (QSs). Bottom black line is the quasi-universal relationship found by [15]. Blue curve in the middle is the relationship for rotating QSs with strangeon star model and the top red line represents the relationship for the MIT bag model. The universal relationship cannot be extended for QSs easily.

2.2. Triaxial Rotating Quark Stars

Rotating QSs in a triaxial Jacobian sequence are another interesting type of QS and occur when the kinetic energy to potential ratio ($T/|W|$) is sufficient large. On the one hand, the post merger product in a BNS merger or a newly born compact star from a core collapse supernova possesses quite a large angular momentum, which might lead to a sufficiently large $T/|W|$ ratio for the bifurcation for a triaxial Jacobian sequence to occur [35–39]. On the other hand, triaxially rotating compact stars is an effective GW radiator itself [40].

Unlike NSs, which are bound by gravity, QSs are self-bound by strong interaction. Consequently, rotating QSs can reach a much larger $T/|W|$ ratio compared with NSs due to the finite surface density. Therefore, the triaxial instability can play a more important role [41–43] for QSs. The triaxial bar mode (Jacobi-like) instability for the MIT bag-model EoS has been investigated in a general relativistic framework [44].

Here, we build quasi-equilibrium constant rest mass sequences (axisymmetric and triaxial) for both the MIT bag-model EoS and the strangeon star model EoS. The surface fit coordinates used in COCAL allows us to treat the surface density discontinuity properly. We begin with the axisymmetric sequence in which we calculate solution sequences with varying parameters, i.e., the central rest-mass density (ρ_c) and the axis ratio (R_z/R_x). We first impose axisymmetry as a separate condition and manage to reach the mass shedding limit for all the sequences. In order to access the triaxial solutions, we recompute the above sequence of solutions but this time without imposing axisymmetry. As the rotation rate increases (R_z/R_x decreases), the triaxial deformation ($R_y/R_x < 1$) is *spontaneously* triggered, since at a large rotation rate the triaxial configuration possesses a lower total energy and is therefore favored over the axisymmetric solution.

Overall, three main properties of triaxially rotating QSs are found according to our calculations. Firstly, QSs generally have triaxial sequences of solutions that are longer than those of NSs. In another words, QSs can see larger triaxial deformations before the sequence is terminated at the mass-shedding limit (c.f. Figures 2 and 3 for an example for the comparison with the $n = 0.3$ NS model), due to the much higher $T/|W|$ ratio attained by rotating QSs. Secondly, when considering similar triaxial configurations, QSs are (slightly) more efficient GW sources because of the finite surface rest-mass density and hence larger mass quadrupole for QSs (c.f. Figure 4). Thirdly, triaxial supramassive solutions can be found for QSs.

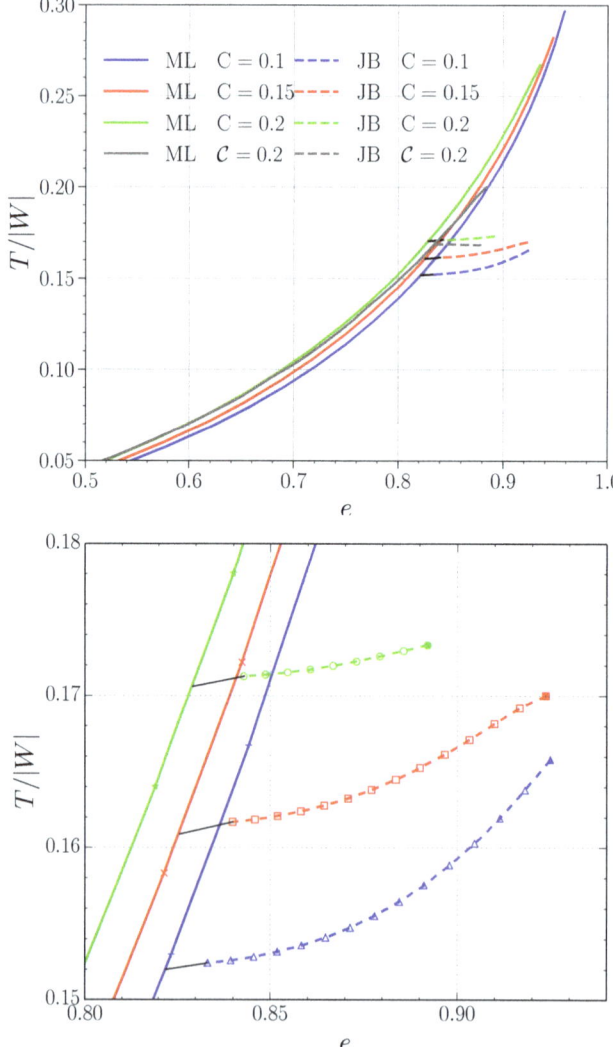

Figure 2. *Upper panel:* $T/|W|$ versus eccentricity $e := \sqrt{1-(z/x)^2}$ (in proper length) for the MIT bag-model Equation of State (EoS) sequences as well as for NSs with $n = 0.3$ EoS reported in [21] (labeled with gray curves). Solid curves are axisymmetric solution sequences, and dashed curves are triaxial solution sequences, which correspond to $C = M/R = 0.2$ (green curves), 0.15 (red curves), and 0.1 (blue curves), respectively. Note that M is the spherical ADM mass. *Lower panel:* Magnification of the region near the onset of the triaxial solutions marked with empty symbols. Filled symbols mark the models at the mass-shedding limit. Solutions labeled with "ML" are axisymmetric solutions (Maclaurin spheroids), while those labeled "JB" are triaxial solutions (Jacobi ellipsoids).

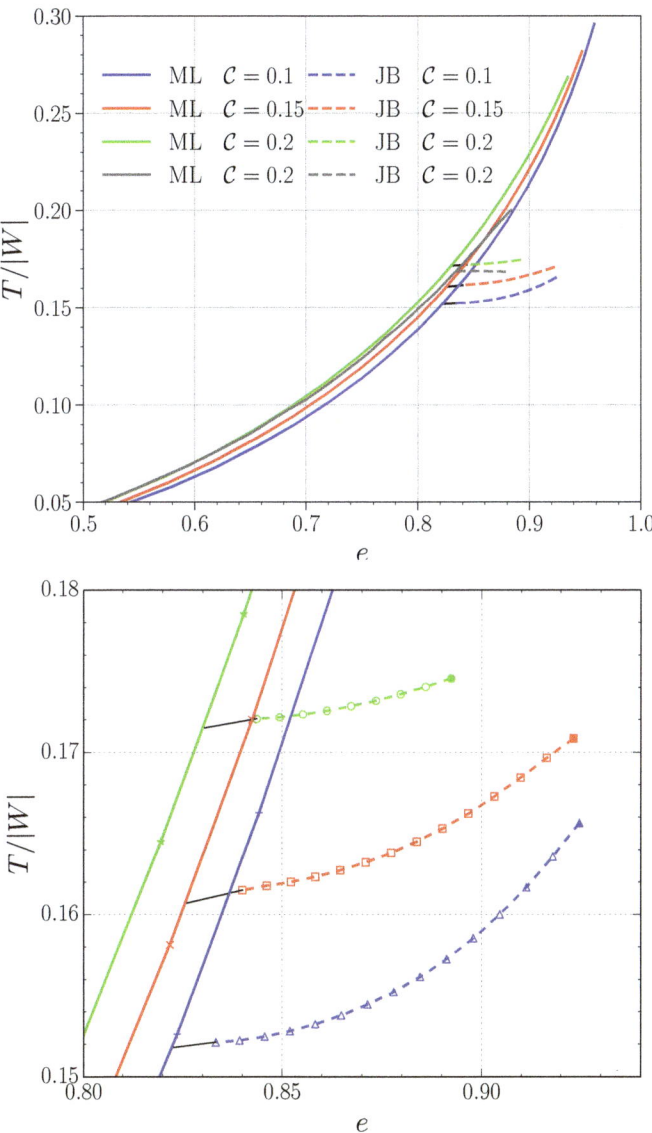

Figure 3. The same as Figure 2 but for the LX EoS sequences.

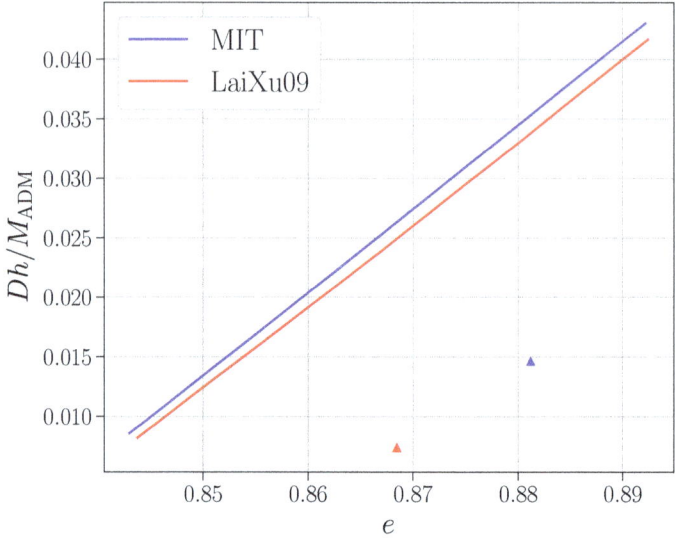

Figure 4. Estimates of the GW strain amplitude according to the quadrupole formula for the $\mathcal{C} = 0.2$ triaxial sequence for both the MIT bag-model EoS (blue solid curve) and the LX EoS (red dashed curve). Shown is the GW strain for the $\ell = m = 2$ mode normalized by the distance and the ADM mass of the source. Triangle markers in blue and red stand for the triaxially rotating NS cases G4C010 and G4C025 models in [40], which indicates that triaxially deformed QSs are more effective gravitational-wave (GW) sources compared with NSs.

3. Discussion

We have built both axisymmetric and triaxial solutions for uniformly rotating QSs. For axisymmetric rotating QSs, we have investigated the critical mass-angular momentum relation and find that it deviates from the universal relationship for rotating NSs. Especially, uniformly rotating QSs can be more massive compared with M_{TOV}, indicating a quite different post-merger phase in binary merger events. For triaxially rotating QSs, we have identified the bifurcation point from axisymmetric solutions. Triaxial solutions have been constructed from the bifurcation point to the mass shedding limit. The GW emission from such a triaxial rotating QS is estimated with a quadrupole formula and is found to be more effective than that of an NS, which can also be tested with future GW observations.

Additionally, since the spin period of a triaxially rotating star increases as the angular momentum increases, the spin frequency at the bifurcation point somehow represents a maximum spin frequency that can be attained by a pulsar when spun-up by accretion. Particularly, a solid QS model is suggested for the strangeon star model [45], which means that r-mode instability could be totally suppressed for such a star and a strangeon star might be spun up to the bifurcation frequency. With the construction of more power radio telescopes such as SKA and FAST, this limit could be tested by searching for faster spinning pulsars and might provide an important clue on the properties of the dense matter in compact stars.

Acknowledgments: It's a pleasure to thank the China Scholarship Council for supporting E. Z. on the joint PhD training in Frankfurt. This work was supported in part by the ERC synergy grant "BlackHoleCam: Imaging the Event Horizon of Black Holes" (Grant No. 610058), by "NewCompStar", COST Action MP1304, by the LOEWE-Program in the Helmholtz International Center (HIC) for FAIR, and by the European Union's Horizon 2020 Research and Innovation Programme (Grant 671698) (call FETHPC-1-2014, project ExaHyPE). R.X. was supported by National Key R&D Program (No.2017YFA0402600) and NNSF (11673002,U1531243). A.T. was supported by NSF Grants PHY-1662211 and PHY-1602536, and NASA Grant 80NSSC17K0070. K.U. was supported by JSPS Grant-in-Aid for Scientific Research(C) 15K05085.

Author Contributions: E.Z. modified the COCAL code to include QS models. E.Z. and A.T. performed the simulations of triaxially rotating QSs and the convergence tests. L. R. has designed those studies. R.X. helped with the EoS part of the code, and K.U. modified the COCAL code, which substantially improved the accuracy and convergence performance for QSs.

Conflicts of Interest: We declare that we do not have any commercial or associative interest that represents a conflict of interest in connection with the work submitted. We declare that we have no financial and personal relationships with other people or organizations that can inappropriately influence our work. The founding sponsors had no role in the design of the study; in the collection, analyses, or interpretation of data; in the writing of the manuscript, and in the decision to publish the results .

References

1. Abbott, B.P.; Abbott, R.; Acernese, F.; Ackley, F.; Adams, C.; Adamsa, T.; Addesso, P.; Adhikari, R.X.; Adya, V.B.; et al. GW170817: Observation of Gravitational Waves from a Binary Neutron Star Inspiral. *Phys. Rev. Lett.* **2017**, *119*, 161101.
2. Abbott, B.P.; Abbott, R.; Acernese, F.; Ackley, F.; Adams, C.; Adamsa, T.; Addesso, P.; Adhikari, R.X.; Adya, V.B.; et al. Multi-messenger Observations of a Binary Neutron Star Merger. *Astrophys. J. Lett.* **2017**, *848*, L12.
3. Abbott, B.P.; Abbott, R.; Acernese, F.; Ackley, F.; Adams, C.; Adamsa, T.; Addesso, P.; Adhikari, R.X.; Adya, V.B.; et al. Gravitational Waves and Gamma-Rays from a Binary Neutron Star Merger: GW170817 and GRB 170817A. *Astrophys. J. Lett.* **2017**, *848*, L13.
4. Abbott, B.P.; Abbott, R.; Acernese, F.; Ackley, F.; Adams, C.; Adamsa, T.; Addesso, P.; Adhikari, R.X.; Adya, V.B.; et al. Estimating the Contribution of Dynamical Ejecta in the Kilonova Associated with GW170817. *Astrophys. J. Lett.* **2017**, *850*, L39.
5. Baiotti, L.; Rezzolla, L. Binary neutron-star mergers: A review of Einstein's richest laboratory. *Rep. Prog. Phys.* **2017**, *80*, 096901.
6. Andersson, N.; Ferrari, V.; Jones, D.I.; Kokkotas, K.D.; Krishnan, B.; Read, J.S.; Rezzolla, L.; Zink, B. Gravitational waves from neutron stars: Promises and challenges. *Gen. Relativ. Gravit.* **2011**, *43*, 409–436.
7. Abramovici, A.A.; Althouse, W.; Drever, R.P.; Gursel, Y.; Kawamura, S.; Raab, F.; Shoemaker, D.; Sievers, L.; Spero, R.; Thorne, K.S.; et al. LIGO: The Laser Interferometer Gravitational-Wave Observatory. *Science* **1992**, *256*, 325–333.
8. Punturo, M.; Abernathy, M.; Acernese, F.; Allen, B.; Andersson, N.; Arun, K.; Barone, F.; Barr, B.; Barsuglia, M.; Beker, M.; et al. The third generation of gravitational wave observatories and their science reach. *Class. Quantum Gravity* **2010**, *27*, 084007.
9. Accadia, T.; Acernese, F.; Antonucci, F.; Astone, P.; Gallardin, G.; Barone, F.; Barsuglia, M.; Basti, A.; Bauer, T.S.; Bebronne, M.; et al. Status of the Virgo project. *Class. Quantum Gravity* **2011**, *28*, 114002.
10. Kuroda, K. Status of LCGT. *Class. Quantum Gravity* **2010**, *27*, 084004.
11. Aso, Y.; Michimura, Y.; Somiya, K.; Ando, M.; Miyakawa, O.; Sekiguchi, T.; Tatsumi, D.; Yamamoto, H. Interferometer design of the KAGRA gravitational wave detector. *Phys. Rev. D* **2013**, *88*, 043007.
12. Chandrasekhar, S. *Ellipsoidal Figures of Equilibrium*; Yale Univ. Press: New Haven, CT, USA, 1969.
13. Meinel, R.; Ansorg, M.; Kleinwächter, A.; Neugebauer, G.; Petroff, D. *Relativistic Figures of Equilibrium*; Cambridge University Press: Cambridge, UK, 2008.
14. Friedman, J.L.; Stergioulas, N. *Rotating Relativistic Stars*; Cambridge University Press: Cambridge, UK, 2013.
15. Breu, C.; Rezzolla, L. Maximum mass, moment of inertia and compactness of relativistic stars. *Mon. Not. R. Astron. Soc.* **2016**, *459*, 646–656.
16. Weih, L.R.; Most, E.R.; Rezzolla, L. On the stability and maximum mass of differentially rotating relativistic stars. *Mon. Not. R. Astron. Soc.* **2017**, *473*, L126–L130.
17. Bozzola, G.; Stergioulas, N.; Bauswein, A. Universal relations for differentially rotating relativistic stars at the threshold to collapse. *Mon. Notices Royal Astron. Soc.* **2017**, *474*, 3557–3564.
18. Yagi, K.; Yunes, N. I-Love-Q. *Science* **2013**, *341*, 365.
19. Yagi, K.; Yunes, N. I-Love-Q relations in neutron stars and their applications to astrophysics, gravitational waves, and fundamental physics. *Phys. Rev. D* **2013**, *88*, 023009.

20. Yagi, K.; Yunes, N. Approximate universal relations for neutron stars and quark stars. *Phys. Rep.* **2017**, *681*, 1–72.
21. Huang, X.; Markakis, C.; Sugiyama, N.; Uryū, K. Quasi-equilibrium models for triaxially deformed rotating compact stars. *Phys. Rev. D* **2008**, *78*, 124023.
22. Uryū, K.; Tsokaros, A.; Galeazzi, F.; Hotta, H.; Sugimura, M.; Taniguchi, K.; Yoshida, S. New code for equilibriums and quasiequilibrium initial data of compact objects. III. Axisymmetric and triaxial rotating stars. *Phys. Rev. D* **2016**, *93*, 044056.
23. Itoh, N. Hydrostatic Equilibrium of Hypothetical Quark Stars. *Prog. Theor. Phys.* **1970**, *44*, 291–292.
24. Alcock, C.; Farhi, E.; Olinto, A. Strange stars. *Astrophys. J.* **1986**, *310*, 261–272.
25. Bodmer, A.R. Collapsed Nuclei. *Phys. Rev. D* **1971**, *4*, 1601–1606.
26. Witten, E. Cosmic separation of phases. *Phys. Rev. D* **1984**, *30*, 272–285.
27. Lai, X.Y.; Yu, Y.W.; Zhou, E.P.; Li, Y.Y.; Xu, R.X. Merging Strangeon Stars. *Res. Astron. Astrophys.* **2018**, *18*, 024.
28. Li, A.; Zhang, B.; Zhang, N.B.; Gao, H.; Qi, B.; Liu, T. Internal x-ray plateau in short GRBs: Signature of supramassive fast-rotating quark stars? *Phys. Rev. D* **2016**, *94*, 083010.
29. Uryū, K.; Tsokaros, A. New code for equilibriums and quasiequilibrium initial data of compact objects. *Phys. Rev. D* **2012**, *85*, 064014.
30. Uryū, K.; Tsokaros, A.; Grandclement, P. New code for equilibriums and quasiequilibrium initial data of compact objects. II. Convergence tests and comparisons of binary black hole initial data. *Phys. Rev. D* **2012**, *86*, 104001.
31. Tsokaros, A.; Uryū, K. Binary black hole circular orbits computed with cocal. *J. Eng. Math.* **2012**, *82*, 133–141.
32. Tsokaros, A.; Uryū, K.; Rezzolla, L. New code for quasiequilibrium initial data of binary neutron stars: Corotating, irrotational, and slowly spinning systems. *Phys. Rev. D* **2015**, *91*, 104030.
33. Rezzolla, L.; Most, E.R.; Weih, L.R. Using Gravitational-wave Observations and Quasi-universal Relations to Constrain the Maximum Mass of Neutron Stars. *Astrophys. J. Lett.* **2018**, *852*, L25.
34. Lai, X.Y.; Xu, R.X. Strangeon and Strangeon Star. *J. Phys. Conf. Ser.* **2017**, *861*, 012027.
35. Lai, D.; Shapiro, S.L. Gravitational radiation from rapidly rotating nascent neutron stars. *Astrophys. J.* **1995**, *442*, 259–272.
36. Bildsten, L. Gravitational Radiation and Rotation of Accreting Neutron Stars. *Astrophys. J. Lett.* **1998**, *501*, L89–L93.
37. Woosley, S.; Janka, T. The physics of core-collapse supernovae. *Nat. Phys.* **2005**, *1*, 147–154.
38. Watts, A.L.; Krishnan, B.; Bildsten, L.; Schutz, B.F. Detecting gravitational wave emission from the known accreting neutron stars. *Mon. Not. R. Astron. Soc.* **2008**, *389*, 839–868.
39. Piro, A.L.; Thrane, E. Gravitational Waves from Fallback Accretion onto Neutron Stars. *Astrophys. J.* **2012**, *761*, 63.
40. Tsokaros, A.; Ruiz, M.; Paschalidis, V.; Shapiro, S.L.; Baiotti, L.; Uryō, K. Gravitational wave content and stability of uniformly, rotating, triaxial neutron stars in general relativity. *Phys. Rev. D* **2017**, *95*, 124057.
41. Gondek-Rosinska, D.; Haensel, P.; Zdunik, J.L.; Gourgoulhon, E. Rapidly rotating strange stars. In *IAU Colloq. 177: Pulsar Astronomy-2000 and Beyond*; Astronomical Society of the Pacific Conference Series; Kramer, M., Wex, N., Wielebinski, R., Eds.; Astronomical Society of the Pacific: San Francisco, CA, USA, 2000; Volume 202, p. 661.
42. Gondek-Rosińska, D.; Bulik, T.; Zdunik, L.; Gourgoulhon, E.; Ray, S.; Dey, J.; Dey, M. Rapidly rotating compact strange stars. *Astron. Astrophys.* **2000**, *363*, 1005–1012.
43. Gondek-Rosińska, D.; Stergioulas, N.; Bulik, T.; Kluźniak, W.; Gourgoulhon, E. Lower limits on the maximum orbital frequency around rotating strange stars. *Astron. Astrophys.* **2001**, *380*, 190–197.
44. Gondek-Rosińska, D.; Gourgoulhon, E.; Haensel, P. Are rotating strange quark stars good sources of gravitational waves? *Astron. Astrophys.* **2003**, *412*, 777–790.
45. Xu, R.X. Solid Quark Stars? *Astrophys. J. Lett.* **2003**, *596*, L59–L62.

 © 2018 by the authors. Licensee MDPI, Basel, Switzerland. This article is an open access article distributed under the terms and conditions of the Creative Commons Attribution (CC BY) license (http://creativecommons.org/licenses/by/4.0/).

Article

Non-Radial Oscillation Modes of Superfluid Neutron Stars Modeled with CompOSE

Prashanth Jaikumar *, Thomas Klähn and Raphael Monroy

Department of Physics and Astronomy, California State University Long Beach, 1250 Bellflower Blvd., Long Beach, CA 90840, USA; thomas.klaehn@googlemail.com (T.K.); raphael.r.monroy@gmail.com (R.M.)
* Correspondence: prashanth.jaikumar@csulb.edu; Tel.: +562-985-5592

Received: 16 January 2018; Accepted: 24 February 2018; Published: 9 March 2018

Abstract: We compute the principal non-radial oscillation mode frequencies of Neutron Stars described with a Skyrme-like Equation of State (EoS), taking into account the possibility of neutron and proton superfluidity. Using the CompOSE database and interpolation routines to obtain the needed thermodynamic quantities, we solve the fluid oscillation equations numerically in the background of a fully relativistic star, and identify imprints of the superfluid state. Though these modes cannot be observed with current technology, increased sensitivity of future Gravitational-Wave Observatories could allow us to observe these oscillations and potentially constrain or refine models of dense matter relevant to the interior of neutron stars.

Keywords: neutron stars; gravitational waves; equation of state

1. Introduction

The Laser Interferometer Gravitational-Wave Observatory (LIGO) recently detected gravitational waves emitted from the merger of two neutron stars, heralding a new era of compact star physics [1]. As the sensitivity of gravitational wave detectors increases, we may expect to directly observe the gravitational waves even from isolated neutron stars. Neutron star oscillations allow us to probe the nature of ultra-dense nuclear matter, since the equation of state (EoS) plays a central role in determining the spectrum of oscillation modes. Thus, in the event that actual oscillation modes can be observed, comparison to theoretical templates based on models of dense matter can help to refine or constrain the EoS. For calculating the spectrum of oscillation modes theoretically, we require tabulated EoS from realistic microscopic models of dense matter. The CompOSE database [2] provides an array of such models with thermodynamically consistent interpolation. In these proceedings, we report results from a calculation of the spectrum of non-radial oscillation modes of neutron stars based on Skyrme-like EoS in the CompOSE database, and compare the results to the polytropic equation of state. We also consider the effects of superfluidity in neutron stars on the oscillation modes, using the same family of EoS. After establishing the background structure of the neutron star in Section 2, we present the main equations and numerical results from solving the oscillation equations in Section 3, including the extension to superfluidity. In Section 4, we summarize our findings and comment on further extensions of this work in Section 5.

2. EoS Models and Superfluid Neutron Star Structure

One of the simplest models of a neutron star assumes a polytropic equation of state $P = K\rho^{(n+1)/n}$ with polytropic index n and a proportionality constant K. In this work, we choose $n = 1$ and $K = 63.66$ km^2 in $G = c = 1$ units, which simplifies the numerics of the Tolman-Oppenheimer-Volkov (TOV) equations for the mass and radius of the neutron star. There is a limitation of the $n = 1$ polytrope: the structure of such a star in General Relativity can only provide a maximum mass of about 1.2 $M_\odot$.

We therefore employ Newtonian structure equations for the polytrope, in which case, for $n = 1$, the mass is independent of the radius [3]. This allows us to study the model for stars as heavy as 2 $M_\odot$ while keeping the radius fixed at a canonical value (e.g., 10 km). We emphasize that the polytropic equation of state is only used here as a simple model for a rough estimate of the non-radial oscillation spectra of neutron stars, and the Skyrme-like model [4,5] is used for more realistic calculations. For the latter case, the CompOSE database provides appropriate tables of thermodynamic quantities based on the baryon number density within the star. While the structure of the star can be solved from the standard TOV equations using the energy and pressure alone, finding the oscillation modes for the normal and superfluid cases requires knowing the chemical potentials and densities of the neutrons, protons, and electrons in the star from center to surface. The CompOSE database provides these for various EoS along with consistent interpolation routines. From this database, we choose three Skyrme models with interaction parameters provided by the KDE0, KDE0v1 and the SK255 dataset, as calculated by Agrawal [4,6]. We refer to these three sets of parameters as Skyrme Model A, B and C respectively. These interaction parameters can be thought of as specifying a nucleon energy density [7]

$$U_N = n_b mc^2 + \frac{\hbar^2}{2m}\tau_b + B_1 n_b^2 + B_2(n_n^2 + n_p^2) + B_3 n_b \tau_b + \\ B_4(n_n \tau_n + n_p \tau_p) + B_5 n_b^{2+\alpha} + B_6 n_b^\alpha (n_n^2 + n_p^2) \quad (1)$$

where n_n, n_p, and n_b refer to neutron, proton, and baryon number densities respectively; the τ are based on these densities; and the B_i and α are specific to the interaction.

A neutron star may be considered to have three major regions: the outer crust, the inner crust, and the core. In the outer crust, the density ranges from 10^4 to 10^{11} g/cm^3. This part of the star consists mostly of nuclei and electrons. The inner crust ranges from 10^{11} to 10^{14} g/cm^3. Here, the temperature is below the critical temperature for neutron pairing, and a neutron superfluid can form. The core of the neutron star has densities greater than 10^{14} g/cm^3, and consists of both neutrons and protons in a superfluid state, with electrons present to satisfy charge neutrality. For this work, we adopt a simplified two-layer model of core and inner crust, applying a Skyrme effective force model to describe the nucleon-nucleon interactions. This is similar to previous works such as [8], where the fluid below a certain critical density is treated as normal (i.e, the inner crust), and above the critical density is treated as a superfluid (i.e, the core). In a more complete treatment, the location of fluid and superfluid regions would be determined by a profile of the density-dependent critical temperature $T_c(\rho)$; however, we limit ourselves to a $T = 0$ calculation. It is important to state the possible consequences of this assumption on our results. It is known that non-zero temperatures can shift the frequency of the superfluid mode by approximately (5–10)%, without affecting the non-superfluid modes [9]. This is because the entrainment matrix depends on the critical temperature profile of the superfluids through its density dependence. We choose a simpler form of entrainment that applies only at $T = 0$. Studying the temperature effects requires models of the critical temperature which are currently not available within the EoS models we use, but this would certainly be interesting to explore with a self-consistent finite temperature model of the neutron star. Although entropy entrainment does not arise in the case of zero temperature [10], entrainment in the cores of superfluid neutron stars between neutrons and protons can still arise at $T = 0$ due to the Andreev-Bashkin effect [11], which is encoded in the ρ_{12} terms in the set of equations in Section 3.2 below.

The plots in Figures 1 and 2 show the run of interior variables such as pressure and density in the star for one of the Skyrme EoS (Model A), as well as the mass-radius curves for all three of the Skymre models.

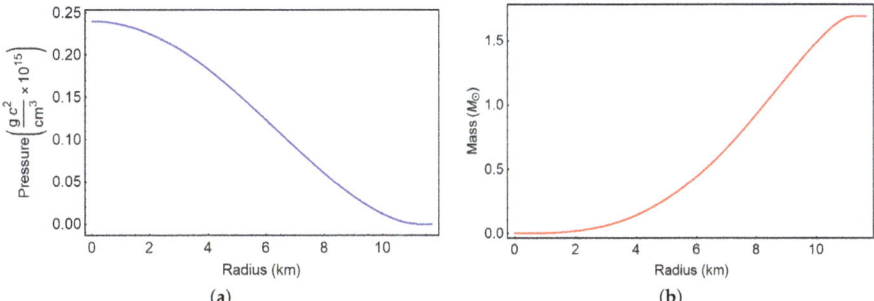

Figure 1. (**a**): Pressure vs. radial distance and (**b**): enclosed mass vs. radial distance for the Skyrme (Model A) equation of state. These curves both correspond to the same star with a central density of 1.102×10^{15} g/cm^3, which yields a mass of 1.69 $M_\odot$ and radius of 11.68 km.

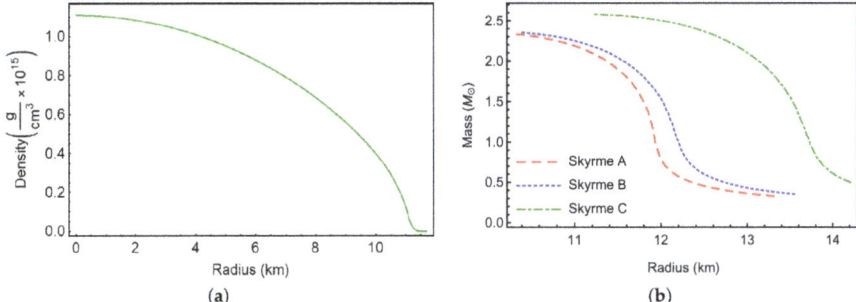

Figure 2. (**a**): Density profile vs. radius of the Skyrme (Model A) equation of state, with a central density of 1.102×10^{15} g/cm^3, corresponding to a mass of 1.69 $M_\odot$ and radius of 11.68 km. The density decrease is not smooth near the surface of the star. This feature requires special care in the solution of the oscillation equations. (**b**): Mass-radius curves for the Skyrme A, B, and C models. All three models can support a maximum mass of more than 2 $M_\odot$.

3. Fluid Oscillation Equations

3.1. Normal Fluid

We assume the neutron and proton fluids to be ideal in nature, neglecting dissipative effects due to viscous drag, which is negligible for a superfluid at $T \ll T_c$. This is the case for cold neutron stars with $T \sim 10^7$ K. However, there can still be dissipation and damping of various oscillation modes due to mutual friction and other microscopic processes such as electron-electron scattering that dominate the shear viscosity when both neutrons and protons are superfluid [12]. Previous works [13] lead us to believe that these effects are not expected to change the real part of the frequency, which is the quantity calculated in this work. For the case of the polytropic EoS, we use a coupled system of differential equations for the radial displacement and pressure perturbation of the fluid [14] to find the oscillation modes of the star. For the polytrope, the stellar structure and the fluid perturbation equations are being described in Newtonian gravity, for the reason explained in Section 2. The structure of the star for the Skyrme models will be treated in General Relativity, although we still employ Newtonian hydrodynamics for the oscillation equations. It is convenient to use the Dziembowski parameterization [14], with non-dimensional variables to efficiently solve for the oscillation eigenfrequencies. The relevant equations in the Cowling approximation, which neglects the back-reaction of the fluid perturbations on the gravitational potential, are:

$$r\frac{dy_1}{dr} = \left(\frac{V}{\Gamma_1} - 3\right)y_1 + \left[\frac{l(l+1)}{c_1\bar{\sigma}^2} - \frac{V}{\Gamma_1}\right]y_2 \qquad (2)$$

$$r\frac{dy_2}{dr} = (c_1\bar{\sigma}^2 + rA)y_1 + (1 - U - rA)y_2 \qquad (3)$$

where y_1, y_2 are related respectively to the perturbations in the radial component of the fluid motion and the fluid pressure; V, U, and A are derived from logarithmic derivatives of the pressure, mass, and density; l is an angular momentum parameter, c_1 is a combination of the mass and radius of the star, and $\bar{\sigma}$ is the normalized frequency of oscillation. For the Skyrme models with superfluidity, we employ a more complicated system of equations given in Section 3.2 that can be applied to the entrained neutron and proton superfluids.

In Figure 3 below, we compare the f- and p-modes for the polytropic EoS and the Skyrme (Model A) EoS. While we have only taken one fixed configuration ($n = 1$, $K = 63.66$ km^2) for the polytrope, the f- and p-mode frequencies for larger n values are even higher [15]. A larger value of K for the same n would make the star less compact and lower the mode frequencies. In any case, the Newtonian polytrope is meant to provide a rough estimate only. The f-modes scale with the mean density $\sqrt{M/R^3}$ in both cases. The p-modes scale with the compactness M/R rather than the mean density, though a general increase with mean density is still evident since increased compactness leads to increased mean density. It follows that the oscillation frequencies of the realistic models is lower than the $n = 1$ polytrope with our chosen value of K, since these stars are less compact and less dense on average than the polytrope. The modes for the polytrope would be lower if the structure was computed in General Relativity, but as previously mentioned, this cannot meet the observational constraint of at least 2 $M_\odot$ maximum mass. The X-axis range is different for the two figures, since we spanned the same mass range for the two EoS, which leads to different ranges for the mean density. The data in the right panel of Figure 3 is for Skyrme Model A.

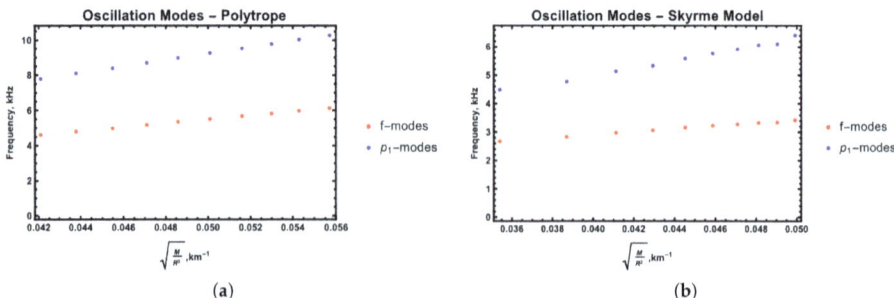

Figure 3. The trend of f- and p-modes with the square root of the mean density of the neutron star, given a polytropic equation of state (**a**) and a Skyrme (Model A) equation of state (**b**).

3.2. Oscillation Equations: Extension to Superfluidity

Since both neutron and proton fluids can undergo pairing, the extension to superfluidity leads to the following equations for determining the oscillation modes, as given in [16]

$$r\frac{dy_1^n}{dr} = \left(-3 - \frac{d\ln\rho_n}{\ln r}\right)y_1^n + \left[\frac{1}{\Delta}\frac{l(l+1)}{c_1\bar{\sigma}^2}\left(\frac{\rho_{22}}{\rho_p} + \frac{m_e}{m_p}\right) - \frac{gr}{\rho_n}Q_{11}\right]y_2^n - \left[\frac{1}{\Delta}\frac{l(l+1)}{c_1\bar{\sigma}^2}\frac{\rho_{12}}{\rho_n} + \frac{gr}{\rho_n}Q_{12}\right]y_2^p + \frac{gr}{\rho_n}(Q_{11} + Q_{12})y_3 \qquad (4)$$

$$r\frac{dy_1^p}{dr} = \left(-3 - \frac{d\ln\rho_p}{\ln r}\right)y_1^p + \left[\frac{1}{\Delta}\frac{l(l+1)}{c_1\bar{\sigma}^2}\frac{\rho_{11}}{\rho_n} - \frac{gr}{\rho_p}Q_{22}\right]y_2^p - \\ \left[\frac{1}{\Delta}\frac{l(l+1)}{c_1\bar{\sigma}^2}\frac{\rho_{12}}{\rho_p} + \frac{gr}{\rho_n}Q_{21}\right]y_2^n + \frac{gr}{\rho_p}(Q_{21} + Q_{22})y_3 \quad (5)$$

$$r\frac{dy_2^n}{dr} = c_1\bar{\sigma}^2\left[\frac{\rho_{11}}{\rho_n}y_1^n + \frac{\rho_{12}}{\rho_n}y_1^p\right] + (1-U)y_2^n \quad (6)$$

$$r\frac{dy_2^p}{dr} = c_1\bar{\sigma}^2\left[\frac{\rho_{12}}{\rho_p}y_1^n + \left(\frac{\rho_{22}}{\rho_p} + \frac{m_e}{m_p}\right)y_1^p\right] + (1-U)y_2^p \quad (7)$$

where the ρ matrix components ($\rho_{11}, \rho_{12} = \rho_{21}, \rho_{22}$) and the parameter Δ are based on the particle densities ρ_p and ρ_n and the nucleon effective masses m_n^* and m_p^*; the Q matrix components ($Q_{11}, Q_{12} = Q_{21}, Q_{22}$) are derivatives of the particle chemical potentials with respect to the particle densities; and g is the gravitational field. As mentioned before, entrainment is included through ρ_{12}. All other parameters take the same definitions as in the normal fluid.

We determine the effective masses self-consistently from the Skyrme EoS model, following Chamel's theoretical treatment of superfluidity [7]. The effective masses and all other required thermodynamic quantities are consistently determined at each grid point in the star using inbuilt interpolation routines in CompOSE. It is important to note that the effective masses are included self-consistently in the calculation of the oscillation spectrum. Though the extension to superfluidity is ultimately straightforward, we find that a minimal fraction of normal fluid is needed to tame the discontinuities in the density profile near the surface of the star, which otherwise cause numerical problems. This is particularly the case for Model A (KDE0 interaction), suggesting that either the model or the interpolation routine needs to be further investigated at low densities. For this work, we confine our calculations to the case of a superfluid core surrounded by a normal envelope with a transition density at nearly half the radius of the star.

4. Results

In this section, we present our numerical results and summarize the main findings from our work. In Table 1, we show the oscillation mode frequencies for neutron stars of mass $\{1.4, 1.8, 2.2\}\,M_\odot$ with normal fluid and superfluid components.

Table 1. Mode Frequency in kHz ($l = 2$).

Mode	Polytrope	Model A	Model B	Model C
		1.4 $M_\odot$		
f	4.77	2.74 (2.73)	2.69 (2.69)	2.41 (2.41)
p	8.34	4.42 (4.40)	4.29 (4.27)	3.98 (3.97)
s	–	3.66	3.59	3.34
		1.8 $M_\odot$		
f	5.67	2.91 (2.88)	2.87 (2.85)	2.55 (2.54)
p	9.52	5.05 (5.02)	4.96 (4.94)	4.76 (4.74)
s	–	4.21	4.18	3.87
		2.2 $M_\odot$		
f	6.26	3.52 (3.50)	3.37 (3.34)	3.05 (3.03)
p	10.55	6.28 (6.24)	6.18 (6.15)	5.66 (5.65)
s	–	5.13	5.07	4.75

The polytropic model has only a normal component, while the Skyrme EoS admits both normal and superfluid components. For the latter case, the numbers in brackets are the mode frequencies if only the normal component exists in the star, showing that these modes (f and p) remain approximately unchanged even in the presence of superfluidity. The last row is the frequency of the superfluid mode. Similar to previous works with other EoS [8,16], we find that the modes of the two-fluid star support f- and p-modes that are very close to the corresponding normal fluid star, with the addition of one or two superfluid modes, the lower one of which is intermediate between the f and p modes and is denoted by "s" in Table 1. The superfluid mode frequency also decreases with decreasing compactness, though no simple scaling with global stellar properties as for normal fluid modes could be determined. It is likely that the superfluid mode is quite sensitive to the details of the EoS such as the effective masses, and the superfluid fraction, which must ideally be determined by the critical temperature profile in the star. The systematic thoeretical study of the superfluid modes with variable superfluid fraction, EoS models and including non-zero temperature certainly deserves more study. Nevertheless, even in our simple framework, the appearance of these distinct modes in the spectrum indicating superfluidity in the cores of neutron stars is an exciting observational possibility for gravitational wave detectors of increased sensitivity. At this stage, it is useful to summarize the main new developments in this work:

- The appearance of a new superfluid mode, while leaving the frequencies of the normal modes almost unchanged, was already noted in [8], but they used a much older EoS [17] based on the Walecka model. We have used the Skyrme EoS which obeys modern constraints from isospin diffusion data and the slope of the symmetry energy. In addition, we employ neutron and proton effective masses that are obtained consistently within the model [7]. This was not the case, for e.g., in all other previous computations of the superfluid oscillation modes [8,16]. This is an important point since the entrainment matrix depends on the effective masses, and different models for the density dependent effective mass yield different numerical values for the modes [8]. In this way, our results are based on an EoS that is built from a unified treatment of terrestrial nuclear experiments and the astrophysics of compact stars, and our results are more consistent from a quantitative standpoint.
- We have reported the mode frequencies for a 1.8 $M_\odot$ and 2 $M_\odot$ configuration, which is a new result since earlier works [8,16] that considered configurations of 1.4 $M_\odot$ predate the discovery of the presently observed heaviest neutron star.
- We demonstrate the utility of the CompOSE database in using modern EoS for studies of neutron star oscillations and gravitational waves. The EoS models taken from CompOSE calculate the nucleon effective masses consistently, which is important in superfluid mode calculations. This database also provides easy-to-use interpolation routines that are necessary since the computational grid for the oscillation equations requires more points than are typically provided in tabulated EoS. Readers interested in using the CompOSE repository for compact star and supernova studies may consult the manual [18] or write to the authors of this manuscript.

5. Future Work

The results presented here are part of an ongoing, more systematic study of the effects of superfluidity in dense matter on the oscillation modes of compact stars. We made some approximations for the structure of the neutron star that can, and should be improved upon. For example, the EoS models adopted here assume that the interior consists primarily of nucleons, with the addition of electrons to satisfy charge neutrality. If hyperons are included, the entrainment matrix must be modified [19] and this will affect the superfluid modes. The normal fluid modes will also change, since the addition of hyperons softens the EoS at high density, making the star more compact. We expect, as shown in previous studies with polytropic models [20], that with increasing compactness (i.e., decreasing R/M), the f and p-mode frequencies will also increase. We are currently working on extensions to include quark matter, which has a relativistic dispersion and additional neutrality conditions. In such a case, the superfluid equations become more complicated: instead of dealing with the two species of nucleon, one would likely have to account for the entrainment between up, down and possibly strange quarks. Thus, not only does the equation of state have to be modified, but the fluid equations must be altered as well. The non-radial oscillations of a superconducting quark fluid have been recently discussed in [21], although its superfluid modes have not been calculated. Further improvements such as extending the analysis to finite temperature and including viscous dissipation that damps the oscillation modes are needed to provide a more complete picture of the spectrum and damping times of oscillations in a neutron star. As shown in this work, the CompOSE database can serve as a useful repository for studies in this direction.

Acknowledgments: P. J. gratefully acknowledges support from the U.S. National Science Foundation as P. I. of Grant No. PHY 1608959. R. M. was supported as a graduate student by the same grant.

Author Contributions: R. M. and P. J. performed the calculations. R. M. also made the figures, P. J. drafted the paper and T. K. provided critical input via CompOSE data and interpretation of results. All three authors approved the final version to be published.

Conflicts of Interest: The authors declare no conflict of interest.

Abbreviations

The following abbreviations are used in this manuscript:

EoS   equation of state

References

1. Abbott, B.P.; Abbott, R.; Abbott, T.D.; Acernese, F.; Ackley, K.; Adams, C.; Adams, T.; Addesso, P.; Adhikari, R.X.; Adya, V.B.; et al. GW170817: Observation of Gravitational Waves from a Binary Neutron Star Inspiral. *Phys. Rev. Lett.* **2017**, *119*, 161101.
2. CompStar Online Supernovae Equations of State. Available online: https://compose.obspm.fr/ (accessed on 6 March 2018).
3. Chandrasekhar, S. *An Introduction to the Study of Stellar Structure*; Dover Publications, Incorporated: Mineola, NY, USA, 1967; ISBN 13: 9780486604138.
4. Agrawal, B.K.; Shlomo, S.; Kim Au, V. Determination of the parameters of a Skyrme type effective interaction using the simulated annealing approach. *Phys. Rev. C* **2005**, *72*, 014310.
5. Gulminelli, F.; Raduta, A.R. Unified treatment of subsaturation stellar matter at zero and finite temperature. *Phys. Rev. C* **2015**, *92*, 055803.
6. Agrawal, B.K.; Shlomo, S.; Kim Au, V. Nuclear matter incompressibility coefficient in relativistic and nonrelativistic microscopic models. *Phys. Rev. C* **2003**, *68*, 031304.
7. Chamel, N. Two-fluid models of superfluid neutron star cores. *Mon. Not. R. Astron. Soc.* **2008**, *388*, 737–752.
8. Lindblom, L.; Mendell, G. The oscillations of superfluid neutron stars. *Astrophys. J.* **1994**, *421*, 689–704.
9. Gualtieri, L.; Kantor, E.M.; Gusakov, M.E.; Chugunov, A.I. Quasinormal modes of superfluid neutron stars. *Phys. Rev. D* **2014**, *90*, 024010.
10. Andersson, N.; Comer, G.L. Entropy entrainment and dissipation in superfluid Helium. *Int. J. Mod. Phys. D* **2011**, *20*, 1215.
11. Andreev, A.F.; Bashkin, E.P. Three-velocity hydrodynamics of superfluid solutions. *Sov. Phys. J. Exp. Theor. Phys.* **1976**, *42*, 164.
12. Andersson, N.; Comer, G.L.; Glampedakis, K. How viscous is a superfluid neutron star core? *Nucl. Phys. A* **2005**, *763*, 212–229.
13. Lindblom, L.; Mendell, G. R-modes in superfluid neutron stars. *Phys. Rev. D* **2000**, *61*, 104003.
14. Mcdermott, P.N.; Van Horn, H.M.; Hansen, C.J. Nonradial oscillations of neutron stars. *Astrophys. J.* **1988**, *325*, 725–748.
15. Asbell, J.L. Non-Radial Fluid Pulsation Modes of Compact Stars. Master's Thesis, California State University, Long Beach, CA, USA, 2016.
16. Lee, U. Nonradial oscillations of neutron stars with the superfluid core. *Astron. Astrophys.* **1995**, *303*, 515–525.

17. Serot, B.D. A relativistic nuclear field theory with π and ρ mesons. *Phys. Lett. B* **1979**, *86*, 146–150.
18. Typel, S.; Oertel, M.; Klähn, T. CompOSE CompStar online supernova equations of state; harmonising the concert of nuclear physics and astrophysics compose.obspm.fr. *Phys. Part. Nucl.* **1979**, *46*, 633–664.
19. Gusakov, M.E.; Kantor, M.; Haensel, P. The relativistic entrainment matrix of a superfluid nucleon-hyperon mixture at zero temperature. *Phys. Rev. C* **2009**, *79*, 055806.
20. Andersson, N.; Kokkotas, K.D. Pulsation modes for increasingly relativistic polytropes. *Mon. Not. R. Astron. Soc.* **1998**, *297*, 493–496.
21. Flores, C.V.; Lugones, G. Constraining color flavor locked strange stars in the gravitational wave era. *Phys. Rev. C* **2017**, *95*, 025808.

© 2018 by the authors. Licensee MDPI, Basel, Switzerland. This article is an open access article distributed under the terms and conditions of the Creative Commons Attribution (CC BY) license (http://creativecommons.org/licenses/by/4.0/).

Article

Anomalous Electromagnetic Transport in Compact Stars

Efrain J. Ferrer * and Vivian de la Incera

Department of Engineering Science and Physics, CUNY-College of Staten Island and CUNY-Graduate Center, New York, NY 10314, USA; vivian.incera@csi.cuny.edu
* Correspondence: efrain.ferrer@csi.cuny.edu; Tel.: +1-718-982-2919

Received: 9 January 2018; Accepted: 13 February 2018; Published: 12 March 2018

Abstract: We study the anomalous electromagnetic transport properties of a quark-matter phase that can be realized in the presence of a magnetic field in the low-temperature/moderate-high-density region of the Quantum Chromodynamics (QCD) phase map. In this so-called Magnetic Dual Chiral Density Wave phase, an inhomogeneous condensate is dynamically induced producing a nontrivial topology, a consequence of the asymmetry of the lowest Landau level modes of the quasiparticles in this phase. The nontrivial topology manifests in the electromagnetic effective action via a chiral anomaly term $\theta F^{\mu\nu}\tilde{F}_{\mu\nu}$, with an axion field θ given by the phase of the Dual Chiral Density Wave condensate. The coupling of the axion with the electromagnetic field leads to several macroscopic effects that include, among others, an anomalous, nondissipative Hall current, an anomalous electric charge, magnetoelectricity, and the formation of a hybridized propagating mode known as an axion polariton. The possible existence of this phase in the inner core of neutron stars opens a window to search for signals of its anomalous transport properties.

Keywords: chiral symmetry; axion QED; quark-hole pairing; cold-dense QCD; magnetic DCDW

1. Introduction

Neutron stars, the remnants of supernova collapse, are very dense objects produced by the gravitational colapse of very massive stars (stars with masses between 10 and 30 solar masses). They can reach densities several times larger than the nuclear density of 4×10^{17} kg/m^3. An interesting question that still remains unsettled is about the state of matter that supports such a dense medium. In this regard, very precise mass measurements for two compact objects, PSR J1614 − 2230 and PSR J0348 + 0432 with $M = 1.97 \pm 0.04 M_\odot$ [1] and $M = 2.01 \pm 0.04 M_\odot$ [2], respectively, where $M_\odot$ is the solar mass, have provided an important clue on the possible candidates for their interior composition: the phase of matter there should have an equation of state (EoS) rather stiff at high densities

The possibility of an interior composition based on a nucleon medium phase, formed mainly by neutrons, faces the difficulty of reproducing the required EoS stiffness. The reason is that, while under terrestrial conditions, hyperons are unstable and decay into nucleons through weak interactions, in neutron stars, the equilibrium conditions at core densities of order $2 - 3\rho_0$ (with $\rho_0 \simeq 0.16$ fm^{-3} the nuclear saturation density) can make the inverse process possible [3–12]. Thus, at large enough baryon chemical potential, the conversion of nucleons into hyperons becomes energetically favorable. This conversion releases the Fermi pressure exerted by the nucleons and makes the EoS soft enough to lead to a significant reduction of the star mass [4,13–20]. Different attempts to overcome this problem exist in the literature, as the inclusion of a repulsive hyperon-hyperon interaction through the exchange of vector mesons [21–25], or the inclusion of repulsive hyperonic three-body forces [26–31]. However, the possibility to reach the $2M_\odot$ with a nucleon inner phase still remains an open question under discussion.

On the other hand, in the highly dense cores of the compact objects, the neutron-rich matter can give rise to new degrees of freedom, by forming quark matter (see e.g., [32,33]). Even more, cold strange quark matter has been argued to be absolutely stable [34]. Thus, a phase transition can take place in the core favoring quark matter in the entire star interior, therefore giving rise to a strange star. Nevertheless, since the existence of $2M_\odot$ stars [35,36] was reported, there were also claims [37,38] that quark matter has too soft an EoS to reach such a high-mass value. Nonetheless, it was later realized that to study quark matter in compact objects, one has to rely on effective models like the Nambu-Jona-Lasinio (NJL) models with parameters matched to nuclear data. The one-gluon exchange interaction of QCD contains a dominant attractive diquark channel that is incorporated as a four-fermion interaction term in NJL models. This attractive interaction gives rise to color superconductivity (CS) [39–42]. NJL models have predicted that the most favored phase of CS at asymptotically high densities is the three-flavor color-flavor-locked (CFL) phase with a significantly large gap. The existence of a large superconducting gap together with a repulsive vector interaction, which is always present in a dense medium [43], can help to make the EoS stiffer. Yet, it was also found more recently that gluons in the color superconducting medium can soften the EoS [44]. Then, it is not clear at present what will be the more significant factor among all these competing effects, since they depend on parameter values that cannot be fixed with total precision.

Moreover, even assuming that neutron stars can realize a quark phase, this does not take place at asymptotically large densities. Their location in the QCD phase map will better correspond to the low-temperature, intermediate-density region. This is a particularly challenging region with the possibility of spatially inhomogeneous phases. To visualize that, we should take into account that coming from the low density region, the energy separation between quarks and antiquarks grows with increasing density up to a point where to excite antiquarks all the way from the Dirac sea in order to pair them with the quarks at the Fermi surface is not energetically favorable anymore. In this case, instead of undergoing a transition to a chirally restored phase, the system prefers to pair quarks and holes with parallel momenta close to the Fermi surface, giving rise to inhomogeneous chiral condensates. Spatially inhomogeneous phases with quark-hole condensates have been found in the large-N limit of QCD [45–47], in quarkyonic matter [48–51], and in NJL models [52–57]. Hence, although most NJL models had predicted a first-order chiral transition with increasing density [58], it turned out that the transition is more likely to occur via some intermediate state(s) characterized by inhomogeneous chiral condensates.

Inhomogeneous phases become favored also in CS [59], when the intermediate density region is approached from the region of low temperatures and asymptotically high density values. As already pointed out, at asymptotic densities, the most favored CS phase is the CFL, a homogeneous phase in which all flavors pair with each other via the strong attractive quark-quark channel. This phase is based on BCS quark-pairing and relies on the assumption that the quarks that pair with equal and opposite momenta can each be arbitrarily close to their common Fermi surface. However, with decreasing density, the combined effect of the strange quark mass, neutrality constraint and beta equilibrium, tends to pull apart the Fermi momenta of different flavors, imposing an extra energy cost on the formation of Cooper pairs. Thus, we conclude that BCS-pairing dominates as long as the energy cost of forcing all species to have the same Fermi momentum is compensated by the pairing energy that is released by the formation of Cooper pairs.Then, with decreasing density, the CFL phase eventually becomes gapless and, most importantly, becomes unstable [60,61]. The instability, known as chromomagnetic instability, manifests itself in the form of imaginary Meissner masses for some of the gluons and indicates an instability towards spontaneous breaking of translational invariance [62–65]. In other words, it indicates the formation of a spatially inhomogeneous phase. Most inhomogeneous CS phases are based on the idea of Larkin and Ovchinnikov (LO) [66] and Fulde and Ferrell (FF) [67], originally applied to condensed matter. In the CS LOFF phases [68–70], quarks of different flavors pair even though they have different Fermi momenta, because they form Cooper pairs with nonzero momentum. CS inhomogeneous phases with gluon condensates that break rotational symmetry [71] have also been considered to remove the chromomagnetic instability. However, to the best of our knowledge, the question of which CS phase is the most favorable in the region of intermediate densities still remains unanswered.

In addition to their high densities, neutron stars typically have strong magnetic fields, which become extremely large in the case of magnetars, with inner values that have been estimated to range from 10^{18} G for nuclear matter [72] to 10^{20} G for quark matter [73]. The fact that strong magnetic fields populate the vast majority of the astrophysical compact objects and that they can significantly affect several properties of the star have served as motivation for many works focused on the study of the EoS of magnetized neutron stars, without considering [73–98], or considering the magnetic-field interaction with the particle anomalous-magnetic-moment [88,99–113]. An important characteristic is that the EoS in a uniform magnetic field becomes anisotropic, with different pressures along the field and transverse to it [73,94,98]. The transverse and longitudinal pressures can be found performing a quantum-statistical average of the energy-momentum tensor, as done in [73] using a path-integral approach based on the partition function of the grand canonical emsemble, or in [94] using the many-particle density matrix.

It has been found that as a consequence of the anisotropy in the EoS, the effects of the two pressures produce opposite contributions to the TOV equations. When the anisotropy becomes significant, the TOV approach is inadequate, since it is based on a medium with spherical symmetry, while the influence of a strong uniform magnetic field makes the geometry cylindrical. Lacking a suitable approach that is compatible with the symmetry of the problem, any conclusion about the effect of the magnetic field on the neutron star mass is in principle unreliable.

In summary, up to now it has been impossible to reliably determine if neutron stars (magnetars included) are formed by neutrons or quarks, or by a hybrid combination of them, and if quark matter is present, what its phase will be. That is why it is so important to look for new signals that can be attributed uniquely to a certain phase of the interior matter with the hope of using observations to pinpoint the inner composition of the star. Transport properties could be a way to reach this goal. This paper aims to advance such a strategy. We present the electromagnetic transport properties [114] of one of the quark phases: the Magnetic Dual Chiral Density Wave (MDCDW) phase, that can take place in the low-temperature/intermediate-density region of quark matter. An important attribute of this phase is its similarities with condensed matter topological materials like Weyl semimetals (WSM) [115,116]. This opens the possibility to take advantage of new understandings within these materials to infer potentially measurable effects in the MDCDW phase of quark matter, and then use that insight to design clever ways to probe the presence of this quark phase in neutron stars.

2. The MCDCW Phase

To study the electromagnetic properties of the MCDCW phase, we should start by modeling QCD + QED with the help of the following Lagrangian density that combines electromagnetism with a two-flavor NJL model of strongly interacting quarks,

$$\mathcal{L} = -\frac{1}{4}F_{\mu\nu}F^{\mu\nu} + \bar{\psi}[i\gamma^{\mu}(\partial_{\mu} + iQA_{\mu}) + \gamma_0\mu]\psi \\ + G[(\bar{\psi}\psi)^2 + (\bar{\psi}i\tau\gamma_5\psi)^2], \tag{1}$$

Here $Q = \mathrm{diag}(e_u, e_d) = \mathrm{diag}(\frac{2}{3}e, -\frac{1}{3}e)$, $\psi^T = (u, d)$; μ is the baryon chemical potential; and G is the four-fermion coupling. The electromagnetic potential A^{μ} is formed by the background $\bar{A}^{\mu} = (0, 0, Bx, 0)$, which corresponds to a constant and uniform magnetic field **B** pointing in the z-direction, with $x^{\mu} = (t, x, y, z)$, plus the fluctuation field $\tilde{A}$. In the presence of electromagnetic interactions the flavor symmetry $SU(2)_L \times SU(2)_R$ of the original NJL model is reduced to the subgroup $U(1)_L \times U(1)_R$.

In the presence of a magnetic field, the formation of a dual chiral density wave (DCDW) condensate with magnitude Δ and modulation $q^\mu = (0,0,0,q)$ along the field direction [117,118] is favored

$$\langle \bar{\psi}\psi \rangle = \Delta \cos q_\mu x^\mu, \quad \langle \bar{\psi} i\tau_3 \gamma_5 \psi \rangle = \Delta \sin q_\mu x^\mu, \tag{2}$$

The mean-field Lagrangian of the MDCDW phase is then,

$$\begin{aligned}\mathcal{L}_{MF} &= \bar{\psi}[i\gamma^\mu(\partial_\mu + iQA_\mu) + \gamma_0\mu]\psi - m\bar{\psi}e^{i\tau_3\gamma_5 q_\mu x^\mu}\psi \\ &\quad - \frac{m^2}{4G} - \frac{1}{4}F_{\mu\nu}F^{\mu\nu},\end{aligned} \tag{3}$$

where $m = -2G\Delta$.

To remove the spatial modulation of the mass, we use a local chiral transformation

$$\psi \to e^{i\tau_3\gamma_5\theta}\psi, \quad \bar{\psi} \to \bar{\psi} e^{i\tau_3\gamma_5\theta}, \tag{4}$$

with $\theta(x) = qz/2$.

After the chiral transformation (4), the mean-field Lagrangian density (3) becomes

$$\begin{aligned}\mathcal{L}_{MF} &= \bar{\psi}[i\gamma^\mu(\partial_\mu + iQA_\mu + i\tau_3\gamma_5\partial_\mu\theta) + \gamma_0\mu - m]\psi \\ &\quad - \frac{m^2}{4G} - \frac{1}{4}F_{\mu\nu}F^{\mu\nu}\end{aligned} \tag{5}$$

The corresponding modified Dirac Hamiltonian of flavor f is

$$H_f = -i\gamma^0\gamma^i(\partial_i + ie_f A_i + i\frac{e_f}{|e_f|}\gamma_5\partial_i\theta) + \gamma^0 m, \tag{6}$$

Here e_f is the flavor electric charge. The single-particle energy spectrum is given by the eigenvalues of H_f. It separates into two sets of energy modes [117], the LLL ($l = 0$) modes

$$E^0 = \epsilon\sqrt{m^2 + k_3^2} + q/2, \quad \epsilon = \pm, \tag{7}$$

and the higher Landau level ($l \neq 0$) modes

$$E^l = \epsilon\sqrt{(\xi\sqrt{m^2 + k_3^2} + q/2)^2 + 2|e_f B|l}, \quad \epsilon = \pm, \xi = \pm, l = 1,2,3,... \tag{8}$$

In (8) $\xi = \pm$ indicates spin projection and $\epsilon = \pm$ particle/antiparticle energies. In contrast, only one spin projection (+ for positively charged and − for negatively charged quarks) contributes to the LLL spectrum. An important feature of this spectrum is that the LLL energies are not symmetric about the zero-energy level. As a consequence, the ± sign in front of the square root should not be interpreted as particle/antiparticle in the LLL case.

A peculiarity of the local chiral transformation (4) is that it does not leave invariant the fermion measure in the path-integral. To take this into consideration, we need to calculate the contribution of the Jacobian $J(\theta(x)) = (\text{Det}U_A)^{-2}$ with $U_A = e^{i\tau_3\gamma_5\theta}$

$$\mathcal{D}\bar{\psi}(x)\mathcal{D}\psi(x) \to (\text{Det}U_A)^{-2}\mathcal{D}\bar{\psi}(x)\mathcal{D}\psi(x), \tag{9}$$

to the effective action [114]. As a consequence, the fermion effective action entering in the partition function after the local chiral transformation is implemented reads

$$S_{eff} = \int d^4x \{\bar{\psi}[i\gamma^\mu(\partial_\mu + iQA_\mu + i\tau_3\gamma_5\partial_\mu\theta) + \gamma_0\mu - m]\psi - \frac{m^2}{4G} + \frac{\kappa}{4}\theta(x)F_{\mu\nu}\tilde{F}^{\mu\nu}\}, \quad (10)$$

on which $\frac{\kappa}{4} = \frac{3(e_u^2 - e_d^2)}{8\pi^2} = \frac{e^2}{8\pi^2} = \frac{\alpha}{2\pi}$ reflects the contribution of all the flavors and colors. The axion term $\frac{\kappa}{4}\theta(x)F_{\mu\nu}\tilde{F}^{\mu\nu}$ denotes the contribution of the Jacobian of the fermionic measure [114].

3. Axion Electrodynamics in the MDCDW Phase

To find the Maxwell equations of the MDCDW phase, we need to find the zero temperature electromagnetic effective action $\Gamma(A)$ corresponding to the effective classical action (10),

$$\Gamma = -i\log Z, \quad (11)$$

with Z the partition function given by

$$Z = e^{i\Gamma_{matter}} = \int \mathcal{D}\bar{\psi}(x)\mathcal{D}\psi(x)e^{iS_{eff}} \quad (12)$$

with S_{eff} given in (10).

After integrating in the fermion fields and carrying out the finite-temperature Matsubara sum to take the zero-temperature limit, we can expand Γ in powers of the fluctuation field $\tilde{A}$ to obtain

$$\Gamma(A) = -V\Omega + \int d^4x \left[-\frac{1}{4}F_{\mu\nu}F^{\mu\nu} + \frac{\kappa}{4}\theta(x)F_{\mu\nu}\tilde{F}^{\mu\nu}\right] \quad (13)$$
$$+ \sum_{i=1}^{\infty}\int dx_1...dx_i \Pi^{\mu_1,\mu_2,...\mu_i}(x_1, x_2,...x_i)\tilde{A}_{\mu_1}(x_1)...\tilde{A}_{\mu_i}(x_i),$$

with V the four-volume, Ω the mean-field thermodynamic potential obtained for this phase in Ref. [117], and $\Pi^{\mu_1,\mu_2,...\mu_i}$ the i-vertex tensors corresponding to the one-loop polarization operators with internal lines of fermion Green functions in the MDCDW phase and i external lines of photons. In (13) we added the pure electromagnetic field contribution $-\frac{1}{4}F_{\mu\nu}F^{\mu\nu}$.

We are interested in the linear response of the MCDCW phase to a small electromagnetic probe $\tilde{A}$. Furthermore, for consistency of the approximation, we can neglect all the radiative corrections of order higher than α, as α is the order of the axion term in (13). These two conditions imply that we shall cut the series in (13) at $i = 1$, which can be shown to provide the medium corrections to the Maxwell equations that are linear in the electromagnetic field and of the desired order in α.

Then, $\Gamma(A)$ becomes

$$\Gamma(A) = -V\Omega + \int d^4x \left[-\frac{1}{4}F_{\mu\nu}F^{\mu\nu} - \kappa\int d^4x \epsilon^{\mu\alpha\nu\beta} A_\alpha \partial_\nu A_\beta \partial_\mu \theta\right]$$
$$- \int d^4x \tilde{A}_\mu(x) J^\mu(x), \quad (14)$$

where we integrated by parts the third term in the r.h.s. of (13). $J^\mu(x) = (J^0, \mathbf{J})$ represents the contribution of the ordinary (non-anomalous) electric four-current, determined by the one-loop tadpole diagrams.

The Euler-Lagrange equations derived from the action (14) give rise to the modified Maxwell equations

$$\nabla \cdot \mathbf{E} = J^0 + \frac{e^2}{4\pi^2} q B, \qquad (15)$$

$$\nabla \times \mathbf{B} - \partial \mathbf{E}/\partial t = \mathbf{J} - \frac{e^2}{4\pi^2} \mathbf{q} \times \mathbf{E}, \qquad (16)$$

$$\nabla \cdot \mathbf{B} = 0, \quad \nabla \times \mathbf{E} + \partial \mathbf{B}/\partial t = 0, \qquad (17)$$

on which we already used that $\theta = \frac{qz}{2}$. These are the equations of axion electrodynamics for the MDCDW phase, which are a particular case of those proposed by Wilczek [119] many years ago for a general axion field θ.

From Equations (15) and (16) we have that the axion term leads to an anomalous electric charge density,

$$J^0_{anom} = \frac{e^2}{4\pi^2} q B, \qquad (18)$$

as well as to an anomalous Hall current density,

$$\mathbf{J}_{anom} = -\frac{e^2}{4\pi^2} \mathbf{q} \times \mathbf{E} \qquad (19)$$

The anomalous Hall current is perpendicular to both the magnetic and the electric field, since $\mathbf{q}$ is aligned with $\mathbf{B}$. Besides, $\mathbf{J}_{anom}$ is dissipationless and as such, it can significantly influence the transport properties of the system.

In (15) and (16), J^0 and $\mathbf{J}$ are the ordinary charge and current densities respectively, which are calculated from the tadpole diagrams of the theory. Of special interest is to check if the ordinary charge so obtained can annihilate the anomalous contribution J^0_{anom}. As discussed in [114], since J^0_{anom} only gets contributions from the LLL, we should compare it with the corresponding tadpole result calculated in the LLL approximation. Then, calculating the tadpole diagrams we obtain

$$J^{1,2,3}_{LLL} = 0 \qquad (20)$$

and

$$J^0_{LLL} = \sum_f J^0_{LLL}(\operatorname{sgn}(e_f)) \qquad (21)$$

$$= \frac{e^2 B}{2\pi^2} \sqrt{(\mu - q/2)^2 - m^2} [\Theta(\mu - q/2 - m) - \Theta(q/2 - \mu - m)]$$

Thus, there is no LLL current density, but only a charge density. Comparing Equations (18) and (21) we can see that they do not cancel out. Only if $m = 0$ in (21), meaning setting the condensate amplitude to zero, the anomalous electric charge density will be cancelled out by the ordinary charge in Equation (15). In such a situation, the resulting LLL contribution to the net electric charge density reduces to $\frac{e^2 B}{2\pi^2} \mu$, a non-anomalous term which, as expected, is independent of q since no physical quantity should depend on q when there is no MDCDW condensate (i.e., when $m = 0$).

Finally, since the Maxwell Equation (16) contains an anomalous Hall current, it is important to investigate if it can be cancelled out by an ordinary Hall current. The Hall conductivity can be easily found from the charge density in the case that it is linearly dependent on the magnetic field [114],

$$\sigma_{xy} = \frac{\partial J^0}{\partial B} \qquad (22)$$

Applying this formula to the ordinary electric charge density (21), the corresponding LLL contribution to the Hall conductivity is

$$\sigma_{xy}^{ord} = \frac{\partial J_{LLL}^0}{\partial B} = \frac{e^2}{2\pi^2}\sqrt{(\mu-q/2)^2-m^2}[\Theta(\mu-q/2-m)-\Theta(q/2-\mu-m)] \quad (23)$$

which leads to the LLL ordinary Hall current $\mathbf{J}_{LLL}^{ord} = (\sigma_{xy}^{ord}E_y, -\sigma_{xy}^{ord}E_x, 0)$. Clearly, $\mathbf{J}_{LLL}^{ord}$ does not cancel out the anomalous current (19).

Likewise, the anomalous Hall conductivity can be found either from the anomalous charge (18),

$$\sigma_{xy}^{anom} = \frac{\partial J_{anom}^0}{\partial B} = \frac{e^2}{4\pi^2}q, \quad (24)$$

or directly from the anomalous Hall current $\mathbf{J}_{anom}$ given in (19). As J_{anom}^0 is due to the LLL, so is σ_{xy}^{anom}, thereby underlining once again the LLL origin of $\mathbf{J}_{anom}$.

4. Anomalous Transport in the MDCDW Phase

The MDCDW phase exhibits some interesting electromagnetic features, such as for example, linear magnetoelectricity [114]. To see this, we should define the **D** and **H** fields as

$$\mathbf{D} = \mathbf{E} - \kappa\theta\mathbf{B}, \quad \mathbf{H} = \mathbf{B} + \kappa\theta\mathbf{E} \quad (25)$$

Then, rewriting the Maxwell Equations (15) and (16) in terms of these fields we obtain,

$$\nabla \cdot \mathbf{D} = J^0, \quad \nabla \times \mathbf{H} - \frac{\partial \mathbf{D}}{\partial t} = \mathbf{J} \quad (26)$$

Equations (25) imply that a magnetic field induces an electric polarization $\mathbf{P} = -\kappa\theta\mathbf{B}$ and an electric field induces a magnetization $\mathbf{M} = -\kappa\theta\mathbf{E}$. Both parameters, **P** and **M**, depend on the theta angle, which evidences the anomalous character of this result. It is important to point out that the appearance of a linear magnetoelectricity in this medium is possible because, as seen from (25) and (26), the MDCDW ground state breaks parity and time-reversal symmetries. We call attention to the fact that the magnetoelectricity here is different from the one found in the magnetic-CFL phase of color superconductivity [120–122], where P was not broken and the effect was a consequence of an anisotropic electric susceptibility [123], so it was not linear.

In this formulation, it follows from (26) that the anomalous Hall current is given by a medium-induced, magnetic current density $\nabla \times \mathbf{M}$, due to the space-dependent anomalous magnetization coming from the axion term.

An interesting property of the anomalous Hall current is its dissipationless character. This is a consequence of the fact that this current (19) is perpendicular to **E** and to the modulation vector **q**, which in turn is parallel to **B**. We already proved that the anomalous Hall conductivity is given by

$$\sigma_{xy}^{anom} = e^2 q/4\pi^2 \quad (27)$$

Its anomalous character is reflected in the fact that it does not depend on the fermion mass m, which is consistent with the nondissipative character of the anomalous Hall current.

The same expression of the anomalous Hall conductivity has been found in WSM [115], where the role of the modulation parameter q is played by the separation in momentum of the Weyl nodes. A similar Hall conductivity can appear also at the boundary between a topological and a normal insulator [124] when there is an electric field in the plane of the boundary. However, in the topological insulator case, the anomalous Hall conductivity is discrete because the axion field θ jumps from 0 to π in the surface of the two insulators. Our results are also connected to optical lattices, as 3D topological insulators have been proposed to exist in 3D optical lattices [125].

It is worth pointing out the relevance of these results for neutron stars. The values of chemical potential and magnetic field needed for the realization of the MDCDW phase are all within the possible ranges of these parameters at the core of hybrid stars or in quark stars. As shown in [117], for baryon chemical potentials between 300 and 600 MeV, and magnetic field strengths $\sqrt{eB} \sim 150 - 500$ MeV ($\sim 10^{18}$–10^{19} G) the MDCDW is very robust, with b and m in the range between 300–500 MeV. The inhomogeneity parameter b actually sets at arbitrarily small values of magnetic field and chemical potential because the anomalous term in the thermodynamic potential drives $b \neq 0$ from the beginning, but b becomes of the order of the dynamical mass when the matter density becomes nonzero, which typically occurs when the chemical potential is comparable to the vacuum dynamical mass. Furthermore, the application to neutron stars requires introducing electrical neutrality (either locally or globally, as the neutrality does not need to be satisfied locally for compact hybrid stars [126]) and beta equilibrium conditions, as well as vector interactions to be able to accommodate the $2M_\odot$ observations. None of these conditions however impede the realization of the MDCDW phase, as shown in detail in [93], where their main consequence was that each flavor acquires its own modulation and dynamical mass, and the MDCDW remains robust at fields of the order of 10^{18} G and chemical potentials between 300 and 600 MeV. A significant result in [93] was the fact that the MDCDW phase is compatible with the $2M_\odot$ observations. It is easy to see that the effect on the anomalous terms will simply be that each flavor still contributes to the anomalous electric charge and Hall current, only now each depends on the flavor's particular modulation parameter. The anomalous currents could serve to resolve the issue with the stability of the magnetic field strength in magnetars [127,128]. It will be worthy to understand if the magneto-transport properties of the MDCDW phase can significantly affect the thermal and electrical conductivity producing a tangible separation between the transport properties of compact stars formed by neutrons or by quarks in this inhomogeneous phase. These and other questions highlight the importance of exploring which observable signatures could be identified and then used as telltales of the presence of the MDCDW phase in the star core.

Another interesting effect can be connected to the fluctuations $\delta\theta$ of the axion field. If one goes beyond the mean-field approximation, there will be mass and kinetic terms of the axion field fluctuation. Besides, due to the background magnetic field, the axion fluctuation couples linearly to the electric field via the term $\kappa\delta\theta \mathbf{E} \cdot \mathbf{B}$, so the field equations of the axion fluctuation and the electromagnetic field will be mixed, giving rise to a quasiparticle mode known as the axion polariton mode [129]. The axion polariton mode is gapped with a gap proportional to the background magnetic field. This implies that electromagnetic waves of certain frequencies will be attenuated by the MDCDW matter, since in this medium they propagate as polaritons. The axion polariton could be useful to probe the presence of the MDCDW phase in different media.

5. Conclusions

In this paper we present the electromagnetic anomalous transport properties of the so called MDCDW phase of quark matter at low temperatures and moderate densities in the presence of a magnetic field. This is a phase that in principle can be realized in neutron hybrid stars with a quark core or in strange stars.

The system under study has a non-trivial topology, which is due to the combined effect of a ground state having an inhomogeneous particle-hole condensate and the dimensional reduction affecting the quasiparticles occupying the LLL. As a consequence, the system exhibits an anomalous charge that depends on the applied magnetic field and the modulation of the particle-hole condensate. The topological nature of the electric charge can be traced back to the spectral asymmetry of the LLL modes. The spectral asymmetry is also responsible for an anomalous non-dissipative Hall current that depends on the modulation parameter.

We should mention that the reported results can also have importance for Heavy Ion Collision (HIC) physics. Future HIC experiments, that will take place at lower temperatures and higher densities, will certainly generate strong magnetic and electric fields in their off-central collisions and will open a much more sensitive window to look into a very challenging region of QCD [130]. For example, the second phase of the RHIC energy scan (BES-II) [131], the planned experiments at the Facility for Antiproton and Ion Research (FAIR) [132] at the GSI site in Germany, and the Nuclotron-based Ion Collider Facility (NICA) [133,134] at JINR laboratory in Dubna, Russia, are all designed to run at unprecedented interaction rates to provide high-precision measures of observables in the high baryon density region. That is why it is so timely and relevant to carry out detailed theoretical investigations of all potential observables of the MDCDW phase. Therefore, we hope that our findings will serve to stimulate quantitative studies to identify signatures of the anomalous effects here discussed in the future HIC experiments.

In addition, we should notice that the anomalous effects of the MDCDW phase share many properties with similar phenomena in condensed matter systems with non-trivial topologies as topological insulators [124], where θ depends on the band structure of the insulator; Dirac semimetals [135–138], a 3D bulk analogue of graphene with non-trivial topological structures; and WSM [115], where the derivative of the angle θ is related to the momentum separation between the Weyl nodes. Countertop experiments with these materials can therefore help us to gain useful insight into the physics governing the challenging region of strongly coupled QCD, thereby inspiring new strategies to probe the presence of the MDCDW phase in neutron stars and HIC.

Acknowledgments: This work was supported in part by NSF grant PHY-1714183, and PSC-CUNY Award 60650-00 48.

Author Contributions: These authors contributed equally to this work.

References

1. Demorest, P.; Pennucci, T.; Ransom, S.M.; Roberts, M.S.E.; Hessels, J.W.T. Shapiro delay measurement of a two solar mass neutron star. *Nature* **2010**, *467*, 1081–1083.
2. Antoniadis, J.; Paulo, C.C.; Freire, P.C.C.; Wex, N.; Tauris, T.M.; Lynch, R.S.; van Kerkwijk, M.H.; Kramer, M.; Bassa, C.; Dhillon, V.S.; et al. A massive pulsar in a compact relativistic binary. *Science* **2013**, *340*, 1233232.
3. Ambartsumyan, V.A.; Saakyan, G.S. The degenerate superdense gas of elementary particles. *Sov. Astron.* **1960**, *4*, 187–201.
4. Glendenning, N.K. The hyperon composition of neutron stars. *Phys. Lett. B* **1982**, *114*, 392–396.
5. Glendenning, N.K. Neutron stars are giant hypernuclei? *Astrophys. J.* **1985**, *293*, 470–493.
6. Weber, F.; Weigel, M.K. Baryon composition and macroscopic properties of neutron stars. *Nucl. Phys. A* **1989**, *505*, 779–822.
7. Knorren, R.; Prakash, M.; Ellis, P.J. Strangeness in hadronic stellar matter. *Phys. Rev. C* **1995**, *52*, 3470–3482.
8. Schulze, H.-J.; Sagawa, H.; Wu, C.-X.; Zhao, E.-G. Hypernuclei in the deformed Skyrme-Hartree-Fock approach. *Phys. Rev. C* **2007**, *76*, 034312.
9. Zhou, X.-R.; Polls, A.; Schulze, H.-J.; Vidaña, I. Lambda hyperons and the neutron drip line. *Phys. Rev. C* **2008**, *78*, 054306.
10. Sammarruca, F. Effects of Lambda hyperons on the nuclear equation of state in a Dirac-Brueckner-Hartree-Fock model. *Phys. Rev. C* **2009**, *79*, 034301.
11. Lonardoni, D.; Pederiva, F.; Gandolfi, S. Accurate determination of the interaction between hyperons and nucleons from auxiliary field diffusion Monte Carlo calculations. *Phys. Rev. C* **2014**, *89*, 014314.
12. Katayama, T.; Saito, K. Hyperons in neutron stars. *Phys. Lett. B* **2015**, *747*, 43–47.
13. Glendenning, N.K. Hyperons in neutron stars. *Z. Phys. A* **1987**, *326*, 57–64.
14. Schulze, H.-J.; Baldo, M.; Lombardo, U.; Cugnon, J.; Lejeune, A. Hypernuclear matter in the Bruckner-Hartree-Fock approximation. *Phys. Lett. B* **1995**, *355*, 21–26.
15. Schulze, H.-J.; Baldo, M.; Lombardo, U.; Cugnon, J.; Lejeune, A. Hyperonic nuclear matter in Bruckner theory. *Phys. Rev. C* **1998**, *57*, 704.
16. Baldo, M.; Burgio, G.F.; Schulze, H.-J. Hyperon stars in the Brueckner-Bethe-Goldstone theory. *Phys. Rev. C* **2000**, *61*, 055801.
17. Vidaña, I.; Polls, A.; Ramos, A.; Engvik, L.; Hjorth-Jensen, M. Hyperon-hyperon interactions and properties of neutron star matter. *Phys. Rev. C* **2000**, *62*, 035801.

18. Schulze, H.-J.; Polls, A.; Ramos, A.; Vidaña, I. Maximum mass of neutron stars. *Phys. Rev. C* **2006**, *73*, 058801.
19. Djapo, H.; Schaefer, B.-J.; Wambach, J. Appearance of hyperons in neutron stars. *Phys. Rev. C* **2010**, *81*, 035803.
20. Schulze, H.-J.; Rijken, T. Maximum mass of hyperon stars with the Nijmegen ESC-08 model. *Phys. Rev. C* **2011**, *84*, 035801.
21. Bednarek, I.; Haensel, P.; Zdunik, L.; Bejger, M.; Mánka, R. Hyperons in neutron-star cores and two-solar-mass pulsar. *Astron. Astrophys.* **2012**, *157*, 543.
22. Weissenborn, S.; Chatterjee, D.; Schaffner-Bielich, J. Hyperons and massive neutron stars: vector repulsion and SU(3) symmetry. *Phys. Rev. C* **2012**, *85*, 065802.
23. Weissenborn, S.; Chatterjee, D.; Schaffner-Bielich, J. Hyperons and massive neutron stars: vector repulsion and SU(3) symmetry (Erratum). *Phys. Rev. C* **2014**, *90*, 019904.
24. Oertel, M.; Providência, C.; Gulminelli, F.; Raduta, A.R. Hyperons in neutron star matter within relativistic mean-field models. *J. Phys. G* **2015**, *42*, 075202.
25. Maslov, K.A.; Kolomeitsev, E.E.; Voskresensky, D.N. Solution of the hyperon puzzle within a relativistic mean-field model *Phys. Lett. B* **2015**, *748*, 369–375.
26. Takatsuka, T.; Nishizaki, S.; Yamamoto, Y. Necessity of extra repulsion in hypernuclear systems: Suggestion from neutron stars. *Eur. Phys. J. A* **2002**, *13*, 213.
27. Takatsuka, T.; Nishizaki, S.; Tamagaki, R. Three-Body force as an extra repulsion suggested from hyperon-mixed neutron stars. *Prog. Theor. Phys. Suppl.* **2002**, *174*, 80–83.
28. Vidaña, I.; Logoteta, D.; Providência, C.; Polls, A.; Bombaci, I. Estimation of the effect of hyperonic three-body forces on the maximum mass of neutron stars. *Eur. Phys. Lett.* **2011**, *94*, 11002.
29. Yamamoto, Y.; Furumoto, T.; Yasutake, B.; Rijken, T.A. Multi-pomeron repulsion and the neutron-star mass. *Phys. Rev. C* **2013**, *88*, 022801.
30. Yamamoto, Y.; Furumoto, T.; Yasutake, B.; Rijken, T.A. Hyperon mixing and universal many-body repulsion in neutron stars. *Phys. Rev. C* **2014**, *90*, 045805.
31. Lonardoni, D.; Lovato, A.; Gandolfi, S.; Pederiva, F. Hyperon puzzle: Hints from Quantum Monte Carlo calculations. *Phys. Rev. Lett.* **2015**, *114*, 092301.
32. Weber, F. Strange quark matter and compact stars. *Prog. Part Nucl. Phys.* **2005**, *54*, 193–288.
33. Lattimer, J.M. Neutron star equation of state. *New Astron. Rev.* **2010**, *54*, 101.
34. Witten, E. Cosmic separation of phases. *Phys. Rev. D* **1984**, *30*, 272.
35. Nice, D.J.; Splaver E.M.; Stairs, I.H.; Loehmer, O.; Jessner, A.; Kramer, M.; Cordes, J.M. A 2.1 solar mass pulsar measured by relativistic orbital decay. *Astrophys. J.* **2005**, *634*, 1242–1249.
36. Barret, D.; Olive, J.F.; Miller, M.C. An abrupt drop in the coherence of the lower kilohertz QPO in 4U 1636-536. *Mon. Not. Roy. Astron. Soc.* **2005**, *361*, 855–860.
37. Trümper, J.E.; Burwitz, V.; Haberl, F.; Zavlin, V.E. The puzzles of RX J1856.5-3754: Neutron star or quark star? *Nucl. Phys. Proc. Suppl.* **2004**, *132*, 560–565.
38. Özel, F. Soft equations of state for neutron-star matter ruled out by EXO 0748-676. *Nature* **2006**, *441*, 1115–1117.
39. Alford, M.G.; Schmitt, A.; Rajagopal, K.; Schäfer, T. Color superconductivity in dense quark matter. *Rev. Mod. Phys.* **2008**, *80*, 1455–1515.
40. Huang, M. QCD phase diagram at high temperature and density. *arXiv* **2010**, arXiv:1001.3216.
41. Schmitt, A. *Dense Matter in Compact Stars, Lecture Notes in Physics v. 811*; Springer: Berlin/Heidelberg, Germany, 2010.
42. Fukushima, K.; Hatsuda, T. The phase diagram of dense QCD. *Rept. Prog. Phys.* **2011**, *74*, 014001.
43. Kitazawa, M.; Koide, T.; Kunihiro, T.; Nemoto, Y. Chiral and color superconducting phase transitions with vector interaction in a simple model. *Prog. Theor. Phys.* **2002**, *108*, 929–951.
44. Ferrer, E.J.; de la Incera, V.; Paulucci, L. Gluon effects on the equation of state of color superconducting strange stars. *Phys. Rev. D* **2015**, *92*, 043010.
45. Deryagin, D.V.; Grigoriev, D.Y.; Rubakov, V.A. Standing wave ground state in high density, zero temperature QCD at large N(c). *Int. J. Mod. Phys. A* **1992**, *7*, 659–681.
46. Shuster, E.; Son, D.T. On finite density QCD at large N(c). *Nucl. Phys. B* **2000**, *573*, 434–446.
47. Park, B.-Y.; Rho, M.; Wirzba, A.; Zahed, I. Dense QCD: Overhauser or BCS pairing? *Phys. Rev. D* **2000**, *62*, 034015.

48. Kojo, T.; Hidaka, Y.; McLerran, L.; Pisarski, R.D. Quarkyonic chiral spirals. *Nucl. Phys. A* **2010**, *843*, 37–58.
49. Kojo, T.; Hidaka, Y.; Fukushima, K.; McLerran, L.D.; Pisarski, R.D. Interweaving chiral spirals. *Nucl. Phys. A* **2012**, *875*, 94–138 .
50. Kojo, T.; Pisarski, R.D.; Tsvelik, A.M. Covering the Fermi surface with patches of quarkyonic chiral spirals. *Phys. Rev. D* **2010**, *82*, 074015.
51. Kojo, T. A (1+1) dimensional example of quarkyonic matter. *Nucl. Phys. A* **2012**, *877*, 70–94.
52. Nickel, D. How many phases meet at the chiral critical point? *Phys. Rev. Lett.* **2009**, *103*, 072301.
53. Nickel, D. Inhomogeneous phases in the Nambu-Jona-Lasino and quark-meson model. *Phys. Rev. D* **2009**, *80*, 074025.
54. Rapp, R.; Shuryak, E.; Zahed, I. A Chiral crystal in cold QCD matter at intermediate densities? *Phys. Rev. D* **2001**, *63*, 034008.
55. Gubina, N.V.; Klimenko, K.G.; Kurbanov, S.G.; Zhukovsky, V.C. Inhomogeneous charged pion condensation phenomenon in the NJL_2 model with quark number and isospin chemical potentials. *Phys. Rev. D* **2012**, *86*, 085011.
56. Carignano, S.; Nickel, D.; Buballa, M. Influence of vector interaction and Polyakov loop dynamics on inhomogeneous chiral symmetry breaking phases. *Phys. Rev. D* **2010**, *82*, 054009.
57. Abuki, H.; Ishibashi, D.; Suzuki, K. Crystalline chiral condensates off the tricritical point in a generalized Ginzburg-Landau approach. *Phys. Rev. D* **2012**, *85*, 074002.
58. Klevansky, S. The Nambu-Jona-Lasinio model of quantum chromodynamics. *Rev. Mod. Phys.* **1992**, *64*, 649–708.
59. Anglani, R.; Casalbuoni, R.; Ciminale, M.; Ippolito, N.; Gatto, R.; Mannarelli, M.; Ruggieri, M. Crystalline color superconductors. *Rev. Mod. Phys.* **2014**, *86*, 509–561.
60. Casalbuoni, R.; Gatto, R.; Mannarelli, M.; Nardulli, G.; Ruggieri, M. Meissner masses in the gCFL phase of QCD. *Phys. Lett. B* **2005**, *605*, 362–368.
61. Fukushima, K. Analytical and numerical evaluation of the Debye and Meissner masses in dense neutral three-flavor quark matter. *Phys. Rev. D* **2005**, *72*, 074002.
62. Reddy, S.; Rupak, G. Phase structure of 2-flavor quark matter: Heterogeneous superconductors. *Phys. Rev. C* **2005**, *71*, 025201.
63. Fukushima, K. Characterizing the Larkin-Ovchinnikov-Fulde-Ferrel phase induced by the chromomagnetic instability. *Phys. Rev. D* **2006**, *73*, 094016.
64. Hashimoto, M. Manifestation of instabilities in Nambu-Jona-Lasinio type models. *Phys. Lett. B* **2006**, *642*, 93–99.
65. Huang, M. Spontaneous current generation in the 2SC phase. *Phys. Rev. D* **2006**, *73*, 045007.
66. Larkin, A.I.; Ovchinnikov, Y.N. Nonuniform state of superconductors. *Sov. Phys. JETP* **1965**, *20*, 762–770.
67. Fulde, P.; Ferrell, R.A. Superconductivity in a strong spin-exchange field. *Phys. Rev.* **1964**, *135*, A550–A563.
68. Alford, M.G.; Bowers, J.A.; Rajagopal, K. Crystalline color superconductivity. *Phys. Rev. D* **2001**, *63*, 074016.
69. Bowers, J.A.; Rajagopal, K. The crystallography of color superconductivity. *Phys. Rev. D* **2001**, *66*, 065002.
70. Casalbuoni, R.; Nardulli, G. Inhomogeneous superconductivity in condensed matter and QCD. *Rev. Mod. Phys.* **2004**, *76*, 263–320.
71. Ferrer, E.J.; de la Incera, V. Chromomagnetic instability and induced magnetic field in neutral two-flavor color superconductivity. *Phys. Rev. D* **2007**, *76*, 114012.
72. Dong, L; Shapiro, S. L. Cold equation of state in a strong magnetic field - Effects of inverse beta-decay. *ApJ* **1991**, *383*, 745–751.
73. Ferrer, E.J.; de la Incera, V.; Keith, J.P.; Portillo, I.; Springsteen, P.L. Equation of state of a dense and magnetized fermion system. *Phys. Rev. C* **2010**, *82*, 065802.
74. Fushiki, I.; Gudmundsson, E.H.; Pethick, C.J. Surface structure of neutron stars with high magnetic fields. *Astrophys. J.* **1989**, *342*, 958–975.
75. Abrahams, A.M.; Shapiro, S.L. Equation of state in a strong magnetic field-Finite temperature and gradient corrections. *Astrophys. J.* **1991**, *374*, 652–667.
76. Fushiki, I.; Gudmundsson, E H.; Yngvason, J.; Pethick, C.J. Matter in a magnetic field in the Thomas-Fermi and related theories. *Ann. Phys.* **1992**, *216*, 29–72.
77. Chakrabarty, S.; Bandyopadhyay, D.; Pal, S. Dense nuclear matter in a strong magnetic field. *Phys. Rev. Lett.* **1997**, *78*, 2898–2901.

78. Bandyopadhyay, D.; Chakrabarty, S.; Dey, P. Rapid cooling of magnetized neutron stars. *Phys. Rev. D* **1998**, *58*, 121301.
79. Broderick, A.; Prakash, M.; Lattimer, J.M. The Equation of state of neutron star matter in strong magnetic fields. *Astrophys. J.* **2000**, *537*, 351–367.
80. Cardall, C.Y.; Prakash, M.; Lattimer, J.M. Effects of strong magnetic fields on neutron star structure. *Astrophys. J.* **2001**, *554*, 322–339.
81. Suh, I.-S.; Mathews, G.J. Cold ideal equation of state for strongly magnetized neutron star matter: Effects on muon production and pion condensation. *Astrophys. J.* **2001**, *546*, 1126–1136.
82. Perez-Martinez, A.; Perez-Rojas, H.; Mosquera-Cuesta, H.J. Magnetic collapse of a neutron gas: Can magnetars indeed be formed? *Eur. Phys. J. C* **2003**, *29*, 111–123.
83. Wei, F.X.; Mao, G.J.; Ko, C.M.; Kisslinger, L.S.; Stoecker, H.; Greiner, W. Effect of isovector-scalar meson on neutron star matter in strong magnetic fields. *J. Phys. G* **2006**, *32*, 47–61.
84. Hardings, A.K.; Lai, D. Physics of strongly magnetized neutron stars. *Rep. Prog. Phys.* **2006**, *69*, 2631.
85. Chen, W.; Zhang, P.-Q.; Liu, L.-G. The influence of the magnetic field on the properties of neutron star matter. *Mod. Phys. Lett. A* **2007**, *22*, 623–630.
86. Rabhi, A.; Providencia, C.; da Providencia, J. Stellar matter with a strong magnetic field within density-dependent relativistic models. *J. Phys. G* **2008**, *35*, 125201.
87. Gonzalez-Felipe, R.; Perez-Martinez, A.; Perez-Rojas, H.; Orsaria, M. Magnetized strange quark matter and magnetized strange quark stars. *Phys. Rev. C* **2008**, *77*, 015807.
88. Yue, P.; Yang, F.; Shen, H. Properties of hyperonic matter in strong magnetic fields. *Phys. Rev. C* **2009**, *79*, 025803.
89. Menezes, D.P.; Benghi Pinto, M.; Avancini, S. S.; Perez Martinez, A.; Providencia, C. Quark matter under strong magnetic fields in the Nambu-Jona-Lasinio Model. *Phys. Rev. C* **2009**, *79*, 035807.
90. Menezes, D.P.; Benghi Pinto, M.; Avancini, S. S.; Providencia, C. Quark matter under strong magnetic fields in the SU(3) Nambu-Jona-Lasinio model. *Phys. Rev. C* **2009**, *80*, 065805.
91. Paulucci, L.; Ferrer, E.J.; de la Incera, V.; Horvath, J.E. Equation of state for the MCFL phase and its implications for compact star models. *Phys. Rev. D* **2011**, *83*, 043009.
92. Fayazbakhsh, S.; Sadooghi, N. Anomalous magnetic moment of hot quarks, inverse magnetic catalysis, and reentrance of the chiral symmetry broken phase. *Phys. Rev. D* **2014**, *90*, 105030.
93. Carignano, S.; Ferrer, E.J.; de la Incera, V.; Paulucci, L. Crystalline chiral condensates as a component of compact stars. *Phys. Rev. D* **2015**, *92*, 105018.
94. Canuto, V.; Chiu, H.Y. Quantum theory of an electron gas in intense magnetic fields. *Phys. Rev.* **1968**, *173*, 1210–1219.
95. Canuto, V.; Chiu, H.Y. Thermodynamic properties of a magnetized fermi gas. *Phys. Rev.* **1968**, *173*, 1220.
96. Canuto, V.; Chiu, H.Y. Magnetic moment of a magnetized fermi gas. *Phys. Rev.* **1968**, *173*, 1229–1235.
97. Chiu, H.Y.; Canuto, V.; Fassio-Canuto, L. Quantum Theory of an Electron Gas with Anomalous Magnetic Moments in Intense Magnetic Fields. *Phys. Rev.* **1968**, *176*, 1438–1442.
98. Chaichian, M.; Masood, S.S.; Montonen, C.; Perez Martinez, A.; Perez Rojas, H. Quantum magnetic and gravitational collapse. *Phys. Rev. Lett.* **2000**, *84*, 5261–5264.
99. Chakrabarty, S. Quark matter in strong magnetic field. *Phys. Rev. D* **1996**, *54*, 1306–1316.
100. Broderick, A.; Prakash, M.; Lattimer, J.M. Effects of strong magnetic fields in strange baryonic matter. *Phys. Lett. B* **2002**, *531*, 167–174.
101. Khalilov, V.R. Macroscopic effects in cold magnetized nucleons and electrons with anomalous magnetic moments. *Phys. Rev. D* **2002**, *65*, 056001.
102. Mao, G.; Kondratyev, V. N.; A. Iwamoto, A.; Li, Z.; Wu, X.; Greiner, W. Neutron star composition in strong magnetic fields. *Chin. Phys. Lett.* **2003**, *20*, 1238–1241.
103. Mao, G.; Iwamoto, A.; Li, Z. Study of the neutron star structure in strong magnetic fields including the anomalous magnetic moments. *Chin. J. Astron. Astrophys.* **2003**, *3*, 359–374.
104. Perez Martinez, A.; Perez Rojas, H.; Mosquera Cuesta, H. J.; Boligan, M.; Orsaria, M. G. Quark stars and quantum-magnetically induced collapse. *Int. J. Mod. Phys. D* **2005**, *14*, 1959.
105. Felipe, R.G.; Martinez, A.P.; Rojas, H.P.; Orsaria, M. Magnetized strange quark matter and magnetized strange quark stars. *Phys. Rev. C* **2008**, *77*, 015807.

106. Perez-Garcia, M.A.; Navarro, J.; Polls, A. Neutron Fermi Liquids under the presence of a strong magnetic field with effective nuclear forces. *Phys. Rev. C* **2009**, *80*, 025802.
107. Rabhi, A.; Pais, H.; Panda, P.K.; Providencia, C.J. Quark-hadron phase transition in a neutron star under strong magnetic fields. *Phys. G* **2009**, *36*, 115204.
108. Dexheimer, V.; Negreiros, R.; Schramm, S. Hybrid Stars in a strong magnetic field. *Eur. Phys. J. A* **2012**, *48*, 189.
109. Strickland, M.; Dexheimer, V.; Menezes, D.P. Bulk properties of a Fermi gas in a magnetic field. *Phys. Rev. D* **2012**, *86*, 125032.
110. Dong, J.; Lombardo, U.; Zuo, W.; Zhang, H. Dense nuclear matter and symmetry energy in strong magnetic fields. *Nucl. Phys. A* **2013**, *898*, 32–42.
111. Casali, R.H.; Castro, L.B.; Menezes, D.P. Hadronic and hybrid stars subject to density dependent magnetic fields. *Phys. Rev. C* **2014**, *89*, 015805.
112. Manreza Paret, D.; Perez Martinez, A.; Ferrer, E.J.; de la Incera, V. Effects of AMM on the EoS of magnetized dense systems. *Astron. Nachr.* **2014**, *335*, 685–690.
113. Ferrer, E.J.; de la Incera, V.; Manreza Peret, D.; Perez Martinez, A.; Sanchez, A. Insignificance of the anomalous magnetic moment of charged fermions for the equation of state of a magnetized and dense medium. *Phys. Rev. D* **2015**, *91*, 085041.
114. Ferrer, E.J.; de la Incera, V. Dissipationless Hall current in dense quark matter in a magnetic field. *Phys. Lett. B* **2017**, *69*, 208–212.
115. Burkov, A.A.; Balents, I. Weyl semimetal in a topological insulator multilayer. *Phys. Rev. Lett.* **2011**, *107*, 127205.
116. Zyuzin, A.A.; Burkov, A.A. Topological response in Weyl semimetals and the chiral anomaly. *Phys. Rev. B* **2012**, *86*, 115133.
117. Frolov, I.E.; Zhukovsky, V.C.; Klimenko, K.G. Chiral density waves in quark matter within the Nambu-Jona-Lasinio model in an external magnetic field. *Phys. Rev. D* **2010**, *82*, 076002.
118. Tatsumi, T.; Nishiyama, K.; Karasawa, S. Novel Lifshitz point for chiral transition in the magnetic field. *Phys. Lett. B* **2015**, *743*, 66–70.
119. Wilczek, F. Two Applications of Axion Electrodynamics. *Phys. Rev. Lett.* **1987**, *58*, 1799–1802.
120. Ferrer, E.J.; de la Incera, V.; Manuel, C. Magnetic color flavor locking phase in high density QCD. *Phys. Rev. Lett.* **2005**, 152002.
121. Ferrer, E.J.; de la Incera, V.; Manuel, C. Color-superconducting gap in the presence of a magnetic field. *Nucl. Phys. B* **2006**, *747*, 88–112.
122. Ferrer, E.J.; de la Incera, V.; Manuel, C. Colour superconductivity in a strong magnetic field. *J. Phys. A* **2006**, *39*, 6349–6355.
123. Feng, B.; Ferrer, E.J.; de la Incera, V. Magnetoelectric effect in strongly magnetized color superconductivity. *Phys. Lett. B* **2011**, *706*, 232–238.
124. Qi, X.-L.; Hughes, T.L.; Zhang, S.-C. Topological field theory of time-reversal invariant insulators. *Phys. Rev. B* **2008**, *78*, 195424.
125. Bermudez, A.; Mazza, L.; Rizzi, M.; Goldman, N.; Lewenstein, M.; Martin-Delgado, M.A. Wilson fermions and axion electrodynamics in optical lattices. *Phys. Rev. Lett.* **2010**, *105*, 190404.
126. Glendenning, N.K. *Compact Stars, Nuclear Physics, Particle Physics, General Relativity*; Springer: New York, NY, USA, 2000.
127. Harding, A.K.; Lai, D. Physics of strongly magnetized neutron stars. *Rept. Prog. Phys.* **2006**, *69*, 2631–2708.
128. Spruit, H.C. Origin of neutron star magnetic fields. *AIP Conf. Proc.* **2008**, *983*, 391–398.
129. Li, R.; Wang, J.-L.; Qi, X.-C.; Zhang, S. Dynamical axion field in topological magnetic insulators. *Nat. Phys.* **2010**, *6*, 284–288.
130. Ferrer, E.J.; de la Incera, V. Exploring dense and cold QCD in magnetic fields. *Eur. Phys. J. A* **2016**, *52*, 266.
131. Odyniec, G. The RHIC beam energy scan program in STAR and what's next ... *J. Phys. Conf. Ser.* **2013**, *455*, 012037.
132. Ablyazimov, T.; Abuhoza, A.; Adak, P.P.; Adamczyk, M.; Agarwal, K.; Aggarwal, M.M.; Ahammed, Z.; Ahmad, F.; Ahmad, N.; Ahmad, S.; et al. [CBM Collaboration] Challenges in QCD matter physics –The scientific program of the Compressed Baryonic Matter experiment at FAIR. *Eur. Phys. J. A* **2017**, *53*, 60.

133. Deng, W.-T.; Huang, X.-G. Event-by-event generation of electromagnetic fields in heavy-ion collisions. *Phys. Rev. C* **2012**, *85*, 044907.
134. Toneev, V.; Rogachevsky, O.; Voronyuk, V. Evidence for creation of strong electromagnetic fields in relativistic heavy-ion collisions. *Eur. Phys. J. A* **2016**, *52*, 264.
135. Young, S.M.; Zaheer, S.; Teo, J.C.Y.; Kane, C.L.; Mele, E.J.; Rappe, A.M. Dirac semimetal in three dimensions. *Phys. Rev. Lett.* **2012**, *108*, 140405.
136. Borisenko, S.; Gibson, Q.; Evtushinsky, D.; Zabolotnyy, V.; Buchner, B.; Cava, R.J. Experimental realization of a three-dimensional Dirac semimetal. *Phys. Rev. Lett.* **2014**, *113*, 027603.
137. Neupane, M.; Xu, S.Y.; Sankar, R.; Alidoust, N.; Bian, G.; Liu, C.; Belopolski, I.; Chang, T.R.; Jeng, H. T.; Lin, H.; Bansil, A; et al. Observation of a three-dimensional topological Dirac semimetal phase in high-mobility Cd_3As_2. *Nat. Commun.* **2014**, *5*, 3786.
138. Liu, Z.K.; Jiang, J.; Zhou, B.; Wang, Z.J.; Zhang, Y.; Weng, H.M.; Prabhakaran, D.; Mo, S.-K.; Peng, H.; Dudin, P.; et al. A stable three-dimensional topological Dirac semimetal Cd_3As_2. *Nat. Mat.* **2014**, *13*, 677–681.

© 2018 by the authors. Licensee MDPI, Basel, Switzerland. This article is an open access article distributed under the terms and conditions of the Creative Commons Attribution (CC BY) license (http://creativecommons.org/licenses/by/4.0/).

Article

Neutrino Emissivity in the Quark-Hadron Mixed Phase

William M. Spinella [1], Fridolin Weber [2,3,*], Milva G. Orsaria [4,5] and Gustavo A. Contrera [4,5,6]

1. Department of Sciences, Wentworth Institute of Technology, 550 Huntington Avenue, Boston, MA 02115, USA; spinellaw@wit.edu
2. Department of Physics, San Diego State University, San Diego, CA 92182, USA
3. Center for Astrophysics and Space Sciences, University of California at San Diego, La Jolla, CA 92093, USA
4. National Scientific and Technical Research Council (CONICET), Godoy Cruz 2290, Buenos Aires 1425, Argentina; morsaria@fcaglp.unlp.edu.ar (M.G.O.); contrera@fisica.unlp.edu.ar (G.A.C.)
5. Grupo de Gravitación, Astrofísica y Cosmología, Facultad de Ciencias Astronómicas y Geofísicas, Universidad Nacional de La Plata, La Plata 1900, Argentina
6. Instituto de Física La Plata, CONICET, Universidad Nacional de La Plata, La Plata 1900, Argentina
* Correspondence: fweber@sdsu.edu or fweber@ucsd.edu

Received: 28 January 2018; Accepted: 4 May 2018; Published: 16 May 2018

Abstract: In this work we investigate the effect a crystalline quark–hadron mixed phase can have on the neutrino emissivity from the cores of neutron stars. To this end we use relativistic mean-field equations of state to model hadronic matter and a nonlocal extension of the three-flavor Nambu–Jona–Lasinio model for quark matter. Next we determine the extent of the quark–hadron mixed phase and its crystalline structure using the Glendenning construction, allowing for the formation of spherical blob, rod, and slab rare phase geometries. Finally, we calculate the neutrino emissivity due to electron–lattice interactions utilizing the formalism developed for the analogous process in neutron star crusts. We find that the contribution to the neutrino emissivity due to the presence of a crystalline quark–hadron mixed phase is substantial compared to other mechanisms at fairly low temperatures ($\lesssim 10^9$ K) and quark fractions ($\lesssim 30\%$), and that contributions due to lattice vibrations are insignificant compared to static-lattice contributions. There are a number of open issues that need to be addressed in a future study on the neutrino emission rates caused by electron–quark blob bremsstrahlung. Chiefly among them are the role of collective oscillations of matter, electron band structures, and of gaps at the boundaries of the Brillouin zones on bremsstrahlung, as discussed in the summary section of this paper. We hope this paper will stimulate studies addressing these issues.

Keywords: quark matter; hadronic matter; quark deconfinement; neutron star matter; nuclear equation of state; phase transition; crystalline structure; neutrino emissivities

1. Introduction

It was shown by Glendenning [1,2] that if electric charge neutrality in a neutron star [3–5] is treated globally rather than locally, the possible first order phase transition from hadronic matter to quark matter in the neutron star core will result in a mixed phase in which both phases of matter coexist. To minimize the total isospin asymmetry energy the two phases will segregate themselves, which results in positively charged regions of hadronic matter and negatively charged regions of quark matter, with the rare phase occupying sites on a Coulomb lattice. The situation is schematically illustrated in Figure 1. Further, the competition between the Coulomb and surface energy densities will cause the matter to arrange itself into energy minimizing geometric configurations [1,2].

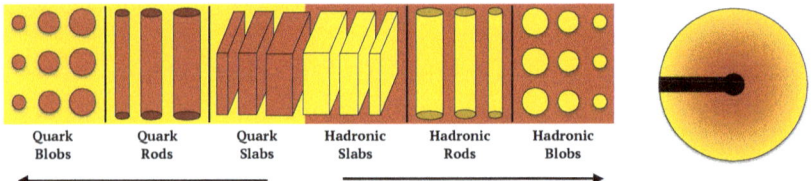

Figure 1. Schematic illustrating the rare phase structures that may form in the quark–hadron mixed phase [6,7]). An increase in the volume fraction of quark matter, described by χ, is accompanied by an increase in baryon number density and depth within a neutron star.

The presence of the Coulomb lattice and the nature of the geometric configurations of matter in the quark–hadron mixed phase may have a significant effect on the neutrino emissivity from the core. More specifically, neutrino-antineutrino pairs will be created by the scattering of electrons from these charged lattice structures,

$$e^- + (Z, A) \to e^- + (Z, A) + \nu + \bar{\nu}, \tag{1}$$

and this will increase the emissivity in the mixed phase. This process is analogous to neutrino-pair bremsstrahlung of electrons in the neutron star crust, where ions exist on a lattice immersed in an electron gas, and for which there exists a large body of work (see, for example [8–14]). The situation is more complicated in the quark–hadron mixed phase, but the operative interaction is still the Coulomb interaction. Thus, to estimate the neutrino-pair Bremsstrahlung of electrons from rare phase structures in the quark–hadron mixed phase we rely heavily on this body of work (particularly [8]). We will refer to this additional mechanism as mixed phase Bremsstrahlung (MPB).

Neutrino emissivity due to the interaction of electrons with a crystalline quark–hadron mixed phase has been previously studied in this manner in [6,15]. In the present work we use a set of nuclear equations of state which are in better agreement with the latest nuclear matter constraints at saturation density than those utilized in [6], and are consistent with the 2.01 $M_\odot$ mass constraint set by PSR J0348 + 0432 [16]. To describe quark matter we use the nonlocal SU(3) Nambu–Jona–Lasinio (n3NJL) model discussed in [6,17–21]. The n3NJL parametrization used is given as "Set I" in [22], and is in better agreement with the empirical quark masses than the parametrization utilized in [6]. We consider three geometries for the range of possible structures in the mixed phase including spherical blobs, rods, and slabs, and calculate the associated static lattice contributions to the neutrino emissivity. Phonon contributions to the emissivity for rod and slab geometries are not considered, though a comparison of the phonon and static lattice contributions for spherical blobs is given and indicates that phonon contributions may not be significant. Finally, the extent of the conversion to quark matter in the core was determined in [7], and this allows for a comparison between emissivity contributions from standard and enhanced neutrino emission mechanisms including the direct Urca (DU), modified Urca (MU), and baryon–baryon and quark–quark Bremsstrahlung (NPB) processes, and contributions from electron–lattice interactions. For a detailed summary including the equations and coefficients used for the calculation of the standard and enhanced neutrino emission mechanisms, see [7].

The results for different parametrizations are numerous and qualitatively similar, so the DD2 parametrization will be presented exclusively in this paper. The results of the other parametrizations can be found in [7].

2. Improved Set of Models for the Nuclear Equation of State

Hadronic matter is modeled in the framework of the relativistic nonlinear mean-field (RMF) approach [23,24], which describes baryons interacting through the exchange of scalar, vector, and isovector mesons (for details, see [6,7,25]). The RMF approach is parametrized to reproduce the following properties of symmetric nuclear matter at saturation density n_0 (see Table 1): the binding energy per nucleon (E_0), the nuclear incompressibility (K_0), the isospin asymmetry energy (J), and the effective mass (m^*/m_N). In addition, the RMF parametrizations used in this work employ a density-dependent isovector–meson–baryon coupling constant that can be fit to the slope of the asymmetry energy (L_0) at n_0. The scalar- and vector-meson–baryon coupling constants of the density-dependent relativistic mean-field models DD2 and ME2 are fit to properties of finite nuclei [7,26,27]. These models are an extension of the standard RMF approach that account for medium effects by making the meson–baryon coupling constants dependent on the local baryon number density [28]. The density-dependence of the meson–baryon coupling constants is given by

$$g_{iB}(n) = g_{iB}(n_0) f_i(x), \qquad (2)$$

where $i \in \{\sigma, \omega, \rho\}$, $x = n/n_0$, and $f_i(x)$ provides the functional form for the density dependence. The most commonly utilized ansatz for $f_i(x)$ are given by [29]

$$f_i(x) = a_i \frac{1 + b_i(x + d_i)^2}{1 + c_i(x + d_i)^2}, \qquad (3)$$

for $i \in \{\sigma, \omega\}$, and

$$f_\rho(x) = \exp\left[-a_\rho(x - 1)\right]. \qquad (4)$$

The nine parameters of the density dependence ($a_\sigma, b_\sigma, c_\sigma, d_\sigma, a_\omega, b_\omega, c_\omega, d_\omega, a_\rho$), the values of the meson–nucleon couplings at n_0 ($g_{\sigma N}(n_0), g_{\omega N}(n_0), g_{\rho B}(n_0)$), and the mass of the scalar meson (m_σ) are all fit to properties of symmetric nuclear matter at n_0 and to the properties of finite nuclei including but not limited to binding energies, charge and diffraction radii, spin–orbit splittings, and neutron skin thickness (see [27,30]).

In addition to the nucleons, hyperons and delta isobars (Δs) are also considered in the composition of hadronic matter. The scalar-meson–hyperon coupling constants are fit to the following hypernuclear potentials at saturation (see [7] and references therein),

$$U_\Lambda^{(N)} = -28\,\text{MeV}, \quad U_\Sigma^{(N)} = +30\,\text{MeV}, \quad U_\Xi^{(N)} = -18\,\text{MeV}. \qquad (5)$$

The vector-meson–hyperon coupling constants are taken to be those given by the ESC08 model in SU(3) symmetry [7,31,32],

$$g_{\omega\Lambda} = g_{\omega\Sigma} \approx 0.79\, g_{\omega N}, \quad g_{\omega\Xi} \approx 0.59\, g_{\omega N}. \qquad (6)$$

The scalar- and vector-meson–Δ coupling constants are given as follows,

$$x_{\sigma\Delta} = x_{\omega\Delta} = 1.1, \quad x_{\rho\Delta} = 1.0. \qquad (7)$$

Finally, the isovector-meson–hyperon and isovector-meson–Δ coupling constants are taken to be universal, with the differences in the baryon isospin accounted for by the isospin operator in the lagrangian.

Table 1. Properties of symmetric nuclear matter at saturation density for the hadronic parametrizations of this work.

Saturation Property	SWL [7]	GM1L [1,7]	DD2 [26]	ME2 [27]
n_0 (fm^{-3})	0.150	0.153	0.149	0.152
E_0 (MeV)	−16.00	−16.30	−16.02	−16.14
K_0 (MeV)	260.0	300.0	242.7	250.9
m^*/m_N	0.70	0.70	0.56	0.57
J (MeV)	31.0	32.5	32.8	32.3
L_0 (MeV)	55.0	55.0	55.3	51.3

3. Crystalline Structure of the Quark–Hadron Mixed Phase

A mixed phase of hadronic and quark matter will arrange itself so as to minimize the total energy of the phase. Under the condition of global charge neutrality, this is the same as minimizing the contributions to the total energy due to phase segregation, which includes the surface and Coulomb energy contributions. Expressions for the Coulomb (ϵ_C) and surface (ϵ_S) energy densities can be written as [1,2]

$$\mathcal{E}_C = 2\pi e^2 \left[q_H(\chi) - q_Q(\chi)\right]^2 r^2 x f_D(x), \tag{8}$$

$$\mathcal{E}_S = D x \alpha(\chi)/r, \tag{9}$$

where q_H (q_Q) is the hadronic (quark) phase charge density, r is the radius of the rare phase structure, and $\alpha(\chi)$ is the surface tension between the two phases. The parameter χ, which varies between 0 and 1, represents the volume fraction of quark matter at a given density. The quantities x and $f_D(x)$ in (8) are defined as

$$x = \min(\chi, 1 - \chi) \tag{10}$$

and

$$f_D(x) = \frac{1}{D+2}\left[\frac{1}{D-2}(2 - D x^{1-2/D}) + x\right], \tag{11}$$

where D is the dimensionality of the lattice. The phase rearrangement process will result in the formation of geometrical structures of the rare phase distributed in a crystalline lattice that is immersed in the dominant phase (see Figure 1). The rare phase structures are approximated for convenience as spherical blobs, rods, and slabs [1,2]. The spherical blobs occupy sites in a three dimensional ($D = 3$) body centered cubic (BCC) lattice, the rods in a two dimensional ($D = 2$) triangular lattice, and the slabs in a simple one dimensional ($D = 1$) lattice [8]. At $\chi = 0.5$ both hadronic and quark matter exist as slabs in the same proportion, and at $\chi > 0.5$ the hadronic phase becomes the rare phase with its geometry evolving in reverse order (from slabs to rods to blobs).

Direct determination of the surface tension of the quark–hadron interface is problematic because of difficulties in constructing a single theory that can accurately describe both hadronic matter and quark matter. Therefore, we employ an approximation proposed by Gibbs where the surface tension is taken to be proportional to the difference in the energy densities of the interacting phases [1,2],

$$\alpha(\chi) = \eta L \left[\mathcal{E}_Q(\chi) - \mathcal{E}_H(\chi)\right], \tag{12}$$

where L is proportional to the surface thickness which should be on the order of the range of the strong interaction (1 fm), and η is a proportionality constant. In this work we maintain the energy density proportionality but set the parameter $\eta = 0.08$ so that the surface tension falls below 70 MeV fm^{-2} for all parametrizations, a reasonable upper limit for the existence of a quark–hadron mixed phase [33]. The surface tension as a function of χ is given in Figure 2 for the nuclear DD2 parametrization, introduced in Section 2.

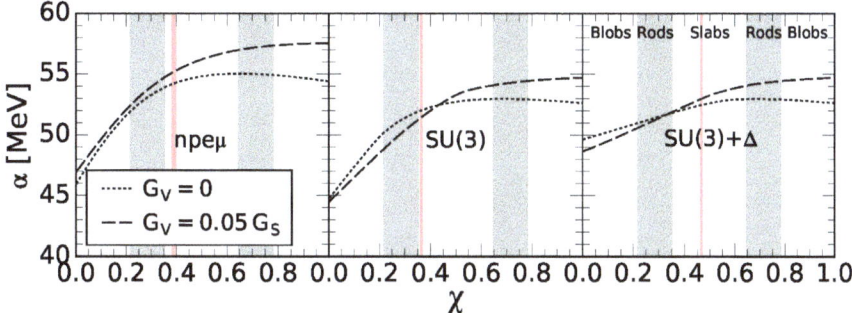

Figure 2. Surface tension α in the quark–hadron mixed phase for the DD2 parametrization [7]. The red shading indicates the range for the maximum quark fraction χ_{max} for the two values of the quark vector coupling constant G_V. (Left panel) Only nucleons and leptons are included in the hadronic phase. (Center panel) Hyperons are included in the hadronic phase. (Right panel) Delta isobars are included in addition to hyperons in the hadronic phase. Similar figures for the SWL, GM1L, and ME2 parametrizations can be found in Reference [7].

We note that, in this work, we restricted ourselves to considering G_V values that are in the range of $0 < G_V < 0.05 G_S$, as this choice leads to gravitational masses of neutron stars with quark-hybrid compositions that satisfy the $2 M_\odot$ constraint. Exploring the possibility of larger G_V values would certainly be worthwhile, but this is beyond the scope of this work.

The size of the rare phase structures is given by the radius (r) and is determined by minimizing the sum of the Coulomb and surface energies, $\partial(\mathcal{E}_C + \mathcal{E}_S)/\partial r$, and solving for r [1,2],

$$r(\chi) = \left(\frac{D\alpha(\chi)}{4\pi e^2 f_D(\chi) \left[q_H(\chi) - q_Q(\chi) \right]^2} \right)^{\frac{1}{3}}. \tag{13}$$

Rare phase structures are centered in the primitive cell of the lattice, taken to be a Wigner–Seitz cell of the same geometry as the rare phase structure. The Wigner–Seitz cell radius R is set so that the primitive cell is charge neutral,

$$R(\chi) = r\chi^{-1/D}. \tag{14}$$

Figure 3 shows r and R as a function of the quark fraction in the mixed phase. Both r and R increase with an increase in the baryonic degrees of freedom, particularly when $\chi \lesssim 0.5$ and the vector interaction is included. Note that the blob radius should vanish for $\chi \in \{0, 1\}$, but does not due to the approximate nature of the geometry function $f_D(\chi)$ [15]. The number density of rare phase blobs will be important for calculating the phonon contribution to the emissivity. Since there is one rare phase blob per Wigner–Seitz cell, the number density of rare phase blobs (n_b) is simply the reciprocal of the Wigner–Seitz cell volume,

$$n_b = (4\pi R^3/3)^{-1}. \tag{15}$$

The density of electrons in the mixed phase is taken to be uniform throughout. Charge densities in both the rare and dominant phases are also taken to be uniform, an approximation supported by a recent study by Yasutake et al. [33]. The uniformity of charge in the rare phase also justifies the use of the nuclear form factor ($F(q)$) presented in Section 4. The total charge number per unit volume ($|Z|/V_{Rare}$) of the rare phase structures is given in Figure 4.

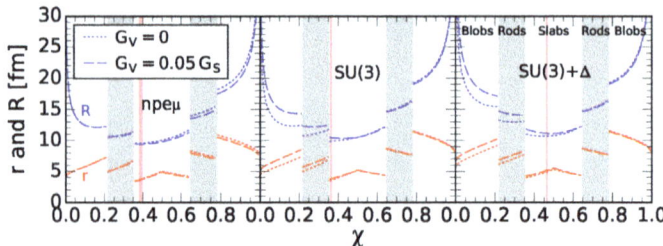

Figure 3. Radius of the rare phase structure r and Wigner–Seitz cell R in the quark–hadron mixed phase for the DD2 parametrization [7]. See Figure 2 for additional details. Similar figures for the SWL, GM1L, and ME2 parametrizations can be found in Reference [7].

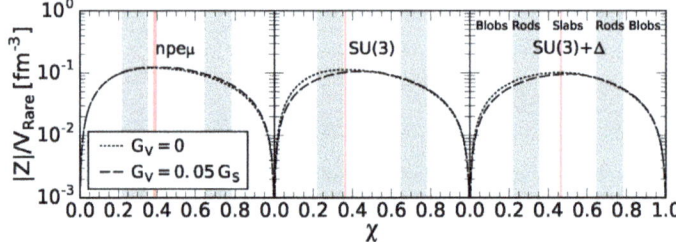

Figure 4. Charge number per unit volume of the rare phase structures for the DD2 parametrization [7]. See Figure 2 for additional details. Similar figures for the SWL, GM1L, and ME2 parametrizations can be found in Reference [7].

4. Neutrino Emissivity Due to a Crystalline Quark–Hadron Lattice

We begin this section with a brief discussion of the neutrino emissivity due to a crystalline quark–hadron lattice [6]. Modeling the complex interactions of electrons with a background of neutrons, protons, hyperons, muons, and quarks is an exceptionally complicated problem. However, to make a determination of the neutrino emissivity that is due to electron–lattice interactions in the quark–hadron mixed phase we need only consider the Coulomb interaction between them. This simplifies the problem greatly, as a significant body of work exists for the analogous process of electron–ion scattering that takes place in the crusts of neutron stars.

To determine the state of the lattice in the quark–hadron mixed phase we use the dimensionless ion coupling parameter given by

$$\Gamma = \frac{Z^2 e^2}{R k_B T}. \tag{16}$$

Below $\Gamma_{melt} = 175$ the lattice behaves as a Coulomb liquid, and above as a Coulomb crystal [34,35]. It was shown in Reference [15] that the emissivity due to electron-blob interactions in the mixed phase was insignificant compared to other contributions at temperatures above $T \gtrsim 10^{10}$ K. Therefore, in this work we consider temperatures in the range $10^7 \text{ K} \leq T \leq 10^{10} \text{ K}$. At these temperatures the value of the ion coupling parameter is well above Γ_{melt}, and so the lattice in the quark–hadron mixed phase is taken to be a Coulomb crystal.

To account for the fact that the elasticity of scattering events is temperature dependent we need to compute the Debye–Waller factor, which is known for spherical blobs only and requires the plasma frequency and temperature given by

$$\omega_p = \sqrt{\frac{4\pi Z^2 e^2 n_b}{m_b}}, \tag{17}$$

$$T_p = \frac{\hbar \omega_p}{k_B}, \tag{18}$$

where m_b is the mass of a spherical blob [8]. The Debye–Waller factor is then given by

$$W(q) = \begin{cases} \frac{aq^2}{8k_e^2}\left(1.399\,e^{-9.1t_p} + 12.972\,t_p\right) & \text{spherical blobs,} \\ 0 & \text{rods and slabs,} \end{cases} \tag{19}$$

where $q = |q|$ is a phonon or scattering wave vector, $a = 4\hbar^2 k_e^2/(k_B T_p m_b)$, and $t_p = T/T_p$ [8,36]. In order to smooth out the charge distribution over the radial extent of the rare phase structure we adopt the nuclear form factor given in [8],

$$F(q) = \frac{3}{(qR^3)}\left[\sin(qR) - qR\cos(qR)\right]. \tag{20}$$

Screening of the Coulomb potential by electrons is taken into account by the static dielectric factor $\epsilon(q,0) = \epsilon(q)$, given in [10]. However, the charge number of the rare phase structures is high and the electron number density is low, so setting this factor to unity has no noticeable effect on the calculated neutrino emissivity. Finally, the effective interaction is given by [8]

$$V(q) = \frac{4\pi e \rho_Z F(q)}{q^2 \epsilon(q,0)} e^{-W(q)}. \tag{21}$$

General expressions for the neutrino emissivity due to the MPB electron–lattice interactions were derived by Haensel et al. [37] for spherical blobs and by Pethick et al. [14] for rods and slabs,

$$\epsilon_{MPB}^{blobs} \approx 5.37 \times 10^{20}\, n T_9^6 Z^2 L \;\text{erg s}^{-1}\,\text{cm}^{-3}, \tag{22}$$

$$\epsilon_{MPB}^{rods,slabs} \approx 4.81 \times 10^{17}\, k_e T_9^8 J \;\text{erg s}^{-1}\,\text{cm}^{-3}, \tag{23}$$

where L and J are dimensionless quantities that scale the emissivities. Both L and J contain a contribution due to the static lattice (Bragg scattering), but we consider the additional contribution from lattice vibrations (phonons) for spherical blobs, so $L = L_{sl} + L_{ph}$. We note that the T^8 temperature dependence in Equation (23) is somewhat deceiving since the J factor also depends on temperature and, for a wide range of parameters, is proportional to $1/T^2$. In effect, the neutrino emissivity $\epsilon_{MPB}^{rods,slabs}$ is therefore proportional to T^6.

4.1. Phonon Contribution to Neutrino Emissivity

The expressions for determining the neutrino emissivity due to interactions between electrons and lattice vibrations (phonons) in a Coulomb crystal, with proper treatment of multi-phonon processes, were obtained by Baiko et al. [38] and simplified by Kaminker et al. [8]. The phonon contribution to the emissivity is primarily due to Umklapp processes in which a phonon is created (or absorbed) by an electron that is simultaneously Bragg reflected, resulting in a scattering vector q that lies outside the first Brillouin zone, $q_0 \gtrsim (6\pi^2 n_b)^{1/3}$ [39,40], where n_b is given by Equation (15).

The contribution to MPB due to phonons is contained in L_{ph} and given by Equation (21) in [8],

$$L_{\text{ph}} = \int_{y_0}^{1} dy \frac{S_{\text{eff}}(q)|F(q)|^2}{y|\epsilon(q,0)|^2}\left(1 + \frac{2y^2}{1-y^2}\ln y\right), \tag{24}$$

where $y = q/(2k_e)$, and the lower integration limit y_0 excludes momentum transfers inside the first Brillouin zone. The structure factor S_{eff} is given by (24) and (25) in [8]),

$$S_{\text{eff}}(q) = 189\left(\frac{2}{\pi}\right)^5 e^{-2W} \int_0^{\infty} d\xi \frac{1 - 40\xi^2 + 80\xi^4}{(1+4\xi^2)^5 \cosh^2(\pi\xi)} \times \left(e^{\Phi(\xi)} - 1\right), \tag{25}$$

$$\Phi(\xi) = \frac{\hbar q^2}{2m_b}\left\langle \frac{\cos(\omega_s t)}{\omega_s \sinh(\hbar\omega_s/2k_B T)}\right\rangle, \tag{26}$$

where $\xi = tk_B T/\hbar$ and $\langle \ldots \rangle$ denotes averaging over phonon frequencies and modes,

$$\langle f_s(\mathbf{k})\rangle = \frac{1}{3V_B}\sum_s \int_{V_B} d\mathbf{k}\, f_s(\mathbf{k}). \tag{27}$$

It is assumed that there are three phonon modes s, two linear transverse and one longitudinal. The frequencies of the transverse modes are given by $\omega_{t_i} = a_i k$, where $i = 1, 2$, $a_1 = 0.58273$, and $a_2 = 0.32296$. The frequency of the longitudinal mode ω_l is determined by Kohn's sum rule, $\omega_l^2 = \omega_p^2 - \omega_{t_1}^2 - \omega_{t_2}^2$ [41].

Umklapp processes proceed as long as the temperature $T_{\text{Umklapp}} \gtrsim T_p Z^{1/3} e^2/(\hbar c)$, below which electrons can no longer be treated in the free electron approximation [39]. This limits the phonon contribution to the neutrino emissivity to only a very small range in temperature for a crystalline quark–hadron mixed phase (see Figure 5), and renders it negligible compared to the static lattice contribution as will be shown in the next section.

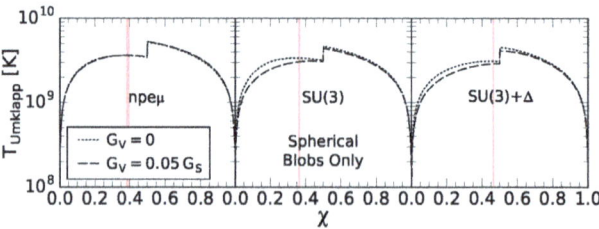

Figure 5. Temperature below which Umklapp processes are frozen out (T_{Umklapp}), and contributions to the neutrino emissivity due to electron–phonon interactions become negligible for the DD2 parametrization [7]. See Figure 2 for additional details. Similar figures for the SWL, GM1L, and ME2 parametrizations can be found in [7].

4.2. Static Lattice Contribution to Neutrino Emissivity

Pethick and Thorsson [14] found that with proper handling of electron band-structure effects the static lattice contribution to the neutrino emissivity in a Coulomb crystal was significantly reduced compared to calculations performed in the free electron approximation. Kaminker et al. [8] presented simplified expressions for calculating the static lattice contribution (L_{sl}) using the formalism developed in [14]. The dimensionless quantities L_{sl} and J that scale the neutrino emissivities for spherical blobs and rods/slabs, respectively, are given by

$$L_{\text{sl}} = \frac{1}{12Z}\sum_{K \neq 0}\frac{(1-y_K^2)}{y_K^2}\frac{|F(K)|^2}{|\epsilon(K)|^2} I(y_K, t_V)\, e^{-2W(K)} \tag{28}$$

246

and

$$J = \sum_{K \neq 0} \frac{y_K^2}{t_V^2} I(y_K, t_V), \qquad (29)$$

where $K = |\mathbf{K}|$ is a scattering vector and restricted to linear combinations of reciprocal lattice vectors, $y_K = K/(2k_e)$, $t_V = k_B T / \left[|V(K)|(1 - y_K^2)\right]$, and $I(y_K, t_V)$ is given by Equation (39) in [8]. The sum over K in (28) and (29) terminates when $K > 2k_e$, prohibiting scattering vectors that lie outside the electron Fermi surface.

5. Neutrino Emissivity Results

The neutrino emissivities due to MPB and the additional emissivity mechanisms are given in Figures 6 and 7 for $G_V = 0$ and $G_V = 0.05\, G_S$ respectively at temperatures between 10^7 K and 10^{10} K. The MPB emissivity is for most of the mixed phase the weakest of the emissivity mechanisms, peaking at low χ (at $\chi \lesssim 0.05$ the MPB emissivity may be overestimated due to the limitations of the dimensionality function), and appears to be slightly larger when hyperons and Δs are included in the composition. Including the vector interaction ($G_V = 0.05\, G_S$) also results in a slight increase in the MPB emissivity. Both additional baryonic degrees of freedom and inclusion of the vector interaction delay the onset of the quark–hadron phase transition, and therefore it may be concluded that the greater the density in the mixed phase, the greater the contribution to the emissivity from MPB. The MPB emissivity is most comparable to the modified Urca emissivity, particularly at 10^8–10^9 K.

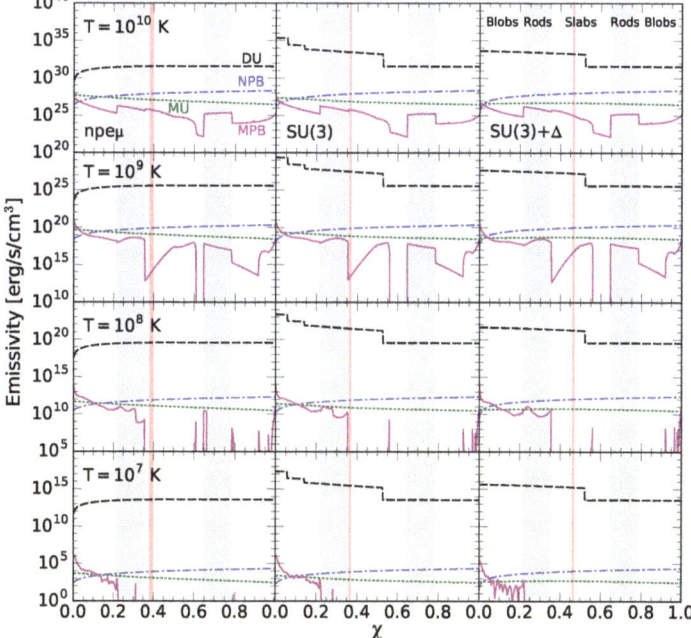

Figure 6. Neutrino emissivity in the quark–hadron mixed phase for the DD2 parametrization with $G_V = 0$ [7]. Contributions due to mixed phase Bremsstrahlung (MPB), nucleon–nucleon and quark–quark neutrino pair Bremsstrahlung (NPB), the nucleon and quark modified Urca processes (MU), and the hyperon and quark direct Urca (DU) processes are included. See Figure 2 for additional details. Similar figures for the SWL, GM1L, and ME2 parametrizations can be found in [7].

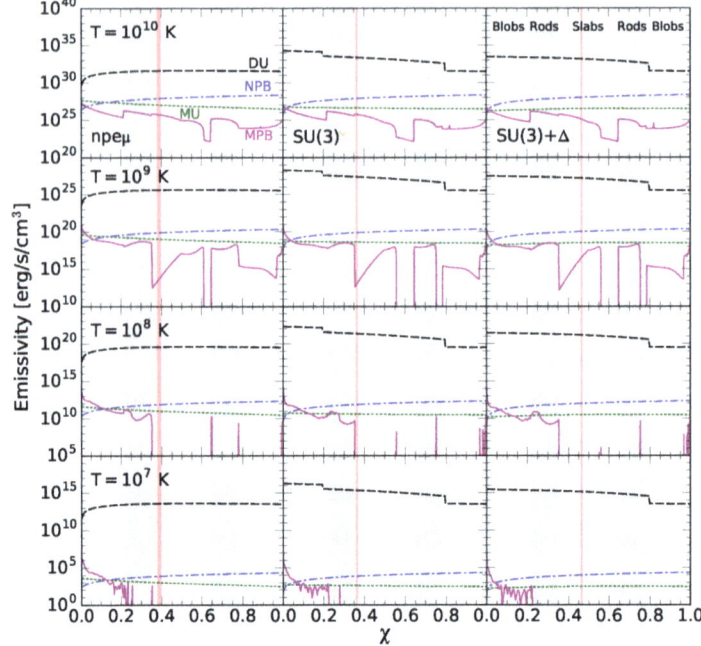

Figure 7. Neutrino emissivity in the quark–hadron mixed phase for the DD2 parametrization with $G_V = 0.05\,G_S$ [7]. Contributions due to mixed phase Bremsstrahlung (MPB), nucleon–nucleon and quark–quark neutrino pair Bremsstrahlung (NPB), the nucleon and quark modified Urca processes (MU), and the hyperon and quark direct Urca (DU) processes are included. See Figure 2 for additional details. Similar figures for the SWL, GM1L, and ME2 parametrizations can be found in [7].

Electron–phonon interactions contribute to the MPB emissivity when the mixed phase consists of spherical blobs ($\chi \lesssim 0.21$ and $\chi \gtrsim 0.79$) and only when $T > T_{\text{Umklapp}}$ (Figure 5), which for the given choices of temperature implies $T = 10^{10}$ K. Figure 8 shows that the static-lattice contribution to the MPB emissivity dominates the phonon contribution rendering it negligible, particularly at quark fractions relevant to the neutron stars of this work ($\chi < 0.5$). Therefore, the MPB emissivity is almost entirely due to the static-lattice contribution (Bragg scattering).

Equations (28) and (29) indicate that the static-lattice contribution to the MPB emissivity is calculated as a sum over scattering vectors $\mathbf{K}$ that satisfy $K < 2k_e$. At the onset of the mixed phase k_e and N_K are at a maximum, but as the quark–hadron phase transition proceeds the negatively charged down and strange quarks take over the process of charge neutralization. Thus, the electron number density and consequently k_e continue to decrease at about the same rate as before the start of the mixed phase. This leads to the steep decline in N_K with increasing χ for $\chi < 0.5$ shown in Figure 9. Further, the rod and slab dimensionality drastically reduces the number of available scattering vectors which contributes to the decrease of the MPB emissivity in those phases, particularly in the slab phase. However, (29) shows that the MPB emissivity from rod and slab phases is dependent on T^8, rather than T^6 for the blob phase, and this explains the dramatic decrease in the MPB emissivity with decreasing temperature.

Direct Urca processes dominate the mixed phase neutrino emissivity at all temperatures, with contributions from the Λ hyperon DU process ($\Lambda \to pe\bar{\nu}$) operating beyond $\chi_{\max}$. Nucleonic DU processes do not operate for any of the parametrizations considered in this work [7]. The hyperon DU process emissivities can be identified as any contribution with an emissivity above that for the quark DU process in the $npe\mu$ composition, and are shown to step down in the mixed phase, vanishing prior to the onset of a pure quark phase. In the absence of the hyperonic DU process, the quark DU process would still dominate the Bremsstrahlung and modified Urca processes unless curtailed by the presence of color superconductivity.

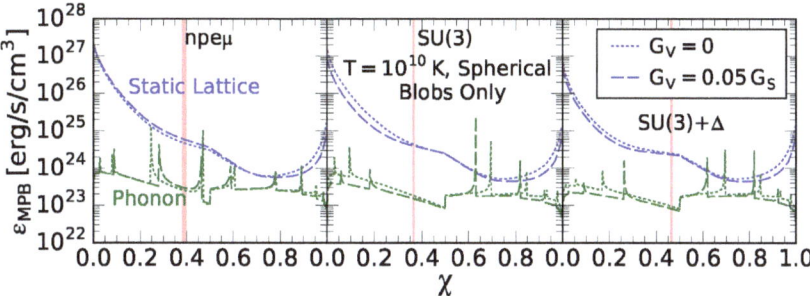

Figure 8. Comparison of the static lattice and phonon contributions to the neutrino emissivity at $T = 10^{10}$ K for the spherical blob geometry only and the DD2 parametrization [7].

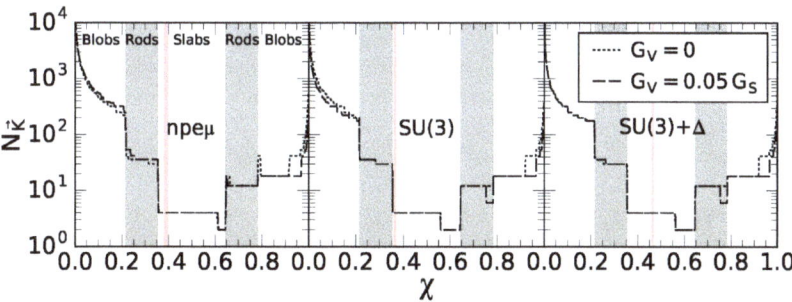

Figure 9. The number of scattering vectors that satisfy the condition $K < 2k_e$ as a function of the quark fraction [7] (see Figure 2 for additional details). Similar figures for the SWL, GM1L, and ME2 parametrizations can be found in [7].

6. Discussion and Summary

In this work we determined that quark blob, rod, and slab structures may exist in a crystalline quark–hadron mixed phase. The study is based on relativistic mean-field equations of state which are used to model hadronic matter and a nonlocal extension of the three-flavor Nambu–Jona–Lasinio model for quark matter. We determined the neutrino emissivities that may result from the elastic scattering of electrons off these quark structures (mixed phase Bremsstrahlung (MBP)), and compared them to standard neutrino emissivity processes that may operate in the mixed phase as well.

We found that the emissivity from the MPB process is comparable to that of the modified Urca process at low volume fractions of quark matter, χ, and in the temperature range of 10^8 K $\lesssim T \lesssim 10^9$ K. The MPB emissivity was found to increase with the inclusion of the vector interaction among quarks and with additional baryonic degrees of freedom in the form of hyperons and Δ baryons [7], both of which lead to an increase in the quark–hadron phase transition density and a higher density core. Further, contributions to the MPB emissivity from phonons were shown to be negligible compared to those from Bragg scattering. Finally, baryonic and quark DU processes were shown to operate in the mixed phase and dominate all other neutrino emissivity mechanisms.

Since it is believed that the hypothetical quark–hadron lattice structures in the core regions of neutron stars are qualitatively reminiscent to the hypothesized structures in the crustal regions of neutron stars [1,2,42–44], we have adopted the Bremsstrahlung formalism developed in the literature for the crustal regions of neutron stars to assess the neutrino emission rates resulting from electron–quark blob (rod, slab) scattering in the cores of neutron stars with quark–hybrid compositions. Because of the complexity of the problem, however, there are several issues that need to be studied further in order to develop refined estimates of the neutrino emission rates presented in this paper. The remaining part of this section is devoted to this topic.

Properties of the sub-nuclear crustal region: The hypothetical structures in the crustal regions of neutron stars range in shape from spheres to rods to slabs at mass densities 10^{14} g cm$^{-3} \lesssim \rho \lesssim 1.5 \times 10^{14}$ g cm^{-3}, which is just below the nuclear saturation density of 2.5×10^{14} g cm^{-3}. At densities where the nuclei are still spherical in such matter, the chemical potential of the electrons is $\mu_e \sim 80$ MeV and the atomic number of the nuclei is $Z \sim 50$ [45]. The corresponding Wigner–Seitz cell has a radius of $R \sim 18$ fm, and the radius of the nucleus inside the cell is $r \sim 9$ fm [45]. The electrons moving in the crystalline lattice formed by the ions are highly relativistic and strongly degenerate. The ion coupling parameter, defined in Equation (16), is $\Gamma \sim 2.3 \times 10^{12}/T$, and the melting temperature $T_{\text{melt}} \sim (Ze)^2/(Rk_B\Gamma_{\text{melt}})$ has a value of $T_{\text{melt}} \sim 1.3 \times 10^{10}$ K.

Properties of the quark–hadron lattice: The size of the Wigner–Seitz cells associated with spherical quark blobs in the crystalline quark–hadron phase is similar to the size of the Wigner–Seitz cells in the crust. (Here, we do not consider the crystalline phases made of quark rods and quark slabs since they contribute much less to Bremsstrahlung because of the much smaller number of electrons in those phases.) For spherical quark blobs at the onset of quark deconfinement, which occurs in our models at densities of around three times nuclear saturation, $3n_0$, the electron chemical potential is $\mu_e = k_e \sim 140$ MeV. Hence, like at sub-nuclear densities, the electrons are ultra-relativistic ($\hbar k_e/mc^2 = 275$) and strongly degenerate. The electron degeneracy temperature is around $T_F \sim 1.6 \times 10^{12}$ K, which is much higher than the temperature range ($\lesssim 10^{10}$ K) considered in this paper. From the results shown in Figure 3, one sees that the radii of the Wigner–Seitz cells containing spherical quark-blobs are around $R \sim 12$ fm and that the quark blobs inside the cells have radii of $r \sim 8$ fm. The density of the Wigner–Seitz cells is $(4\pi R^3/3)^{-1} \sim 1.4 \times 10^{-4}$ fm^{-3} and the atomic number of the quark blobs inside the Wigner–Seitz cell is around $Z \sim 200$.

Plasma temperature and melting temperature: The ion (quark blob) coupling parameter $\Gamma = (Ze)^2/(Rk_BT)$ is given by $\Gamma = 6.7 \times 10^{13}/T$ and the melting temperature of the ion crystal is $T_{\text{melt}} = (Ze)^2/(Rk_B 172) \sim 4 \times 10^{11}$ K. Here we have used $\Gamma_{\text{melt}} = 174$ for which a solid is expected to form [34,35]. Since the melting temperature of the quark crystal exceeds 10^{11} K the quark blobs are expected to be in the crystalline phase at all temperatures ($\lesssim 10^{10}$ K) considered in our study. The plasma temperature of the system follows from $T_P = 7.83 \times 10^9 \sqrt{ZY_e \rho_{12}/A_i}$, where $Y_e = n_e/n_b$ is the number of electrons per baryon, n_e the number density of electrons, n_b the number density of baryons, and ρ_{12} the mass density in units of 10^{12} g/cm^3. For quark blobs with mass numbers of $A \sim 2000$, atomic number $Z \sim 200$, and $Y_e \sim 0.06$ one obtains a plasma temperature of $T_P \sim 2 \times 10^{10}$ K.

Electron–phonon scattering and Umklapp processes: In an Umklapp process, the electron momentum transfer in a scattering event, $\hbar \vec{q}$, lies outside the first Brillouin zone, that is, $\hbar q \gtrsim \hbar q_0$. This is in contrast to the normal processes where $\hbar \vec{q}$ remains in the first Brillouin zone and $\hbar q \lesssim \hbar q_0$, where $q_0 \approx (6\pi^2 n_{\text{Blob}})^{1/3}$. For the quark–blob phase we find $\hbar q_0 \sim 30$ MeV so that $q_0/(2k_e) \sim 0.13$ for the quark-blob lattice, which is of the same order of magnitude as for the crust where $q_0/(2k_e) = (4Z)^{-1/3} \sim 0.01$ [8]. The temperature below which the Umklapp processes are frozen out is $T_{\text{Umklapp}} \sim T_P Z^{1/3} e^2 \sim 8 \times 10^8$ K, with the plasma temperature T_P given just above. We find that the temperatures obtained for T_{Umklapp}, T_P, and T_{melt} in the quark–blob phase are rather similar to their counterparts in the nuclear lattice just below nuclear saturation density, namely $T_{\text{Umklapp}} \sim 10^8$ K, $T_P \sim 10^9$ K, and $T_{\text{melt}} \sim 10^{10}$ K. In our study, both the Umklapp process and the normal process are taken into account since temperatures in the range of 10^6 K $< T < 10^{10}$ K are considered.

Debye–Waller factor: The effective interaction between electrons and quark blobs depends on the thermal quark–blob lattice vibrations which effectively smear-out the quark blob charges. This feature is taken into account via the Debye-Waller factor given in Equation (19). Since estimates for the Debye–Waller factor are only known for spherical blob structures, the Debye–Waller may be the largest source of uncertainty in our study.

Role of electron band structure effects: It has been shown in [46] that gaps in the electron dispersion relation at the boundaries of Brillouin zones can noticeably reduce the static lattice contribution. For point-like quark blobs with atomic number Z and for the smallest reciprocal lattice vector in a bcc lattice, we estimate the electron band splitting from $0.018(Z/60)^{2/3}k_e$ [46]. This leads to a splitting of $\sim$6 MeV for the quark-blob phase, which is around 1 MeV or more for the nuclear lattice case [46].

Author Contributions: The authors contributed equally to the theoretical and numerical aspects of the work presented in this paper.

Funding: This research was supported by CONICET and UNLP, grant numbers PIP 0714, G 157 and G 140, and by the National Science Foundation (USA), grant numbers PHY-1411708 and PHY-1714068.

Acknowledgments: M.G.O. and G.A.C. thank Consejo Nacional de Investigaciones Científicas y Técnicas (CONICET) and the Universidad Nacional de La Plata (UNLP) for financial support. F.W. is supported by the National Science Foundation (USA).

Conflicts of Interest: The authors declare no conflict of interest.

References

1. Glendenning, N.K. First-order phase transitions with more than one conserved charge: Consequences for neutron stars. *Phys. Rev. D* **1992**, *46*, 1274–1287. [CrossRef]
2. Glendenning, N.K. Phase transitions and crystalline structures in neutron star cores. *Phys. Rep.* **2001**, *342*, 393–447. [CrossRef]
3. Page, D.; Reddy, S. Dense Matter in Compact Stars: Theoretical Developments and Observational Constraints. *Ann. Rev. Nucl. Part. Sci.* **2006**, *56*, 327–374. [CrossRef]
4. Becker, W. (Ed.) *Neutron Stars and Pulsars*; Astrophysics and Space Science Library Series 357; Springer: Berlin, Germany, 2009.
5. Buballa, M.; Dexheimer, V.; Drago, A.; Fraga, E.; Haensel, P.; Mishustin, I.; Pagliara, G.; Schaffner-Bielich, J.; Schramm, S.; Sedrakian, A.; et al. EMMI rapid reaction task force meeting on quark matter in compact stars. *J. Phys. G Nucl. Part. Phys.* **2014**, *41*, 123001. [CrossRef]
6. Spinella, W.M.; Weber, F.; Contrera, G.A.; Orsaria, M.G. Neutrino emissivity in the quark–hadron mixed phase of neutron stars. *Eur. Phys. A* **2016**, *52*, 1–12. [CrossRef]
7. Spinella, W.M. A Systematic Investigation of Exotic Matter in Neutron Stars. Ph.D. Thesis, Claremont Graduate University, Claremont, CA, USA, San Diego State University, San Diego, CA, USA, 2017.
8. Kaminker, A.D.; Pethick, C.J.; Potekhin, A.Y.; Yakovlev, D.G. Neutrino-pair bremsstrahlung by electrons in neutron star crusts. *Astron. Astrophys.* **1999**, *343*, 1009–1024.
9. Flowers, E. Neutrino-pair emission in dense matter: A many-body approach. *Astrophys. J.* **1973**, *180*, 911–936. [CrossRef]
10. Itoh, N.; Kohyama, Y. Neutrino-pair bremsstrahlung in dense stars. I—Liquid metal case. *Astrophys. J.* **1983**, *275*, 858–866.
11. Itoh, N.; Matsumoto, N.; Seki, M.; Kohyama, Y. Neutrino-pair bremsstrahlung in dense stars. II—Crystalline lattice case. *Astrophys. J.* **1984**, *279*, 413–418. [CrossRef]
12. Itoh, N.; Kohyama, Y.; Matsumoto, N.; Seki, M. Neutrino-pair bremsstrahlung in dense stars. III—Low-temperature quantum corrections in the liquid metal phase. *Astrophys. J.* **1984**, *280*, 787–791. [CrossRef]
13. Itoh, N.; Kohyama, Y.; Matsumoto, N.; Seki, M. Neutrino-pair bremsstrahlung in dense stars. IV—Phonon contributions in the crystalline lattice phase. *Astrophys. J.* **1984**, *285*, 304–311. [CrossRef]

14. Pethick, C.J.; Thorsson, V. Effects of electron band structure on neutrino pair bremsstrahlung in neutron star crusts. *Phys. Rev. D* **1997**, *56*, 7548–7558. [CrossRef]
15. Na, X.; Xu, R.; Weber, F.; Negreiros, R. On the Transport Properties of a Quark-Hadron Coulomb Lattice in the Cores of Neutron Stars. *Phys. Rev. C* **2012**, *86*, 123016. [CrossRef]
16. Antoniadis, J.; Freire, P.C.C.; Wex, N.; Tauris, T.M.; Lynch, R.S.; Kerkwijk, M.H.; Kramer, M.; Bassa, C.; Dhillon, V.S.; Driebe, T.; et al. A Massive Pulsar in a Compact Relativistic Binary. *Science* **2013**, *340*, 1233232. [CrossRef] [PubMed]
17. Scarpettini, A.; Gómez Dumm, D.; Scoccola, N.N. Light pseudoscalar mesons in a nonlocal SU(3) chiral quark model. *Phys. Rev. C* **2004**, *69*, 114018.
18. Contrera, G.A.; Gómez Dumm, D.; Scoccola, N.N. Nonlocal SU(3) chiral quark models at finite temperature: The role of the Polyakov loop. *Phys. Lett. B* **2008**, *661*, 113–117. [CrossRef]
19. Contrera, G.A.; Gómez Dumm, D.; Scoccola, N.N. Meson properties at finite temperature in a three flavor nonlocal chiral quark model with Polyakov loop. *Phys. Rev. C* **2010**, *81*, 054005. [CrossRef]
20. Orsaria, M.; Rodrigues, H.; Weber, F.; Contrera, G.A. Quark-hybrid matter in the cores of massive neutron stars. *Phys. Rev. D* **2013**, *87*, 023001. [CrossRef]
21. Orsaria, M.; Rodrigues, H.; Weber, F.; Contrera, G.A. Quark deconfinement in high-mass neutron stars. *Phys. Rev. C* **2014**, *89*, 015806. [CrossRef]
22. Ranea-Sandoval, I.F.; Han, S.; Orsaria, M.G.; Contrera, G.A.; Weber, F.; Alford, M.G. Constant-sound-speed parametrization for Nambu–Jona–Lasinio models of quark matter in hybrid stars. *Phys. Rev. C* **2016**, *93*, 045812. [CrossRef]
23. Boguta, J.; Bodmer, A.R. Relativistic calculation of nuclear matter and the nuclear surface. *Nucl. Phys. A* **1977**, *292*, 413–428. [CrossRef]
24. Boguta, J.; Stöcker, H. Systematics of nuclear matter properties in a non-linear relativistic field theory. *Phys. Lett. B* **1983**, *120*, 289–293. [CrossRef]
25. Mellinger, R.D., Jr.; Weber, F.; Spinella, W.; Contrera, G.A.; Orsaria, M.G. Quark Deconfinement in Rotating Neutron Stars. *Universe* **2017**, *3*, 5. [CrossRef]
26. Typel, S.; Ropke, G.; Klahn, T.; Blaschke, D.; Wolter, H.H. Composition and thermodynamics of nuclear matter with light clusters. *Phys. Rev. C* **2010**, *81*, 015803. [CrossRef]
27. Lalazissis, G.A.; Niksic, T.; Vretenar, D.; Ring, P. New relativistic mean-field interaction with density-dependent meson-nucleon couplings. *Phys. Rev. C* **2005**, *71*, 024312. [CrossRef]
28. Fuchs, C.; Lenske, H.; Wolter, H.H. Density dependent hadron field theory. *Phys. Rev. C* **1995**, *52*, 3043. [CrossRef]
29. Typel, S.; Wolter, H.H. Relativistic mean field calculations with density-dependent meson-nucleon coupling. *Nucl. Phys. A* **1999**, *656*, 331–364. [CrossRef]
30. Typel, S. Relativistic model for nuclear matter and atomic nuclei with momentum-dependent self-energies. *Phys. Rev. C* **2005**, *71*, 064301. [CrossRef]
31. Rijken, T.A.; Nagels, M.M.; Yamamoto, Y. Baryon-Baryon Interactions: Nijmegen Extended-Soft-Core Models. *Prog. Theor. Phys. Suppl.* **2010**, *185*, 14–71. [CrossRef]
32. Miyatsu, T.; Cheoun, M.; Saito, K. Equation of state for neutron stars in SU(3) flavor symmetry. *Phys. Rev. C* **2013**, *88*, 015802. [CrossRef]
33. Yasutake, N.; Lastowiecki, R.; Benic, S.; Blaschke, D.; Maruyama, T.; Tatsumi, T. Finite-size effects at the hadron-quark transition and heavy hybrid stars. *Phys. Rev. C* **2014**, *89*, 065803. [CrossRef]
34. Ogata, S.; Ichimaru, S. Critical examination of N dependence in the Monte Carlo calculations for a classical one-component plasma. *Phys. Rev. A* **1987**, *36*, 5451. [CrossRef]
35. Haensel, P.; Potekhin, A.Y.; Yakovlev, D.G. *Neutron Stars 1: Equation of State and Structure*; Springer: New York, NY, USA, 2007.
36. Baiko, D.A.; Yakovlev, D.G. Thermal and electrical conductivities of Coulomb crystals in neutron stars and white dwarfs. *Astron. Lett.* **1995**, *21*, 702–709.
37. Haensel, P.; Kaminker, A.D.; Yakovlev, D.G. Electron neutrino-antineutrino bremsstrahlung in a liquid phase of neutron star crusts. *Astron. Astrophys.* **1996**, *314*, 328–340.
38. Baiko, D.A.; Kaminker, A.D.; Potekhin, A.Y.; Yakovlev, D.G. Ion Structure Factors and Electron Transport in Dense Coulomb Plasmas. *Phys. Rev. Lett.* **1998**, *81*, 5556–5559. [CrossRef]

39. Raikh, M.E.; Yakovlev, D.G. Thermal and electrical conductivities of crystals in neutron stars and degenerate dwarfs. *Astrophys. Space Sci.* **1982**, *87*, 193–203. [CrossRef]
40. Ziman, J.M. *Principles of the Theory of Solids*; Cambridge University Press: Cambridge, UK, 1972.
41. Mochkovitch, R.; Hansen, J.P. Fluid-solid coexistence curve of dense coulombic matter. *Phys. Lett.* **1979**, *73A*, 35–38. [CrossRef]
42. Lamb, D.Q.; Lattimer, J.M.; Pethick, C.J.; Ravenhall, D.G. Physical properties of hot, dense matter: The bulk equilibrium approximation. *Nucl. Phys. A* **1981**, *360*, 459–482. [CrossRef]
43. Ravenhall, D.G.; Pethick, C.J.; Wilson, J.R. Structure of Matter below Nuclear Saturation Density. *Phys. Rev. Lett.* **1983**, *50*, 2066–2069. [CrossRef]
44. Williams, R.D.; Koonin, S.E. Sub-saturation phases of nuclear matter. *Nucl. Phys. A* **1985**, *435*, 844–858. [CrossRef]
45. Lorenz, C.P.; Ravenhall, D.G.; Pethick, C.J. Neutron star crusts. *Phys. Rev. Lett.* **1993**, *70*, 379–382. [CrossRef] [PubMed]
46. Pethick, C.J.; Thorsson, V. Neutrino pair bremsstrahlung in neutron star crusts: A reappraisal. *Phys. Rev. Lett.* **1994**, *72*, 1964–1967. [CrossRef] [PubMed]

© 2018 by the authors. Licensee MDPI, Basel, Switzerland. This article is an open access article distributed under the terms and conditions of the Creative Commons Attribution (CC BY) license (http://creativecommons.org/licenses/by/4.0/).

Conference Report

Strangeness Production in Nucleus-Nucleus Collisions at SIS Energies

Vinzent Steinberg [1,2,*], Dmytro Oliinychenko [3], Jan Staudenmaier [1,2] and Hannah Petersen [1,2,4]

1. Institute for Theoretical Physics, Goethe University, Max-von-Laue-Strasse 1, 60438 Frankfurt, Germany; staudenmaier@fias.uni-frankfurt.de (J.S.); petersen@fias.uni-frankfurt.de (H.P.)
2. Frankfurt Institute for Advanced Studies, Ruth-Moufang-Strasse 1, 60438 Frankfurt, Germany
3. Lawrence Berkeley National Lab, 1 Cyclotron Road, Berkeley, CA 94720, USA; oliiny@fias.uni-frankfurt.de
4. GSI Helmholtzzentrum für Schwerionenforschung, Planck Street 1, 64291 Darmstadt, Germany
* Correspondence: steinberg@fias.uni-frankfurt.de; Tel.: +49-69-798-47677

Received: 30 November 2017; Accepted: 15 January 2018; Published: 13 February 2018

Abstract: Simulating Many Accelerated Strongly-interacting Hadrons (SMASH) is a new hadronic transport approach designed to describe the non-equilibrium evolution of heavy-ion collisions. The production of strange particles in such systems is enhanced compared to elementary reactions (Blume and Markert 2011), providing an interesting signal to study. Two different strangeness production mechanisms are discussed: one based on resonances and another using forced canonical thermalization. Comparisons to experimental data from elementary collisions are shown.

Keywords: relativistic heavy-ion collisions; monte carlo simulations; transport theory; strangeness

1. Introduction

Relativistic heavy-ion collisions provide a unique opportunity to study matter under extreme conditions. These experiments allow the creation of high temperatures similar to the universe during the first few microseconds after the Big Bang, yielding insights into the equation of state of nuclear matter, which is crucial for understanding the high-density physics of neutron stars.

Nuclear matter exclusively consists of up and down quarks, therefore, the newly produced strange quarks during heavy-ion collisions are a particularly interesting probe for studying the evolution of the reaction, see [1] for a recent overview. Strangeness enhancement in heavy-ion reactions compared to elementary proton-proton collisions has been observed some time ago [2]. The High-Acceptance Di-Electron Spectrometer (HADES) collaboration measured surprisingly high multiplicities of ϕ and Ξ hadrons in heavy-ion collisions below the production threshold [3,4]. In the intermediate energy range between the threshold and $\sqrt{s} = 10A$ GeV, the multiplicities of multi-strange particles are still unknown and of high interest to understand this effect.

Overall, there are still a lot of open questions about how strangeness is produced: what role do kaon-nucleon potentials play; how strongly are cross sections affected by the medium; and what are the production mechanisms in and out of equilibrium?

In the following, we focus on heavy-ion reactions at Schwerionen-Synchrotron (SIS) energies ($E_{kin} = 0.5 - 3.5$ GeV) and explore how the hadronic transport approach, Simulating Many Accelerated Strongly-interacting Hadrons (SMASH) [5], models strangeness production out of equilibrium via resonances and in equilibrium via forced thermalization. The aim is to provide a baseline founded on vacuum properties and low-energy physics that can be extended to higher energies and larger systems. Comparisons to elementary cross sections and dilepton measured in experiments are shown, verifying that the resonance approach successfully describes such vacuum properties and providing the foundations for studying the questions raised above within this approach.

2. Model Description

The results presented here are obtained from simulations using SMASH, a microscopic transport approach that solves the relativistic Boltzmann equation

$$p^\mu \partial_\mu f_i(x,p) = C_{\text{coll}}(f_i) \tag{1}$$

using the test particle ansatz and the geometric collision criterion

$$d_{\text{trans}} < d_{\text{int}} = \sqrt{\frac{\sigma_{\text{tot}}}{\pi}}, \tag{2}$$

where d_{trans} is the distance of closest approach between two particles and d_{int} is the interaction distance given by the total cross section σ_{tot}. SMASH has been tested against exact solutions of the Boltzmann equation within an Friedmann-Lemaître-Robertson-Walker expanding metric to verify the numerical implementation of the collision criterion [6]. The $2 \leftrightarrow 2$ and $2 \leftrightarrow 1$ reactions are included, implementing all 106 hadrons species made of up, down and strange quarks that are considered experimentally established by the Particle Data Group (PDG) [7]. This results in tens of thousands of possible reaction pairs, for most of which the cross sections are not measured. In SMASH, these interactions are mostly modeled via resonances. This approach has the advantage that the cross sections can be calculated from the resonance properties, for which the available experimental data has been compiled by the PDG. It can be extended to $m \to n$ reactions by approximating them with a cascade of $1 \leftrightarrow 2$ reactions, maintaining a detailed balance. However, this approach is limited in energy (by the highest mass of the known resonances) and some cross sections are not resonant and have to be parametrized. Furthermore, many resonance properties are only sparsely constrained by experimental data.

SMASH can be used to simulate nuclear collisions (as in this work) or infinite matter and as an afterburner for hydrodynamic simulations of the quark-gluon plasma. It is also able to generate dileptons [8,9] and photons in heavy-ion collisions. The current goal is to test physics at SIS energies, establishing a baseline that can be extended for predictions at Nuclotron-based Ion Collider Facility (NICA) and Facility for Antiproton and Ion Research (FAIR) energies. See [5] for a detailed description of the model and results on pion and proton production in heavy-ion collisions compared to experimental data.

The implemented strange particle species are kaons, 11 kaonic resonances, $\Lambda, \Sigma, \Xi, \Omega$ baryons and 28 baryonic resonances, plus antiparticles. In nucleus-nucleus collisions, hyperons ($Y \in \{\Lambda, \Sigma\}$) and kaons are primarily produced by nucleon resonances

$$NN \to NN^*/\Delta^* \to NYK, \tag{3}$$

while antikaons are produced from strangeness exchange and ϕ decays:

$$NN \to NN^*/\Delta^* \to NYK \quad \pi Y \to Y^* \to \bar{K}N \quad N^* \to \phi N \tag{4}$$

The non-resonant contribution to the strangeness exchange $\pi Y \leftrightarrow \bar{K} N$ is parametrized similar to ref. [10].

3. Results

Let us first concentrate on strangeness production out of equilibrium by resonance excitation. The production of hyperons and ϕ mesons are discussed separately because they are constrained very differently by the available experimental data. Finally, strangeness production via a different approach forcing local thermal equilibrium in hadronic transport is briefly presented and compared to a more traditional hybrid approach of modeling heavy-ion collisions.

3.1. Hyperon Production via Resonances

While the masses and decay widths of the resonances are well-established, their branching ratios are only sparsely known. This can be alleviated by studying elementary cross sections. For example, the reaction $p\pi^- \to \Lambda K^0$ is dominated by the $N^* \to \Lambda K$ branching ratios. Taking the middle of the range given by the PDG results in a cross section as shown in Figure 1a: The cross section reconstructed from SMASH output (lines) overestimates the experimental data (circles) at the threshold and at $\sqrt{s} > 1.8\,\text{GeV}$. Looking at the contributions of the individual resonances reveals that $N(1650)$, $N(2080)$, $N(2190)$, $N(2220)$ and $N(2250)$ overshoot the data and that their branching ratios should be decreased, while the $N(1710)$ and $N(1720)$ branching ratios can be increased to compensate. Adapting the branching ratios in the range given by the PDG data results in Figure 1b: now the cross section is reproduced rather well, especially at the threshold and at high energies. There might be a slight underestimation at $\sqrt{s} = 1.7\,\text{GeV}$ and $\sqrt{s} = 1.8\,\text{GeV}$, but it is hard to tell, due to large uncertainties in the experimental data.

Another constraint on the $N^* \to \Lambda K$ branching ratio is given by the $pp \to \Lambda p K^+$ cross section. Potential conflicts with the constraints posed by the $p\pi^- \to \Lambda K^0$ cross section can usually be resolved by adapting the $N^* \to \pi N$ branching ratios. This has to be done under careful consideration of the effect on pion production. Similarly, the $N^*, \Delta^* \to \Sigma K, \pi N$ branching ratios are constrained by the $N\pi \to \Sigma K$ and $NN \to \Sigma NK$ cross sections. Due to the different possible charge combinations, there are more constraints given by the measured cross sections than for the Λ.

Having constrained the branching ratios using only elementary data, it is now possible to compare strangeness production in SMASH to heavy-ion experiments in the future. Discrepancies in large systems might hint at in-medium effects.

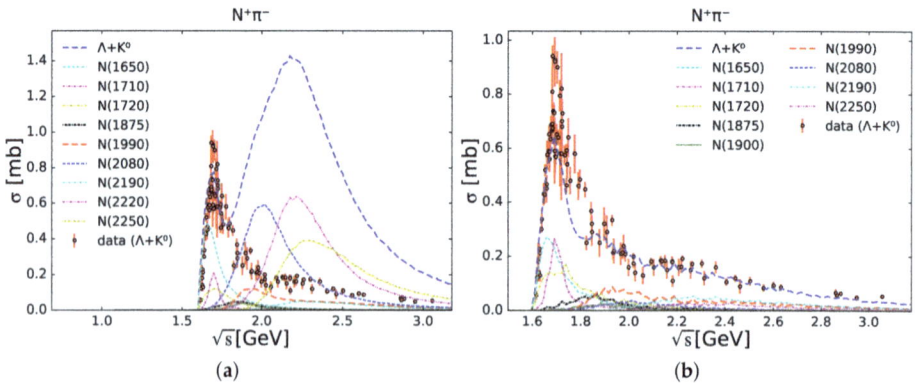

Figure 1. $p\pi^- \to \Lambda K^0$ cross section reconstructed from SMASH output (lines) compared to experimental data [11] (circles). (**a**) Using N^* branching ratios by choosing the middle of the range given by the PDG [7]. (**b**) After tuning the branching ratios to fit the experimental cross section.

3.2. ϕ Production via Resonances

The HADES collaboration concluded that at low energies a significant fraction of the K^- are produced by ϕ decays [3]. However, none of the baryonic resonances given by the PDG decay into any ϕ [7]. To be still able to produce ϕ in a resonance approach, it was proposed to use the experimental uncertainty of the N^* beyond $2\,\text{GeV}$ to introduce a small ϕN branching ratio [12]. Independent experimental data constraining this arbitrary branching ratio is required.

A potential candidate is the $pp \to pp\phi$ cross section shown in Figure 2a. Unfortunately, it has only been measured close to the threshold, so it does not constrain the ϕ production very well. In our resonance approach, the largest contribution stems from higher energies where no data is available. Other observables are the invariant mass spectra of dileptons measured by the HADES collaboration at $E_{kin} = 3.5$ GeV in pNb collisions. As shown in Figure 2b, the experimental data resolves the ϕ peak well enough, constraining the $N^* \to \phi N$ branching ratios. By choosing a branching ratio of 0.5%, SMASH is able to reproduce the dilepton spectra and the cross section shown in Figure 2. It remains to be seen how this approach compares to experimental data from larger systems, where so the far neglected in-medium effects on the ϕ may be important.

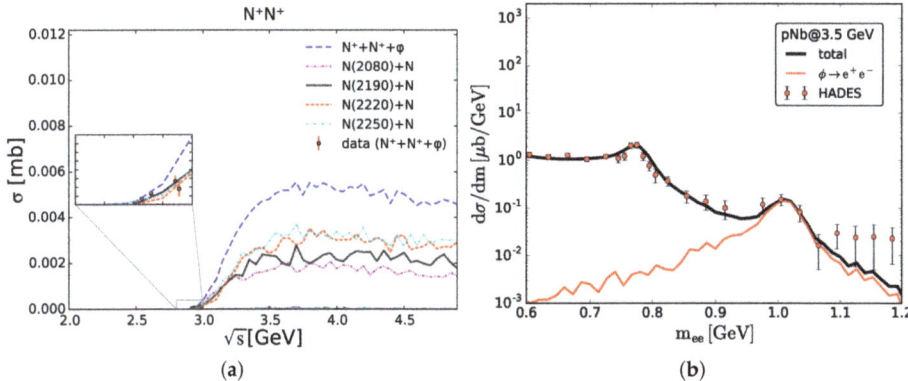

Figure 2. (a) $pp \to pp\phi$ cross section reconstructed from SMASH output (lines) compared to experimental data [13,14] (circles). (b) Dielectron mass spectrum in proton-niobium collisions at $E_{kin} = 3.5$ GeV in SMASH (lines) compared to HADES data (circles). Unlike the elementary cross section, the pNb dilepton spectrum constrains the ϕ production.

3.3. Strangeness Production via Thermalization

Traditionally, hybrid models have been successfully used to simulate high-energy heavy-ion collisions: a hydrodynamical model for the partonic phase and a microscopic model (like SMASH) for the hadronic phase. However, it is not clear how to extend them to intermediate energies relevant for the beam energy scan program at the Relativistic Heavy-Ion Collider (RHIC) and future measurements at NICA and FAIR. In [15] it was proposed to use a different approach based on hadronic transport: if there is a region beyond some critical energy density ϵ_{th}, force thermalization in that region by resampling all particles according to a canonical thermal distribution while conserving all relevant quantum numbers. This has similarities to a thermal model but it assumes local instead of global equilibrium. Effectively, it mimics many-particle scattering and interpolates dynamically between two limits of kinetic theory: the dilute gas and the ideal fluid.

As shown in Figure 3, the forced thermalization does barely affect the pion multiplicity but enhances strangeness similar to a hybrid approach. Note that no mean-field potentials were applied because the collision energy is high enough that they are not so important ($\sqrt{s} = 3A$ GeV). The amount of produced strangeness in the forced thermalization approach is regulated by the parameter ϵ_{th} (the threshold energy density above which thermalization is performed). For low ϵ_{th}, as for instance twice the nuclear ground-state energy density as in Figure 3, strangeness is strongly enhanced and might be too high compared to experiment. High ϵ_{th} leads to a transport simulation without forced thermalization where strangeness is usually underestimated. It remains for future studies to fix ϵ_{th} versus collision energy and test if different strange particles can be described simultaneously.

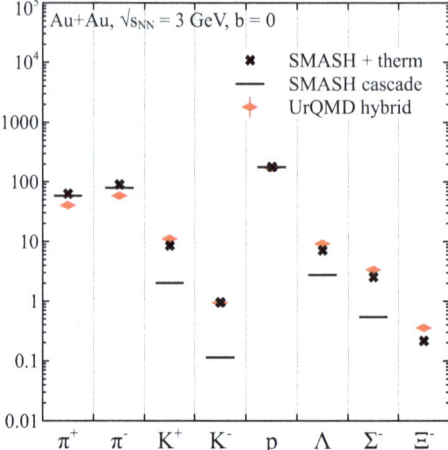

Figure 3. SMASH with and without forced canonical thermalization compared to a UrQMD hybrid model [15]. The energy density threshold for thermalization in SMASH (ϵ_{th}) and particlization in the hybrid approach was set to twice the nuclear ground-state energy density. The strangeness enhancement due to the forced thermalization is comparable to the hybrid approach.

4. Conclusions and Outlook

In this work, it was shown how elementary K, $\bar{K}$, Λ, Σ and ϕ production at low energies can be modeled by via resonances. The PDG data on branching ratios was complemented with constraints from elementary, exclusive cross sections. The ϕ production was modeled by introducing small ϕN branching ratios to heavy N^* resonances, which were successfully constrained by dilepton spectra in pNb collisions. This non-equilibrium strangeness production at low energies via resonances provides a baseline that can be extended to higher energies by including production from strings. Future comparisons to experimental results for larger systems might hint at in-medium effects required to describe the data, such as kaon-nucleon potentials, in-medium cross sections and kaon self-energies [16]. As a mechanism for strangeness production in equilibrium, effective many-particle interactions by forced thermalization were discussed. Promising results of how such an approach enhances strangeness production in heavy-ion collisions similar to a more traditional hybrid approach were shown. It remains to be seen how well this approach can describe the experimental data.

Acknowledgments: The authors thank J. Steinheimer for fruitful discussions. Computational resources have been provided by the Center for Scientific Computing (CSC) at the Goethe-University of Frankfurt and by the Green IT Cube at Gesellschaft für Schwerionenforschung (GSI). The authors acknowledge funding of a Helmholtz Young Investigator Group VH-NG-822 from the Helmholtz Association and GSI. This work was supported by the Helmholtz International Center for the Facility for Antiproton and Ion Research (HIC for FAIR) within the framework of the Landes-Offensive zur Entwicklung Wissenschaftlich-Ökonomischer Exzellenz (LOEWE) program launched by the State of Hesse. D. O. acknowledges support by the Deutsche Telekom Stiftung. D. O. and V. S. acknowledge support by the Helmholtz Graduate School for Hadron and Ion Research (HGS-HIRe). J. S. acknowledges funding by the Deutsche Forschungsgemeinschaft (DFG) grant CRC-TR-2111.

Author Contributions: V.S. implemented strangeness production in SMASH; J.S. implemented dileptons and provided the corresponding plot; D.O. and H.P developed forced thermalization; D.O. implemented it and provided the corresponding plot; H.P. supervised the work; V.S. wrote the paper.

Conflicts of Interest: The authors declare no conflict of interest.

References

1. Blume, C. Open Questions in the Understanding of Strangeness Production in HIC -Experiment Perspective. In Proceedings of the 17th International Conference on Strangeness in Quark Matter (SQM 2017), Utrecht, The Netherlands, 10–15 July 2017.
2. Blume, C.; Markert, C. Strange hadron production in heavy ion collisions from SPS to RHIC. *Prog. Part. Nucl. Phys.* **2011**, *66*, 834–879.
3. Adamczewski-Musch, J.; Arnold, O.; Behnke, C.; Belounnas, A.; Belyaev, A.; Berger-Chen, J.C.; Biernat, J.; Blanco, A.; Blume, C.; Böhmer, M.; et al. Deep sub-threshold ϕ production and implications for the K^+/K^- freeze-out in Au+Au collisions. *arXiv* **2017**, arXiv:1703.08418.
4. Agakishiev, G.; Balanda, A.; Bassini, R.; Belver, D.; Belyaev, A.V.; Blanco, A.; Böhmer, M.; Boyard, J.L.; Braun-Munzinger, P.; Cabanelas, P.; et al. Deep sub-threshold Ξ^- production in Ar+KCl reactions at 1.76A-GeV. *Phys. Rev. Lett.* **2009**, *103*, 132301.
5. Weil, J.; Steinberg, V.; Staudenmaier, J.; Pang, L.G.; Oliinychenko, D.; Mohs, J.; Kretz, M.; Kehrenberg, T.; Goldschmidt, A.; Bäuchle, B.; et al. Particle production and equilibrium properties within a new hadron transport approach for heavy-ion collisions. *Phys. Rev. C* **2016**, *94*, 054905.
6. Tindall, J.; Torres-Rincon, J.M.; Rose, J.B.; Petersen, H. Equilibration and freeze-out of an expanding gas in a transport approach in a Friedmann-Robertson-Walker metric. *Phys. Lett. B* **2017**, *770*, 532–538.
7. Patrignani, C.; Particle Data Group. Review of Particle Physics. *Chin. Phys. C* **2016**, *40*, 100001.
8. Staudenmaier, J.; Weil, J.; Petersen, H. Non-equilibrium dilepton production in hadronic transport approaches. *J. Phys. Conf. Ser.* **2017**, *832*, 012037.
9. Weil, J.; Staudenmaier, J.; Petersen, H. Dilepton production with the SMASH model. *J. Phys. Conf. Ser.* **2016**, *742*, 012034.
10. Graef, G.; Steinheimer, J.; Li, F.; Bleicher, M. Deep sub-threshold Ξ and Λ production in nuclear collisions with the UrQMD transport model. *Phys. Rev. C* **2014**, *90*, 064909.
11. Baldini, A. Numerical Data and Functional Relationships In Science and Technology. Grp. 1: Nuclear and Particle Physics. Vol. 12: Total cross-sections for reactions of high-energy particles (including elastic, topological, inclusive and exclusive reactions). Subvol. In *Landolt-Börnstein, New Series, 1/12B*; Springer: Berlin, Germany, 1988.
12. Steinheimer, J.; Bleicher, M. Sub-threshold ϕ and Ξ^- production by high mass resonances with UrQMD. *J. Phys. G Nucl. Part. Phys.* **2016**, *43*, 015104.
13. Maeda, Y.; Hartmann, M.; Keshelashvili, I.; Barsov, S.; Büscher, M.; Drochner, M.; Dzyuba, A.; Hejny, V.; Kacharava, A.; Kleber, V.; et al. Kaon Pair Production in Proton-Proton Collisions. *Phys. Rev. C* **2008**, *77*, 015204.
14. Balestra, F.; Bedfer, Y.; Bertini, R.; Bland, L.C.; Brenschede, A.; Brochard, F.; Bussa, M.P.; Choi, S.; Debowski, M.; Dressler, R.; et al. ϕ and ω meson production in pp reactions at $p_{lab} = 3.67\,\text{GeV}/c$. *Phys. Rev. C* **2001**, *63*, 024004.
15. Oliinychenko, D.; Petersen, H. Forced canonical thermalization in a hadronic transport approach at high density. *J. Phys. G Nucl. Part. Phys.* **2017**, *44*, 034001.
16. Hartnack, C.; Oeschler, H.; Leifels, Y.; Bratkovskaya, E.L.; Aichelin, J. Strangeness Production close to Threshold in Proton-Nucleus and Heavy-Ion Collisions. *Phys. Rep.* **2012**, *510*, 119–200.

© 2018 by the authors. Licensee MDPI, Basel, Switzerland. This article is an open access article distributed under the terms and conditions of the Creative Commons Attribution (CC BY) license (http://creativecommons.org/licenses/by/4.0/).

MDPI
St. Alban-Anlage 66
4052 Basel
Switzerland
Tel. +41 61 683 77 34
Fax +41 61 302 89 18
www.mdpi.com

Universe Editorial Office
E-mail: universe@mdpi.com
www.mdpi.com/journal/universe

www.ingramcontent.com/pod-product-compliance
Lightning Source LLC
LaVergne TN
LVHW071940080526
838202LV00064B/6643